AF361758

Optimal Control of Hybrid Systems and Renewable Energies

Optimal Control of Hybrid Systems and Renewable Energies

Special Issue Editors

Michela Robba
Mansueto Rossi

MDPI • Basel • Beijing • Wuhan • Barcelona • Belgrade • Manchester • Tokyo • Cluj • Tianjin

Special Issue Editors

Michela Robba
University of Genova
Italy

Mansueto Rossi
University of Genova
Italy

Editorial Office
MDPI
St. Alban-Anlage 66
4052 Basel, Switzerland

This is a reprint of articles from the Special Issue published online in the open access journal *Energies* (ISSN 1996-1073) (available at: https://www.mdpi.com/journal/energies/special_issues/ Control_Hybrid_Systems_Renewable_Energies).

For citation purposes, cite each article independently as indicated on the article page online and as indicated below:

LastName, A.A.; LastName, B.B.; LastName, C.C. Article Title. *Journal Name* **Year**, *Article Number*, Page Range.

ISBN 978-3-03928-897-7 (Hbk)
ISBN 978-3-03928-898-4 (PDF)

Contents

About the Special Issue Editors

Michela Robba is Associate Professor of Systems Engineering at University of Genoa, where she received her Degree in Environmental Engineering in 2000 and PhD in Electronic and Computer Engineering in 2004. Her main research activities are in the field of optimization and control of smart grids, renewable energy resources, and natural resources management. She is the Italian representative of EU ESFRI (European Strategy Forum on Research Infrastructures) for the Energy area and is a member of the scientific board of the Italian Energy Cluster and Liguria Region Innovation Pole on Energy, Environment and Sustainable Development. She has been Editor of the Journal of Control Science and Engineering since 2014 and has participated in numerous international program committees of conferences in the field of control and optimization. Additionally, she is Lecturer for the courses "Simulation of Energy and Environmental Systems" and "Models and Methods for Energy Engineering" at Savona Campus Polytechnic School, University of Genoa. She is author of more than 120 publications.

Mansueto Rossi is Researcher at the University of Genoa, where he received his graduate (cum laude) degree in Electrical Engineering and the Ph.D. in 2000 and 2004, respectively. From April 2004 to April 2007, he was a Temporary Research Assistant with the University of Genoa, where he worked on research projects concerning channel allocation, optimization for cellular base stations, and the finite formulation method applied to the computation of electromagnetic fields. He began here as Researcher in the Department of Electrical Engineering in October 2008. He has coauthored more than 60 papers published in international journals and proceedings of international conferences. His research interests include lightning modeling, transmission line analysis and lightning-induced over-voltages, smart grids, microgrids, and electromagnetic transient analysis of power lines.

Preface to "Optimal Control of Hybrid Systems and Renewable Energies"

International policies for sustainable development have led to an increase in distributed power production based on renewable resources. This, in turn, leads to the necessity of defining new technological solutions that can reduce costs as well as new control strategies to optimally manage renewable resources which are increasingly characterized by the close interaction between different energy vectors and their networks and by a transition from a centralized structure to a decentralized one (both in terms of sources and controls). Since countries all over the world are implementing low-carbon and energy efficiency policies, there has been a rapid increase in the installations of distributed generation (DG) technologies using mainly renewable resources and fossil fuels in the case of co(tri)generation applications. The presence of intermittent, non-programmable renewables as "standalone" entities connected to the distribution networks can result in difficulties in balancing the power of the transmission grid. This problem can be positively mitigated by new solutions, such us bundling distributed energy resources (DERs) into intelligent microgrids (MGs), which can, in an aggregate way, manage all power sources and loads, thereby playing a major role in providing balancing services for the safe operation of the power system and, at the same time, meeting environmental targets. Microgrids can alleviate the management and monitoring burden for the distribution system operator (DSO) by clustering several DERs in a single entity, but they require flexible and reliable energy management systems (EMSs) which, on the basis of simulation and optimization models, automatically schedule the plants according to economic and environmental criteria. Moreover, new actors are now entering the energy market, such as aggregators, and new roles are foreseen for distribution and transmission system operators. This creates new challenges for the integration of new optimization models and new control techniques. In this context, the main aim of this book is to collect papers in the field of the optimal control of power and energy production from renewable resources (wind, PV, biomass, hydrogen, etc.). In the first part of the book, attention is focused both on the optimal control of local technologies (such as solar cooling plants, wind turbines, and hybrid plants and storage systems). Then, the integration of distributed production plants and storage systems in buildings, microgrids, and the energy market is presented. Specifically, optimization and simulation models within SCADA and Energy Management Systems are considered. Finally, the management of buildings and microgrids, equipped with renewables, by an aggregator in the energy market is presented.

Michela Robba, Mansueto Rossi
Special Issue Editors

Article

Hybrid Nonlinear MPC of a Solar Cooling Plant

Eduardo F. Camacho *, Antonio J. Gallego *, Juan M. Escaño * and Adolfo J. Sánchez *

Departamento de Ingeniería de Sistemas y Automática, Universidad de Sevilla, Camino de los Descubrimientos s/n., 41092 Sevilla, Spain
* Correspondence: efcamacho@us.es or eduardo@esi.us.es (E.F.C.); gallegolen@hotmail.com (A.J.G.); jescano@us.es (J.M.E.); adolfo.spf@gmail.com (A.J.S.); Tel.: +34-954-487-347 (E.F.C.)

Received: 6 June 2019; Accepted: 12 July 2019; Published: 16 July 2019

Abstract: Solar energy for cooling systems has been widely used to fulfill the growing air conditioning demand. The advantage of this approach is based on the fact that the need of air conditioning is usually well correlated to solar radiation. These kinds of plants can work in different operation modes resulting on a hybrid system. The control approaches designed for this kind of plant have usually a twofold goal: (a) regulating the outlet temperature of the solar collector field and (b) choosing the operation mode. Since the operation mode is defined by a set of valve positions (discrete variables), the overall control problem is a nonlinear optimization problem which involves discrete and continuous variables. This problems are difficult to solve within the normal sampling times for control purposes (around 20–30 s). In this paper, a two layer control strategy is proposed. The first layer is a nonlinear model predictive controller for regulating the outlet temperature of the solar field. The second layer is a fuzzy algorithm which selects the adequate operation mode for the plant taken into account the operation conditions. The control strategy is tested on a model of the plant showing a proper performance.

Keywords: solar energy; Fresnel collector; model predictive control; fuzzy algorithm; hybrid systems

---

1. Introduction

The need of reducing the impact of fossil energies such as coal or petroleum has led to a great interest in the renewable energies such as wind or solar. In particular, solar energy has experienced a great impulse over the last 30 years. One of the most important advantages of solar energy compared to other renewable energies is the possibility of using cost efficient heat energy storage systems [1,2].

Many solar power plants have been built around the world making use of multiple technologies such as parabolic trough, solar power tower, solar dish, Fresnel collector etc. For example, the experimental solar plant of ACUREX in Almería (Spain), the 50 MW commercial parabolic trough plants Helios 1 and 2 in Castilla la Mancha (Spain) [3], owned by Atlantica Yield, in Écija (Spain), can be mentioned. In the United States the large scale parabolic trough plants of Mojave of 280 MW [4] and SOLANA [5] can also be found.

The use of solar energy for cooling systems has been increasing for several decades spurred by the fact that the need for air conditioning is usually well correlated to high levels of solar radiation [6–8]. The plant used in this paper as a test-bench for control purposes is the solar cooling plant located on the roof of the Engineering School (ESI) of Seville [9,10]. This plant was commissioned in 2008, consisting of a Fresnel collector field, a double effect LiBr+ water absorption chiller and a storage tank. The Fresnel collector delivers pressurized water at 140–170 °C to the absorption machine for producing air conditioning. If solar radiation is not high enough for heating the water up to the required temperature, the storage tank can be used. If neither the solar field nor the storage tank are able to heat the water up to the operation temperature, the absorption machine uses natural gas [11].

The previous developed works for another solar cooling plant installed at the ESI of Seville as well, have shown that the design of control algorithms for this kind of systems is hindered by two facts [12,13]: firstly, the primary energy source, the sun, cannot be manipulated. Secondly, the environmental conditions and cooling demand may change substantially. This previous plant was used in the framework of the Network of Excellence HYCON and served as a benchmark for testing control technologies of hybrid systems.

The differences between the HYCON solar cooling plant and the one described in this paper are as follows:

- The solar field of the plant described here is a Fresnel collector field, whereas the HYCON plant uses a set of flat collectors.
- The storage system is a phase change material (PCM) storage tank, where as the accumulation system of the HYCON plant is composed of two tanks storing water.
- The absorption machine is a double effect LiBr+ water absorption chiller with a theoretical cooling power of 174 kW. The chiller of the HYCON plant was a simple effect absorption machine with a cooling power of 35 kW.

Solar cooling plants may work in multiple operation modes as is pointed out in [14]. In order to ensure an efficient operation of the plant, a model of the plant for control purposes is needed. The control approaches designed for this kind of plant usually have a twofold goal: (a) regulating the outlet temperature of the solar collector field and (b) choose the operation mode. Since the operation mode is defined by a set of valves positions (discrete variables), the overall control problem is a nonlinear optimization problem which involves discrete and continuous variables. These problems are difficult to solve within the normal sampling times for control purposes (around 20–30 s) [15].

In this paper, a different approach is proposed. The control strategy uses two independent algorithms. The first one is a nonlinear model predictive control (MPC) which regulates the outlet temperature of the Fresnel collector field. The main control objective of this kind of plant is to regulate the outlet temperature of the solar collector field around a desired value [16–18]. However, as stated above, the plant can be working in different operation modes which involve the position of different valves and the activation of a particular subsystem. The decision-making process to make a transition between modes of operation is imposed as a restriction and has been designed by the experience of the operators of the plant. The information accumulated by experience comes, on the one hand, clearly defined in operating rules, determined by the limit values of certain variables and the transitions allowed in each of the modes.

However, the operators of the plant have established decision rules that handle variables with limits where certain activation thresholds are taken into account, causing the information to present some undefined limits. Fuzzy logic can handle information closer to the human way, that is, uncertain, vague or inaccurate. The second algorithm is based on a fuzzy logic to decide in which operation mode the plant has to work. Fuzzy logic has been widely used in classification, matching and decision making techniques (see [19–21]). Based on the theory of fuzzy systems and the idea of an expert judgment, the proper mode of operation for the plant can be decided at any time. The results obtained show that the proposed control strategy presents a good performance when applied in a hybrid system with different modes of operation and under different conditions of radiation and demand. The main advantage of this algorithm with respect to a non-linear control algorithm is the speed of computation when the mode of operation is chosen and its easy implementation in systems of low computational capacity such as Programmable Logic controllers (PLCs). On the contrary, its disadvantage is that the global optimum is not guaranteed.

This paper is organized as follows. In Section 2, a brief description of the solar cooling plants is presented. In Section 3, the modeling of each subsystem is carried out. In Section 4, the nonlinear model predictive controller for regulating the outlet temperature of the field is presented. In Section 5, the different operation modes are explained. In Section 6, the fuzzy algorithm developed to select the adequate operation mode is presented. In Section 7, some simulation results are presented and discussed. Finally, the paper ends with concluding remarks.

2. Solar Cooling Plant Description

The solar cooling plant was commissioned in 2008 and consists of three subsystems: the double-effect LiBr+ water absorption chiller of 174 kW nominal cooling capacity. The solar Fresnel collector field aims at heating up the pressurized water used by the absorption machine. The PCM storage tank supplies energy to the water if there is not enough energy reaching the solar field to reach the required temperature. Figure 1 shows the scheme of the whole plant.

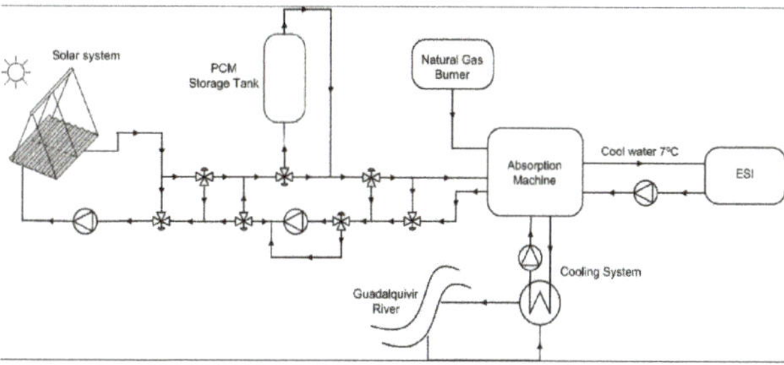

Figure 1. Plant general scheme.

Water absorption chiller: this is a double-effect cycle LiBr+ absorption machine with 174 kW and a theoretical COP of 1.34, which transforms the thermal energy (hot water at 140–170 °C) coming from the Fresnel solar field or the PCM storage tank, into cold water to be used by the ESI of Seville [9]. Apart from the hot water, a cooling fluid for the condenser is needed in the absorption machine. This is obtained from the water catchment of the Guadalquivir river.

Solar field: the solar field consists of a set of Fresnel solar collectors (see Figure 2) which concentrate solar radiation onto a line where an absorption tube is located. The energy is transferred to a heat transfer fluid (in our case, pressurized water).

PCM storage tank: PCM storage is a is a shell-tube heat exchanger 18 m long and 1.31 m in diameter (Figure 3). It consists of a series of tubes containing a heat transfer fluid and PCM fills up the space between the tubes and the shell.

The storage tank uses a hydroquinone as a PCM because the melting temperature is about 170 °C, which is suitable for the water absorption chiller operational range (145–170 °C).

Figure 2. Fresnel collector field.

Figure 3. Phase change material (PCM) storage tank.

3. Modeling of the Plant Subsystems

In this section, the model of each subsystem comprising the whole plant is described. The solar cooling plant has four subsystems: the Fresnel solar field, the storage tank, the absorption machine and the piping system connecting them. The equations governing the dynamics of each subsystem are presented. The models are validated and compared to real data taken from the plant.

Since the main goal of these models is to be used in control algorithms, the balance equations of each subsystem should be as simple as possible and an adequate trade-off between precision and complexity is pursued.

3.1. Fresnel Solar Field

In this subsection, the mathematical model of the Fresnel collector field is presented. Two approaches are usually considered in this kind of systems: the lumped parameter model (developed in [22]) and the distributed parameter model (described in [23]). In this paper, a distributed parameter model has been used. The distributed parameter model is governed by the following two PDE equations [24,25]:

$$\rho_m C_m A_m \frac{\partial T_m}{\partial t} = I K_{opt} noG - H_l G(T_m - T_a) - LH_t(T_m - T_f) \tag{1}$$

$$\rho_f C_f A_f \frac{\partial T_f}{\partial t} + \rho_f C_f q \frac{\partial T_f}{\partial x} = LH_t(T_m - T_f), \tag{2}$$

where m subindex refers to metal and f subindex refers to a fluid. In Table 1, parameters and their units are shown. The same system of equations is used to model the piping system. In this case, the radiation reaching the tube is null and the thermal losses coefficient takes a different value.

The PDE system is solved by dividing the metal and fluid in 64 segments of 1 m long. The integration step is chosen of 0.5 s.

As has been mentioned before, the heat transfer fluid is pressurized water whose density and specific heat have been obtained as polynomial functions of the segment temperature using thermodynamical data of pressurized water. The heat transfer coefficient depends on the segment temperature and the water flow [26]. As far as the thermal losses coefficient is concerned, it was obtained using experimental data from the collector field [10,22]. Figures 4 and 5 show a comparison between the model and the real plant evolution in two different days (in October and June). The model evolution behavior is very similar to the real solar field as can be seen.

Table 1. Parameter description.

Symbol	Description	Units
t	Time	s
x	Space	m
ρ	Density	kgm^{-3}
C	Specific heat capacity	$\mathrm{JK}^{-1}\mathrm{kg}^{-1}$
A	Cross sectional area	m^2
$T(x,y)$	Temperature	K,°C
$q(t)$	Oil flow rate	$\mathrm{m}^3\mathrm{s}^{-1}$
$I(t)$	Solar radiation	Wm^{-2}
no	geometric efficiency	Unitless
K_{opt}	Optical efficiency	Unitless
G	Collector aperture	m
$T_a(t)$	Ambient temperature	K,°C
H_l	Global coefficient of thermal loss	$\mathrm{Wm}^{-2}\mathrm{°C}^{-1}$
H_t	Coefficient of heat transmission metal-fluid	$\mathrm{Wm}^{-2}\mathrm{°C}^{-1}$
L	Length of pipe line	m

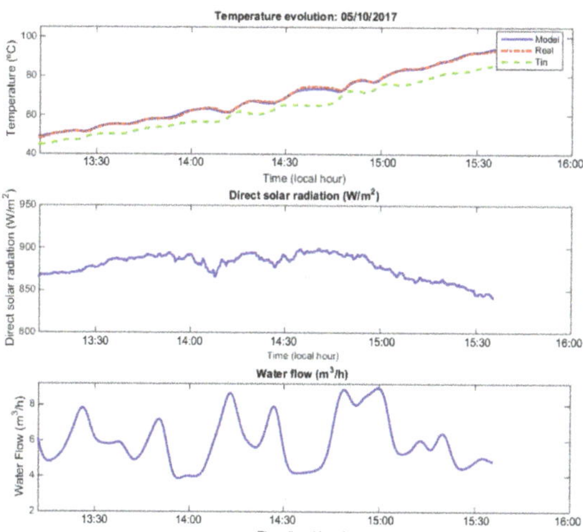

Figure 4. Solar field evolution model vs. real: 5 October 2017.

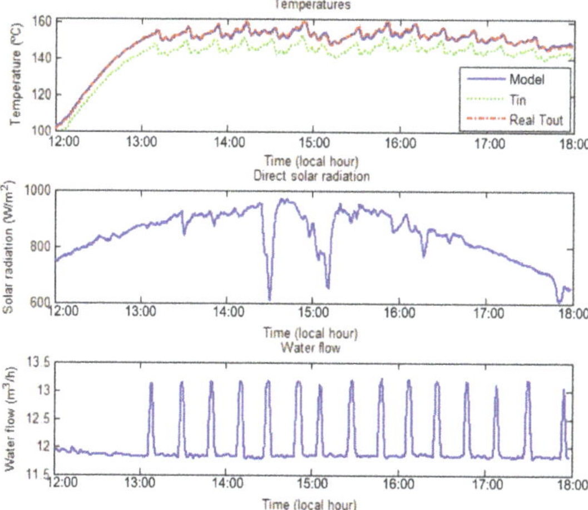

Figure 5. Solar field evolution model vs. real: 29 June 2009.

3.2. PCM Storage Tank

In order to model the storage tank dynamics, two stages are considered: when the PCM is in the sensible heat transmission state, the PCM behavior is modeled by a double-capacity model. In the phase change stage, the solution is based on the Stephan solution. In [27], a more complete description of the PCM storage tank is carried out. The subsection provides a brief description of the model. The model was published in [28].

3.2.1. Double Capacity Model

In this stage, the model consists of two different capacitive zones, with a thermal resistance between both of them. The r_e and r_i radius denote exterior and interior radius respectively, r_m is the separation radius between the two capacities zones, which is a parameter that has to be identified. T_1 and T_2 represent temperatures of zones 1 and 2, h is the convective heat transfer coefficient, K the conductivity, C_p the specific heat and T_∞ stands for the hot water temperature [27].

The model has two differential equations, one per zone:

Zone 1:

$$\rho C_p \pi (r_m^2 - r_i^2)\frac{dT_1}{dt} = 2h\pi r_i (T_\infty - T_1) - \frac{2\pi K(T_1 - T_2)}{ln(r_e/ri)}. \tag{3}$$

Zone 2:

$$\frac{2\pi K(T_1 - T_2)}{ln(r_e/ri)} = \rho C_p \pi (r_e^2 - r_m)\frac{dT_2}{dt}. \tag{4}$$

The aforementioned equations are valid for the liquid and solid phase. However, parameters such as the conductivity K and the density of the hydroquinone ρ may have different values.

3.2.2. Stefan Solution for Phase Change

When the PCM reaches a temperature of 170 °C, the hydroquinone reaches the melting point and the phase change stage starts. To model this stage, the liquid and solid phases are considered to be stable. The dynamics of the hydroquinone temperature is governed by the Stefan solution. This solution establishes an inferior limit of stored energy in a phase change phenomenon as well as a velocity limit for its evolution. Since the Stefan number given by the expression (5) is very small, finding a solution supposing a semi-infinite medium and that all the material is initially at the phase change temperature is possible [29,30].

$$S_T = \frac{C_p(T_f - T(r_i))}{L}. \tag{5}$$

The final expressions of the Stefan solution are given by Equations (6) and (7):

$$T(r) = T_f + \frac{h\,r_i}{K}\left(\frac{T_f - T_\infty \frac{h\,r_i}{K}ln(r_i/R)}{1 - \frac{h\,r_i}{K}ln(r_i/R)} - T_\infty\right)ln(r/R) \tag{6}$$

$$t_{st} = C\left(r_i\left[\frac{r_i h}{K} - 2\right] - R^2\left(\frac{2h}{K}ln(r_i/R) + \frac{h}{K} - \frac{2}{r_i}\right)\right)$$

$$C = \frac{\rho L^*}{4h(T_\infty - T_f)}, \tag{7}$$

where $R = R(t_{st})$ is the interface position which depends on t_{st} (Stefan time), T_f is the melting temperature, $T(r)$ is the PCM temperature which depends on the radius and L^* is the corrected latent heat of hydroquinone. In [27,28], all the modeling details and parameters estimation are better explained.

Figure 6 shows a comparison between the model and the real temperature of the storage tank evolution. At time 14.45 h the inlet valve opens and the hot water heats up the storage tank. As can be observed, the model matches well the real data with a maximum error of a 2.5%.

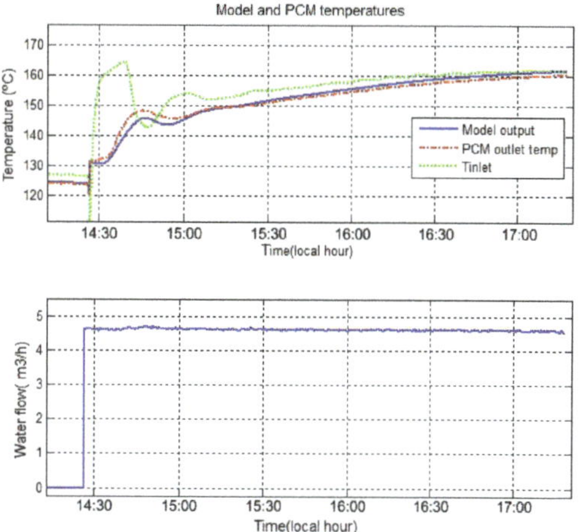

Figure 6. PCM temperature evolution: model vs. real.

3.3. Absorption Machine Model

The water absorption chiller consists of three parts, a high temperature generator, a refrigeration system and an evaporator [31,32]. Each component is modeled as a black-box using input-output data.

More complex thermodynamical models for absorption chillers exist in literature [33], but they are too complicated to be used for control purposes. The model developed in this paper is a simplified control model.

The three subsystems are described by a lumped parameter model with constant heat capacities. All the coefficients involved in the following equations were obtained using data from the real absorption machine. Figures 7 and 8 show a comparison between the model and the real water chiller evolution.

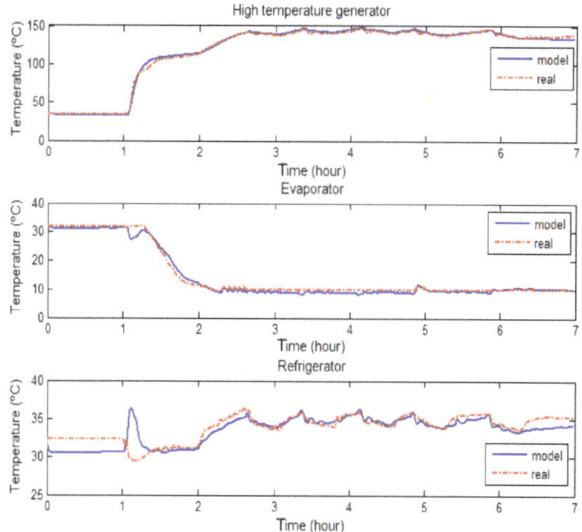

Figure 7. Absorption machine model vs. real: 19 July 2010.

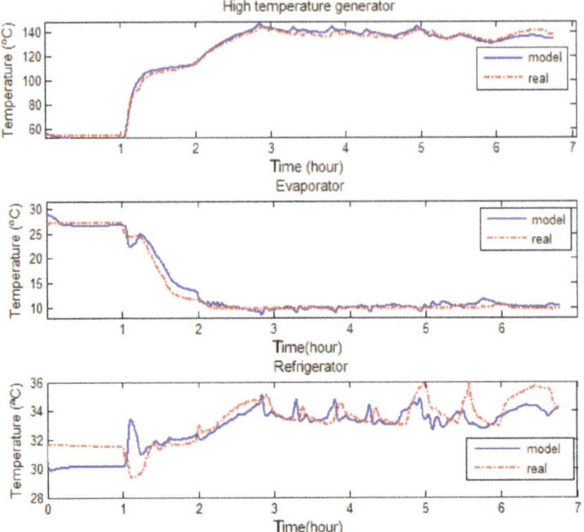

Figure 8. Absorption machine model vs. real: 21 July 2010.

3.3.1. High Temperature Generator

The energy balance equation describing the output temperature of the generator is given by Equation (8):

$$C_g \frac{dT_{ogen}}{dt} = Q_{cald} - Q_{gloss} + \rho_w \, q_h C_w (T_{ogen} - T_{igen}), \tag{8}$$

where T_{ogen} and T_{igen} represent the outlet and the inlet temperature of the high temperature generator, Q_{cald} is the thermal power supplied by the absorption machine burner (W), Q_{gloss} are the thermal losses (W), C_g is the generator heat capacity, ρ_w and C_w are the density and the specific heat of the water, and q_h is the generator water flow.

3.3.2. Evaporator

The energy balance equation describing the output temperature of the evaporator is given by Equation (9):

$$C_{ev} \frac{dT_{oevap}}{dt} = -Q_{ev} - Q_{ev_{loss}} + \rho_w \, q_{ev} C_w (T_{iev} - T_{oev}), \tag{9}$$

where T_{oev} and T_{iev} represent the outlet and the inlet temperature of the evaporator, Q_{ev} is the cooling power supplied by the absorption machine (W), $Q_{ev_{loss}}$ are the thermal losses (W), C_{ev} is the evaporator heat capacity, q_{ev} is the evaporator water flow.

3.3.3. Refrigerator

The energy balance equation describing the output temperature of the refrigerator is given by Equation (10):

$$C_{refr} \frac{dT_{orefr}}{dt} = Q_{refr} - Q_{refr_{loss}} + \rho_w \, q_{ref} C_w (T_{irefr} - T_{orefr}), \tag{10}$$

where T_{orefr} and T_{irefr} represent the outlet and the inlet temperature of the refrigerator, Q_{refr} is the thermal power dissipated by the refrigerator (W), $Q_{refr_{loss}}$ are the thermal losses (W), C_{refr} is the refrigerator heat capacity, q_{ref} is the refrigerator water flow.

4. Solar Field Outlet Temperature Regulator: Nonlinear Model Predictive Control Strategy

As previously mentioned in the introduction, one of the control objectives in solar plants is the regulation of the solar field outlet temperature around a desired set-point. The value of the set-point can be decided by the operator or by optimal criteria as explained in [34].

The regulation of the outlet temperature of a solar collector field around a set-point is hindered by the effect of multiple disturbance sources and its dynamics are greatly affected by the operating conditions [35]. This fact means that conventional linear control strategies do not perform well throughout the entire range of operations. In general, adaptive, robust or nonlinear schemes are needed to cope with the highly nonlinear dynamics of a solar field, especially at low flow levels. In [36] a practical nonlinear MPC is developed. In [37] a nonlinear continuous time generalized predictive control (GPC) is presented and simulation results are shown. In [38], an improvement of a gain scheduling model predictive controller is proposed and tested on a model of the ACUREX solar field.

Concerning the application of control strategies to Fresnel collector fields, there are several works in the literature. For example, in [39] a sliding model predictive control based on a feedforward compensation is developed for a Fresnel collector field and tested on a nonlinear model of a Fresnel collector field. In [10] a gain scheduling generalized predictive controller was developed and tested for a Fresnel collector field.

In this paper, a nonlinear model predictive control strategy is implemented to control the outlet temperature of the Fresnel collector field.

Model Predictive Control Strategy

An MPC control strategy consists of the following three steps [40,41]: in the first place, a model to predict the process evolution depending on a given control sequence is used. Then, it computes the control sequence by minimizing an objective function. Finally, only the first element of the computed control sequence is applied (receding horizon strategy).

The difference in MPC control strategies is usually given by the model used to predict the future evolution of the system. If linear models are used, the resulting MPC problem is a quadratic programming problem which is easily solvable and the optimum is ensured. If the model used is nonlinear, the resulting nonlinear optimization problem is computationally harder to solve and reaching the global optimum is not, in general, ensured [42].

In this paper, the model used is nonlinear. The model is a simplification of the Equation (2), considering four segments (four segments for fluid and metal temperatures) instead of 64 segments used in the full distributed parameter model. This simplification is required to alleviate the computational burden of the resulting problem although a precision loss is produced.

In general, the mathematical expression of the MPC problem can be posed as follows:

$$\min_{\Delta u} J = \sum_{t=1}^{N_p} \left(y_{k+t|k} - y_{k+t}^{ref} \right)^{\mathsf{T}} \left(y_{k+t|k} - y_{k+t}^{ref} \right) + R_u \sum_{t=0}^{N_c-1} \Delta u_{k+t|k}^{\mathsf{T}} \Delta u_{k+t|k} \tag{11}$$

such that

$$
\begin{aligned}
y_{k+t|k} &= f(\Delta u, y_{k+t-1}, y_{k+t-2}, \ldots) \\
u_{k+t|k} &= u_{k+t-1|k} + \Delta u_{k+t|k} \\
u_{\min} &\leq u_{k+t|k} \leq u_{\max} \\
t &= 0, \ldots, N_p - 1
\end{aligned}
\tag{12}
$$

where N_p and N_c are the prediction and the control horizons respectively. The parameter λ penalizes the control effort. Then $u_k \equiv u_{k|k}$ is applied to the system. In this case, constraints in the amplitude of the water flow and the maximum increment per iteration are considered. The sampling time of the control strategy is chosen as 20 s.

Due to the fact that only the inlet and outlet temperature are measurable, the intermediate fluid temperatures and the four metal segments temperature have to be estimated. Since the mathematical model used to predict the future evolution is nonlinear, an unscented Kalman filter is used to estimate the value of the non-measurable states.

The Kalman filter is a widely used tool for state estimation in linear systems [43]. The Kalman filter has been extended to nonlinear state estimation through algorithms such as extended Kalman filter (EFK) or unscented Kalman filter (UKF) [44]. The UKF is based on the unscented transformation, which represents a method for calculating the mean and covariance of a random variable that undergoes a nonlinear transformation [45,46]. For further information the reader is referred to [44].

5. Hybrid Model of the Solar Cooling Plant: Operation Modes

In this section the hybrid model of the solar cooling plant described in this paper, is presented.

As has been mentioned previously, the solar cooling plant can work in different operation modes. The selection of multiple operation modes will be done by a fuzzy algorithm. A particular operation mode can be defined as a particular configuration of the subsystems, valves and pumps which compose the plant (Figure 1). The choice of working in a particular operation mode is determined by the cooling demand, the state of the plant and weather conditions.

The state of the plant is described by integer (for instance, valves position) and real variables (temperatures of subsystems, water flow...). This fact leads to the need of a hybrid description of the system. In order to implement advanced control strategies, a mixed integer programming will be required if the control algorithm has to decide the operation mode [47]. Although there are efficient mixed integer programming algorithms in literature [48], they require more computational resources than those available in plant controllers. Furthermore, the complexity of solving these optimization problems increases greatly with the number of integer variables. If a high number of variables is considered, the optimization problem may be very difficult to solve in real time.

An alternative approach is selecting the operation modes by a fuzzy algorithm based on a series of membership functions obtained using data corresponding to typical operation days. The algorithm is tuned in order to fulfill two conditions: (a) the transition between modes has to be smooth, that is, fast commutations between modes are not allowed and (b) only feasible transitions between the modes are possible. Those transitions are imposed as a constraints and have been designed by the experience of the plant operators.

The operation modes are listed and briefly described below:

- Mode 0: the solar field charges the PCM storage tank. There is no cooling demand and the water chiller is off.
- Mode 1: the solar field is connected to the absorption machine through the PCM tank. This mode is useful in two situations: the storage tank can provide energy if the solar field is not able to heat the water up to the required temperature for powering the chiller. The second situation occurs when there is an excess of solar radiation: the solar field can feed the absorption machine and load the PCM tank.
- Mode 2: the PCM storage tank powers the water chiller while the solar field is heated up by recirculating water.
- Mode 3: the tank is used for heating up the piping system. It can be useful early in the morning for a quick start-up.
- Mode 4: the tank supplies hot water to the water chiller if solar collectors cannot work. For instance, in a cloudy day when solar radiation is nil and the storage tank is charged.
- Mode 5: the water chiller is powered only by the solar collectors.
- Mode 6: water is recirculated in the solar field circuit. The absorption machine uses natural gas.

Although, there are many possible transitions among the modes, only a small group of transitions are feasible. For example, supposing that the plant is working in mode 5 and a sudden decrease of radiation levels cools down the solar field temperature below the minimum required by the water chiller. In this case, if the tank has energy available, the most reasonable decision is operating in mode 1. In the next section, the transition between different operation modes is explained.

6. Transition between Operation Modes

In this section, transitions between the operation modes are described. Only modes 0–2, 5 and 6 are considered to be selected automatically by the hybrid algorithm. The initial mode is considered to be mode 6. Modes 3 and 4 are selected by the operator and deal with special cases in the plant operation. The preheating of the pipes is a operation needed before the operation starts, and it is not a possible state in a normal operation.

Not every transition among operation modes are allowed. The criterion for choosing the target operation mode is that the opening and closing of the plant valves should be smooth: including or excluding more than one subsystem in each transition is not desirable. the possible transitions are showed in Table 2:

Table 2. Allowed transitions between operation modes.

Current \ Target	M. 0	M. 1	M. 2	M. 5	M. 6
M.0	X	X	T_{0-2}	T_{0-5}	T_{0-6}
M.1	T_{1-0}	X	T_{1-2}	T_{1-5}	X
M.2	T_{2-0}	T_{2-1}	X	X	T_{2-6}
M.5	T_{5-0}	T_{5-1}	X	X	T_{5-6}
M.6	T_{6-0}	T_{6-1}	X	T_{6-5}	X

The transitions depend on the plant variables: outlet temperature of the solar field, energy stored in the PCM tank, cooling demand, environmental conditions etc.

Several points are worth pointing out:

- The minimum inlet temperature allowed for correct operation is about 135 °C. This should be taken into account because thermal losses exist in the pipe connecting the solar field to the water chiller.
- The storage tank can proportionate energy for 1 h approximately when it is fully charged. The storage tank should be used only in the case when the amount of stored energy is high or the solar field is able to charge it.
- When solar radiation is high and the solar field is able to charge the storage tank and power the absorption machine, it should be done only if the temperature of the tank is high enough to ensure that the inlet temperature of the absorption machine is adequate for correct operation. For instance, if the outlet temperature of the solar field is 160 °C and the tank temperature is 120 °C, the use of mode 1 is not correct, because the temperature of the tank is smaller than that required for the absorption machine.
- The maximum absolute difference between the outlet temperature of the solar field and the storage tank temperature must be inferior to 30 °C, if the storage tank is to be used.
- In order to avoid very fast commutations of the modes, the fuzzy algorithm computes the adequate operation mode every 120 s.

The experience of operation of the plant establishes that there are conditions for some transitions, clearly defined by rules that handle limit values of variables that would be the following, presented below, where the temperature of the outlet temperature of the solar field as T_{capt}, the temperature of the storage tank is denoted by T_{tank}, the current mode is CM, which takes values between 0 and 6, and the cooling demand is CD, which takes the value of 1 if there is cooling demand and 0 if there is not. $T_{tankOUT}$ is the storage tank outlet temperature and $T_{tankMAX} = 180\,°C$, the storage tank maximum temperature.

Set of rules 1:

- IF $CM = 6$ AND $CD = 0$ AND $T_{tank} < T_{tankMAX}$ AND $6\,°C < T_{capt} - T_{tank}$ AND $T_{capt} - T_{tank} < 30\,°C$ THEN select mode 0;
- IF $CM = 5$ AND $CD = 0$ THEN select mode 6;
- IF $CM = 5$ AND $CD = 1$ AND $139\,°C < T_{capt}$ AND $T_{capt} < 140\,°C$ AND $150\,°C < T_{tank}$ AND $T_{tank} < 155\,°C$ AND $T_{capt} - T_{tank} \leq 30\,°C$ THEN select mode 1;
- IF $CM = 5$ AND $CD = 1$ AND $T_{capt} < 139\,°C$ AND $165\,°C \leq T_{tank}$ THEN select mode 1;
- IF $CM = 0$ AND $CD = 0$ AND $T_{capt} - T_{tank} < -30\,°C$ THEN select mode 6;
- IF $CM = 0$ AND $CD = 0$ AND $T_{capt} - T_{tank} > 30\,°C$ THEN select mode 6;
- IF $CM = 0$ AND $CD = 0$ AND $T_{tank} > T_{tankMAX}$ THEN select mode 6;
- IF $CM = 0$ AND $CD = 0$ AND $T_{capt} - T_{tank} < 0\,°C$ THEN Select mode 6;
- IF $CM = 0$ AND $CD = 1$ AND $T_{capt} - T_{tank} < 0\,°C$ THEN Select mode 6;
- IF $CM = 1$ AND $CD = 0$ THEN select mode 0;
- IF $CM = 1$ AND $CD = 1$ AND $T_{tank} < 135\,°C$ THEN select mode 5;
- IF $CM = 1$ AND $CD = 1$ AND $-3\,°C < T_{capt} - T_{tank} < 3\,°C$ THEN select mode 5;
- IF $CM = 2$ AND $CD = 0$ THEN select mode 0;
- IF $CM = 2$ AND $CD = 1$ AND $T_{tankOUT} < 135\,°C$ THEN Select mode 6;

These rules work with variables whose limits are well defined. However, on the other hand, operators work with other rules that are activated with values of some variables which require a certain threshold, depending on the transition, such as:

Set of rules 2:

- IF $CM = 0$ AND $CD = 1$ AND $T_{capt} > 140\,°C + U_1$, THEN select mode 5;
- IF $CM = 1$ AND $CD = 1$ AND $T_{capt} < 140\,°C + U_1$ AND $T_{tank} < T_{tankMAX}$ AND $T_{capt} - T_{tank} < 30\,°C$ OR $T_{tank} < 135\,°C$ OR $|T_{capt} - T_{tank}| < 3\,°C$, THEN select mode 5;
- IF $CM = 6$ AND $CD = 1$ AND $T_{capt} > 140\,°C + U_2$, THEN select mode 5;
- IF $CM = 6$ AND $CD = 1$ AND $130\,°C < T_{capt} < 140\,°C + U_2$ AND $T_{tank} > 140\,°C + U_3$ AND $T_{capt} - T_{tank} < 30\,°C$, THEN select mode 1;
- IF $CM = 5$ AND $[CD = 1$ AND $T_{capt} - T_{tank} <= 30\,°C$ AND $T_{capt} > 140\,°C + U_3$ AND $[[T_{tank} < T_{tankMAX}$ AND $T_{tank} < T_{capt}$ AND $T_{tank} \geq 135\,°C$ AND $|T_{capt} - T_{tank}| \geq 5]$ OR $[T_{capt} < 140\,°C$ AND $T_{capt} > 139\,°C$ AND $150\,°C < T_{tank} < 155\,°C]$, THEN Select mode 1;
- IF $CM = 5$ AND $CD = 1$ AND $T_{capt} < 139\,°C$ AND $T_{tank} > 140\,°C + U_4$ AND $T_{tank} \geq 165\,°CC$, THEN select mode 1;
- IF $CM = 5$ AND $CD = 1$ AND $T_{capt} < 135\,°C$ AND $T_{tank} < 140\,°C + U_4$, THEN select mode 6;
- IF $CM = 0$ AND $CD = 1$ AND $T_{capt} > 140\,°C + U_5$ AND $|T_{capt} - T_{tank}| > 30\,°C$, THEN select mode 1;
- IF $CM = 0$ AND $CD = 1$ AND $T_{capt} < 135\,°C$ AND $T_{tank} > 140\,°C + U_6$, THEN select mode 2;
- IF $CM = 1$ AND $CD = 1$ AND $T_{capt} < 135\,°C$ AND $T_{tank} < 140\,°C + U_7$, THEN select mode 0;
- IF $CM = 1$ AND $CD = 1$ AND $T_{capt} < 135\,°C$ AND $T_{tank} > 140\,°C + U_7$, THEN Select mode 2;
- IF $CM = 2$ AND $CD = 1$ AND $T_{capt} > 140\,°C + U_8$ AND $T_{tank} < T_{tankMAX}$ AND $|T_{capt} - T_{tank}| < 30\,°C$, THEN select mode 1;

where $U_1, ..., U_8$ are thresholds used by operators. A fuzzy system that integrates these last rules has been implemented. To measure the membership of any value x to the sets $x > A$ and $x < A$, spline-based S-shaped and Z-shaped membership functions have been chosen respectively:

$$S(x,a,b) = \begin{cases} 0 & x \leq a \\ 2\left(\frac{x-a}{b-a}\right)^2 & a \leq x \leq \frac{a+b}{2} \\ 1 - 2\left(\frac{x-b}{b-a}\right)^2 & \frac{a+b}{2} \leq x \leq b \\ 1 & x \geq b \end{cases}$$

$$Z(x,a,b) = \begin{cases} 1 & x \leq a \\ 1 - 2\left(\frac{x-a}{b-a}\right)^2 & a \leq x \leq \frac{a+b}{2} \\ 2\left(\frac{x-b}{b-a}\right)^2 & \frac{a+b}{2} \leq x \leq b \\ 0 & x \geq b \end{cases}$$

To measure the membership of the set $A < x < B$, trapezoidal membership functions $T(x,a,b,c,d)$ have been used (Figure 9):

$$T(x,a,b,c,d) = max\left(min\left(\frac{x-a}{b-a},1,\frac{d-x}{d-c}\right),0\right)$$

Membership functions have been chosen without taking into account the aforementioned thresholds and an adjustment method, based on genetic algorithms [49], has been used to position and shape the functions. A vector k has been established, whose coordinates, added to each one of the parameters of the membership functions, will give new shape and position to them. The evolutionary algorithm will find the vector k, using plant operation data for this.

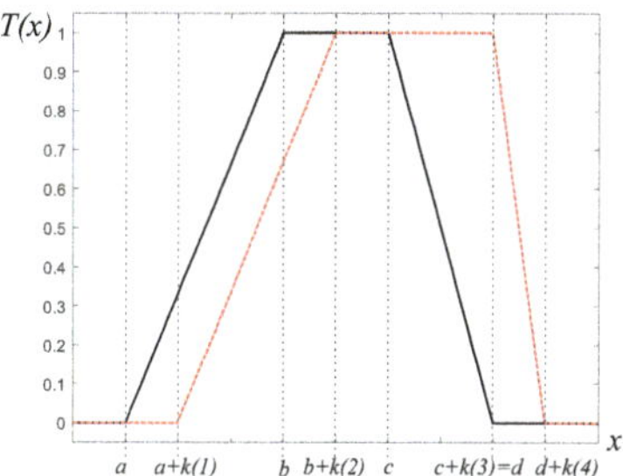

Figure 9. Trapezoidal membership function adjustment.

Five fuzzy inference systems (FIS) with the same antecedent structure have been designed and adjusted to evaluate whether modes 0–2, 5 and 6 are activated or not (FIS0, FIS1, FIS2, FIS5 and FIS6). The systems have five inputs: CM, CD, T_{capt}, T_{tank} and $T_{capt} - T_{tank}$ and the membership functions, after the genetic algorithm result are shown in Figures 10–14. The output variable of each FIS has one membership function with a single-valued constant (singleton membership function at 1).

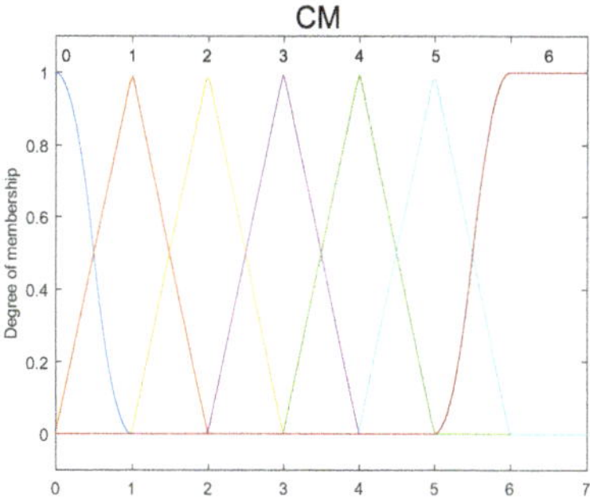

Figure 10. Inputs memberships functions of variable current mode (CM).

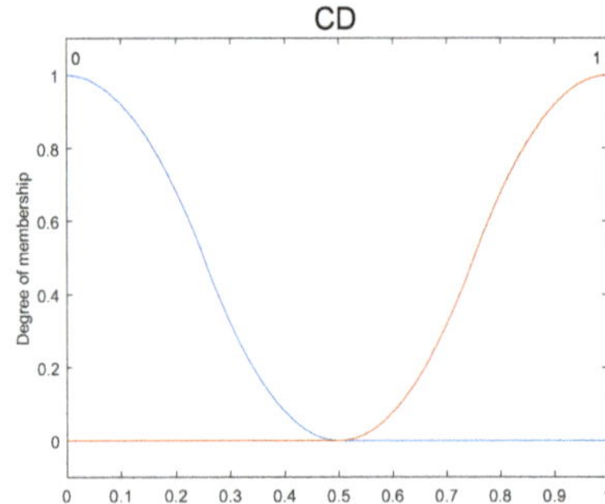

Figure 11. Inputs memberships functions of variable cooling demand (CD).

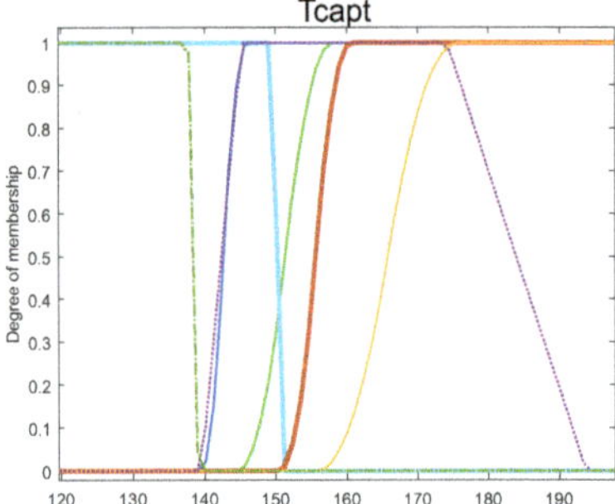

Figure 12. Inputs memberships functions of variable T_{capt}.

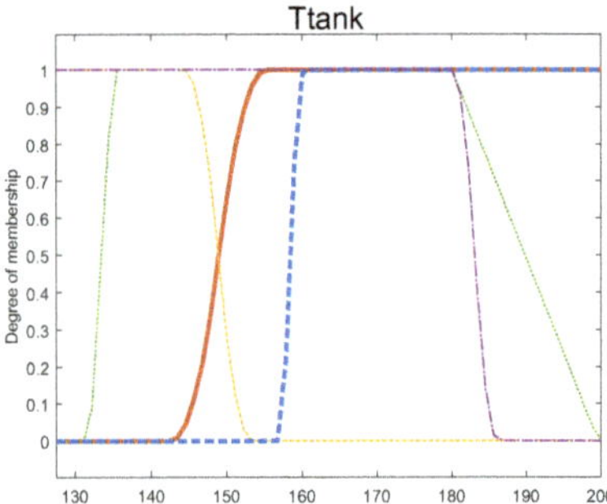

Figure 13. Inputs memberships functions of variable T_{tank}.

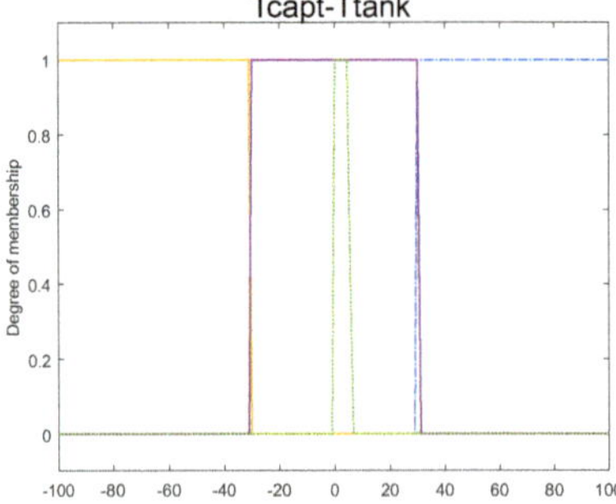

Figure 14. Inputs memberships functions of variable $T_{capt} - T_{tank}$.

The algorithm is the following (Algorithm 1)

Algorithm 1: Fuzzy algorithm

```
Loop
   input=[CM,CD, Tcapt, Ttank, Tcapt-Ttank]; modef(1)=evaluateFIS(input,FIS0);
   modef(2)=evaluateFIS(input,FIS1);
   modef(3)=evaluateFIS(input,FIS2);
   modef(4)=evaluateFIS(input,FIS3);
   modef(5)=evaluateFIS(input,FIS4);
   if (CM=6)&(CD=0) then
      if modef(1)=1 then
         mode=0;
      end
   end
   if (CM=6)&(CD=1) then
      if modef(2)=1 then
         mode=1;
         else if modef(4)=1 then
            mode=5;
         end
      end
   end
   if (CM=5)&(CD=0) then
      if (modef(5)=1) then
         mode=6;
      end
   end
   if (CM=5)&(CD=1) then
      if (modef(3)=1) then
         mode=2;
      end
      else if modef(2)=1 then
         mode=1;
      end
      else if modef(5)=1 then
         mode=6;
      end
   end
   if (CM=0)&(CD=0) then
      if (modef(5)=1) then
         mode=6;
      end
   end
```

Algorithm 1: *Cont.*

```
  if (CM=0)&(CD=1) then
      if (modef(2)=1) then
          mode=1;
      end
      else if modef(3)=1 then
          mode=2;
      end
      else if modef(4)=1 then
          mode=5;
      end
      else if modef(5)=1 then
          mode=6;
      end
  end
  if (CM=1)&(CD=0) then
      if (modef(1)=1) then
          mode=0;
      end
      if (CM=1)&(CD=1) then
          if (modef(1)=1) then
              mode=0;
          end
          else if modef(3)=1 then
              mode=2;
          end
          else if modef(4)=1 then
              mode=5;
          end
      end
  end
  if (CM=2)&(CD=0) then
      if (modef(1)=1) then
          mode=0;
      end
  end
  if (CM=2)&(CD=1) then
      if (modef(2)=1) then
          mode=1;
      end
      else if modef(5)=1 then
          mode=6;
      end
  end
  Evaluate(set of rules);
end
```

7. Simulations and Results

In this section, simulations testing the hybrid algorithm proposed in the previous section are presented. The data corresponding to solar radiation was taken from the real solar field.

Figure 15 shows a clear day where the storage tank had a temperature about 110 °C. The initial operation mode was mode 6 in order to recirculate and heat up the water. The initial temperature reference was about 145 °C. Once the outlet temperature of the solar field reaches an adequate value to feed the absorption machine (higher than 135 °C), the fuzzy algorithm commuted to mode 5, where the solar field was able to power the absorption machine. From 14 h onwards, the temperature reference consisted of a series of increasing steps. As can be seen, the temperature regulator properly tracked these references. This was the normal operation in clear days, where the solar field was sufficient to power the water chiller.

Figure 16 depicts the results of a day where some clouds affect the solar field. From 12.5 h to 12.9 h approximately, the solar field is in recirculation mode (mode 6) accumulation energy and increasing the outlet temperature. The reference temperature for the outlet temperature was 10 °C above the inlet temperature. This was done to maintain an approximately constant thermal jump without decreasing much the water flow. As can be seen, the MPC controller tracks the reference properly in spite of the radiation disturbances.

At 12.9 h the outlet temperature reached a value higher than 135 °C and the fuzzy algorithm commuted to mode 5. At 12.95 h the solar field reaches a temperature about 150 °C which makes not only powering the water chiller possible, but also accumulating energy in the storage tank. From 12.95 h onwards, the outlet temperature of the solar field was increased and the storage tank was charged (dashed green line). The fuzzy algorithm commutes between modes 5 and 1 depending on if the outlet temperature of the solar field and the storage temperature were close enough or not (approximately 4 °C). The main reason to do this was that the temperature reaching the PCM tank was about 4 °C lesser than the outlet temperature of the solar field due to the thermal losses in the pipe. Only when the outlet temperature of the solar field was higher than 4 °C than that of the storage tank, it makes sense charging it.

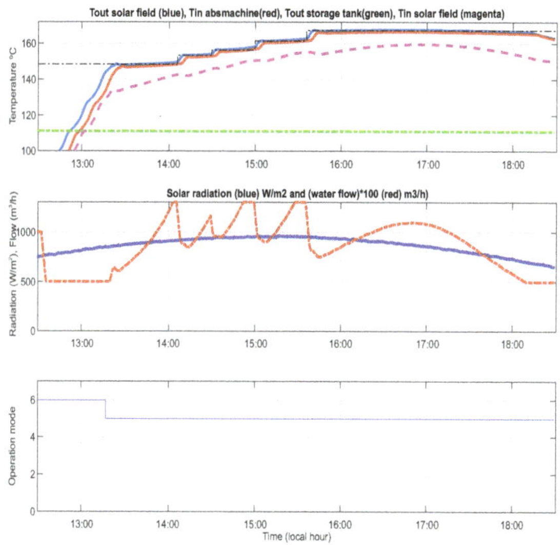

Figure 15. Operation modes in a clear day.

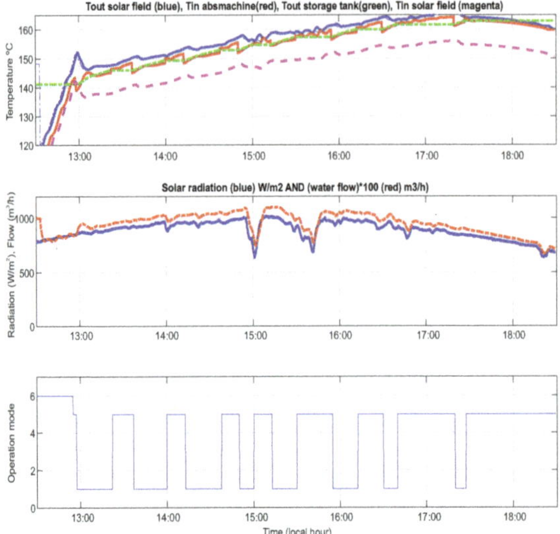

Figure 16. Operation modes in a clear day: storage tank charging.

Figure 17 shows a day with scattered clouds. At the beginning of the day, the absorption machine functions burning natural gas and the operation mode was mode 6. The water recirculates through the solar collectors to be heated. At 12.9 h the water reached 145 °C and the operation mode commuted to mode 5. The solar field powered the absorption machine. At 12.95 h the mode commutes to mode 1 in order not only to power the absorption machine but to accumulate energy in the storage tank. From 13 h to 14.4 h, the temperature reference increases to accumulate more energy in the storage tank.

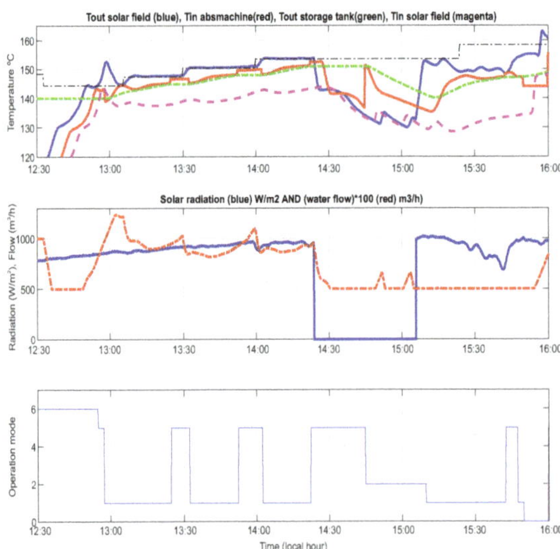

Figure 17. Operation modes in a cloudy day.

At 14.4 h clouds affecting the solar field appear. From 14.4 h to 14.7 h, the water flow is decreased to maintain the temperature of the solar field and the operation mode is 5 because the outlet temperature is higher than 135 °C. At 14.7 h, since there was no energy in the solar field but the storage tank possessed energy, the fuzzy algorithm commuted to mode 2: the solar field was in recirculation and the water chiller is powered by the tank only. This mode is maintained until 15.15 h, when the solar field outlet temperature increases again above 135 °C and it is capable to powers the absorption machine and charging the PCM tank. Thus, the fuzzy algorithm commutes to mode 1.

Finally, in order to test the fuzzy algorithm, from 15.8 to 16 h it was considered that no cooling demand exists. Since no cooling demand existed, the operation mode commuted to 0 and the solar field was in recirculation mode.

As can be seen, the set MPC+fuzzy algorithm has performed adequately in all the tests. This approach of using two independents algorithms has demonstrated to be very effective dealing with the operation of a hybrid system such as the solar cooling plant. The final outcome can be considered satisfactory.

8. Conclusions

Solar energy for cooling systems has been widely used to fulfill the growing air conditioning demand. The advantage of this approach is based on the fact that the need of air conditioning is usually well correlated to solar radiation.

This kind of plants can work in different operation modes resulting on a hybrid system. The control approaches designed for this kind of plant have usually a twofold goal: (a) regulating the outlet temperature of the solar collector field and (b) choose the operation mode. Since the operation mode is defined by a set of valves positions (discrete variables), the overall control problem is a nonlinear optimization problem which involves discrete and continuous variables. This problems are difficult to solve within the normal sampling times for control purposes (around 20–30 s).

A two layer control strategy is proposed in this paper. The first layer is a nonlinear model predictive controller for regulating the outlet temperature of the solar field. The second layer is a fuzzy algorithm which selects the adequate operation mode for the plant taken into account the operation conditions. The set MPC+fuzzy algorithm has performed adequately in all the tests. This approach of using two independents algorithms has demonstrated to be very effective dealing with the operation of a hybrid system such as the solar cooling plant. For the simulations, the data corresponding to solar radiation has been taken from the real solar field. Three different scenarios have been considered: clear day, clear day where some clouds affect the solar field and cloudy day. The control strategy is able to perform the normal operation in clear days, where the solar field is sufficient to power the water chiller. Where some clouds affect the solar field, the MPC controller tracks the reference properly in spite of the radiation disturbances. The system commutes between operating modes using the appropriate logic, contained in the operating rules of the fuzzy algorithm. The final outcome can be considered satisfactory.

Author Contributions: Conceptualization: E.F.C. and A.J.G.; Methodology: E.F.C. and A.J.G.; Software: A.J.G. and J.M.E.; Validation: A.J.G., J.M.E. and A.J.S.; Investigation: A.J.G., J.M.E. and A.J.S. Supervision: E.F.C. Writings and Review: A.J.G., J.M.E. and A.J.S.

Funding: The authors want to thank the European Commission for funding this work under the Advanced Grant OCONTSOLAR (Project ID: 789051).

Conflicts of Interest: The authors declare no conflict of interest.

Abbreviations

The following abbreviations are used in this manuscript:

MPC	Model predictive controller
GPC	Generalized predictive control
GS-GPC	Gain-scheduling generalized predictive controller
UKF	Unscented Kalman filter
PCM	Phase change material
PDE	Partial differential equations
FIS	Fuzzy inference system
PLC	Programmable Logic controller

References

1. Camacho, E.F.; Berenguel, M.; Rubio, F.; Martínez, D. *Control of Solar Energy Systems*; Springer: London, UK, 2012.
2. Camacho, E.F.; Samad, T.; Garcia-Sanz, M.; Hiskens, I. *Control for Renewable Energy and Smart Grids*; Technical Report; IEEE Control Systems Society: New York, NY, USA, 2011.
3. Helios I. 2018. Available online: https://solarpaces.nrel.gov/helios-i (accessed on 7 December 2019).
4. Mojave Solar Project. 2018. Available online: https://solarpaces.nrel.gov/mojave-solar-project (accessed on 7 December 2019).
5. Solana Generating Station. 2018. Available online: https://solarpaces.nrel.gov/solana-generating-station (accessed on 7 December 2019).
6. Sonntag, C.; Ding, H.; Engell, S. Supervisory Control of a Solar Air Conditioning Plant with Hybrid Dynamics. *Eur. J. Control* **2008**, *6*, 451–463. [CrossRef]
7. Kima, D.; Infante-Ferreira, C. Solar refrigeration options a state-of-the-art review. *Int. J. Refrig.* **2008**, *31*, 3–15. [CrossRef]
8. Nkwetta, D.N.; Sandercock, J. A state-of-the-art review of solar air-conditioning systems. *Renew. Sustain. Energy Rev.* **2016**, *60*, 1351–1366. [CrossRef]
9. Bermejo, P.; Pino, F.J.; Rosa, F. Solar absorption cooling plant in Seville. *Sol. Energy* **2010**, *84*, 1503–1512. [CrossRef]
10. Gallego, A.J.; Monguio, G.; Berenguel, M.; Camacho, E.F. Gain-scheduling model predictive control of a Fresnel collector field. *Control Eng. Pract.* **2019**, *82*, 1–13. [CrossRef]
11. Camacho, E.F.; Gallego, A.J.; Sánchez, A.J.; Berenguel, M. Incremental State-Space Model Predictive Control of a Fresnel Solar Collector Field. *Energies* **2018**, *12*, 3. [CrossRef]
12. Zambrano, D.; Bordons, C.; Garcia-Gabin, W.; Camacho, E.F. A solar cooling plant: A benchmark for hybrid systems control. *IFAC Proc. Vol.* **2006**, *39*, 199–204. [CrossRef]
13. Zambrano, D.; Bordons, C.; Garcia-Gabin, W.; Camacho, E.F. Model development and validation of a solar cooling plant. *Int. J. Refrig.* **2008**, *31*, 315–327. [CrossRef]
14. Pasamontes, M.; Álvarez, J.D.; Guzmán, J.L.; Berenguel, M. Hybrid Modeling of a Solar Cooling System. *IFAC Proc. Vol.* **2009**, *42*, 26–31. [CrossRef]
15. Pasamontes, M.; Álvarez, J.D.; Guzmán, J.L.; Berenguel, M.; Camacho, E.F. Hybrid modeling of a solar-thermal heating facility. *Sol. Energy* **2013**, *97*, 577–590. [CrossRef]
16. Witheephanich, K.; Escano, J.M.; Gallego, A.J.; Camacho, E.F. Pressurized water temperature of a Fresnel collector field type cooling system using explicit model predictive control. In Proceedings of the IASTED Conference, Phuket, Thailand, 10–12 April 2013; doi:10.2316/P.2013.800-128.
17. Witheephanich, K.; Escano, J.M.; Bordóns, C. Control strategies of a solar cooling plant with Fresnel collector: A case study. In Proceedings of the 2014 International Electrical Engineering Congress (iEECON), Chonburi, Thailand, 19–21 March 2014.
18. Lemos, J.M.; Neves-Silva, R.; Igreja, J.M. *Adaptive Control of Solar Energy Collector Systems*; Springer: Basel, Switzerland, 2014.
19. Ross, T. *Fuzzy Logic with Engineering Applications*; John Wiley & Sons: Chichester, UK, 2004.
20. Yang, Z.; Wang, Z.; Su, W.; Zhang, J. Multi-mode control method based on fuzzy selector in the four wheel steering control system. In Proceedings of the IEEE ICCA 2010, Xiamen, China, 9–11 June 2010; pp. 1221–1226. [CrossRef]

21. Huchang, L.; Xungie, G.; Zeshui, X. A survey on decision making theory and methodologies of hesitant fuzzy linguistic term set. *Syst. Eng.-Theory Pract.* **2017**, *37*, 35–48.

22. Robledo, M.; Escano, J.M.; Núnez, A.; Bordons, C.; Camacho, E.F. Development and Experimental Validation of a Dynamic Model for a Fresnel Solar Collector. *IFAC Proc. Vol.* **2011**, *44*, 483–488. [CrossRef]

23. Spoladore, M.; Camacho, E.F.; Valcher, M.E. Distributed Parameters Dynamic Model of a Solar Fresnel Collector Field. *IFAC Proc. Vol.* **2011**, *44*, 14784–14790. [CrossRef]

24. Camacho, E.F.; Rubio, F.R.; Berenguel, M. *Advanced Control of Solar Plants*; Springer: London, UK, 1997.

25. Gallego, A.J.; Camacho, E.F. Estimation of effective solar radiation in a parabolic trough field. *Sol. Energy* **2012**, *86*, 3512–3518. [CrossRef]

26. Kreith, F.; Manglik, R.M.; Bohn, M.S. *Principles of Heat Transfer*, 7th ed.; Cengage Learning: Boston, MA, USA, 2011.

27. Ruíz-Pardo, A.; Salmerón, J.M.; Cerezuela-Parish, A.; Gil, A.; Álvarez, S.; Cabeza, L.F. Numerical simulation of a thermal energy storage system with PCM in a shell and tube tank. In Proceedings of the 12th International Conference on Energy Storage (InnoStock 2012), Lleida, Spain, 16–18 May 2012.

28. Gallego, A.J.; Ruíz-Pardo, A.; Martín-Macareno, C.; Cabeza, L.F.; Camacho, E.F.; Oró, E. Mathematical modeling of a PCM storage tank in a solar cooling plant. *Sol. Energy* **2013**, *93*, 1–10. [CrossRef]

29. Lunardini, V.J. *Heat Transfer in Cold Climates*; Van Nostrand Reinhold Company: New York, NY, USA, June 1981.

30. Naterer, G.F. *Heat Transfer in Single and Multiphase Systems*; CRC Press: Boca Raton, FL, USA, 2002; ISBN 978-0-8493-1032-4.

31. Prasartkaew, B. Mathematical Modeling of an Absorption Chiller System Energized by a Hybrid Thermal System: Model Validation. *Energy Procedia* **2013**, *34*, 159–172. [CrossRef]

32. Karimi, M.N.; Ahmad, A.; Aman, S.; Jamshed-Khan, M.D. A review paper on Vapor absorption system working on LiBr/H2O. *Int. Res. J. Eng. Technol.* **2018**, *5*, 857–864.

33. Grossman, G.; Wilk, M. Advanced modular simulation of absorption systems. *Int. J. Refrig.* **1994**, *17*, 231–244. [CrossRef]

34. Camacho, E.F.; Gallego, A.J. Optimal Operation in Solar Trough Plants: A case study. *Sol. Energy* **2013**, *95*, 106–117. [CrossRef]

35. Pin, G.; Falchetta, M.; Fenu, G. Adaptative time-warped control of molten salt distributed collector solar fields. *Control Eng. Pract.* **2007**, *16*, 813–823. [CrossRef]

36. Andrade, G.A.; Pagano, D.J.; Álvarez, J.D.; Berenguel, M. A practical NMPC with robustness of stability applied to distributed solar power plants. *Sol. Energy* **2013**, *92*, 106–122. [CrossRef]

37. Khoukhi, B.; Tadjine, M.; Boucherit, M.S. Nonlinear continuous-time generalized predictive control of solar power plant. *Int. J. Simul. Multidiscip. Design Optim.* **2015**, *A3*, 1–12. [CrossRef]

38. Alsharkawi, A.; Rossiter, J.A. Towards an improved gain scheduling predictive control strategy for a solar thermal power plant. *IET Control Theory Appl.* **2017**, *11*, 1938–1947. [CrossRef]

39. Lu, X.-J.; Dong, H.-Y. Application Research of Sliding Mode Predictive Control Based on Feedforward Compensation in Solar Thermal Power Generation Heat Collecting System. *Int. J. Hybrid Inf. Technol.* **2016**, *9*, 211–220.

40. Camacho, E.F.; Bordons, C. *Model Predictive Control*, 2nd ed.; Springer: London, UK, 2004.

41. Rawlings, J.; Mayne, D. *Model Predictive Control: Theory and Design*; Nob Hill Publishing, LLC: Madison, WI, USA, 2009.

42. Findeisen, R.; Allgöwer, F. *An Introduction to Nonlinear Model Predictive Control*; Technical Report; Institute for Systems Theory in Engineering, University of Stuttgart: Stuttgart, Germany, 2006.

43. Kalman, R. A new approach to linear filtering and predictions problem. *Trans. ASME J. Basic Eng.* **1960**, *82*, 35–60. [CrossRef]

44. Haykin, S. *Kalman Filtering and Neural Networks*; A Wiley-Interscience Publication: New York, NY, USA, 2001.

45. Romanenko, A.; Castro, J.A. The unscented Kalman filter as an alternative to the EKF for nonlinear state estimation: A simulation case study. *Comput. Chem. Eng.* **2004**, *28*, 347–355. [CrossRef]

46. Wang, Y.; Qiu, Z.; Qu, X. An Improved Unscented Kalman Filter for Discrete Nonlinear Systems with Random Parameters. *Discret. Dyn. Nat. Soc.* **2017**, *2017*, 1–10. [CrossRef]

47. Zambrano, D. Modelado y Control Predictivo Híbrido de una Planta de Refrigeración Solar. Ph.D. Thesis, Escuela Superior de Ingenieros de Sevilla, Sevilla, Spain, 2007.

48. Bertsekas, D.P. *Convex Analisys and Optimization*, 1st ed.; Athena Scientific: Nashua, NH, USA, April 2003.
49. Goldberg, D.E. *Genetic Algorithms in Search, Optimization, and Machine Learning*; Addison-Wesley: New York, NY, USA, 1989.

Article

A Feedback Control Loop Optimisation Methodology for Floating Offshore Wind Turbines

Joannes Olondriz [1,*], Josu Jugo [2], Iker Elorza [1], Santiago Alonso-Quesada [2] and Aron Pujana-Arrese [1]

[1] Ikerlan Technology Research Centre, Control and Monitoring Area. P°. J. M. Arizmendiarrieta 2, 20500 Arrasate-Mondragón, Spain; ielorza@ikerlan.es (I.E.); apujana@ikerlan.es (A.P.-A.)
[2] University of the Basque Country UPV/EHU, Electricity and Electronics Area. Bo. Sarriena s/n, 48940 Leioa, Spain; josu.jugo@ehu.es (J.J.); santiago.alonso@ehu.es (S.A.-Q.)
* Correspondence: jolondriz@ikerlan.es; Tel.: +34 943-712-400

Received: 1 July 2019; Accepted: 6 September 2019; Published: 10 September 2019

Abstract: Wind turbines usually present several feedback control loops to improve or counteract some specific performance or behaviour of the system. It is common to find these multiple feedback control loops in Floating Offshore Wind Turbines where the system perferformance is highly influenced by the platform dynamics. This is the case of the Aerodynamic Platform Stabiliser and Wave Rejection feedback control loops which are complementaries to the conventional generator speed PI control loop when it is working in an above rated wind speed region. The multiple feedback control loops sometimes can be tedious to manually improve the initial tuning. Therefore, this article presents a novel optimisation methodology based on the Monte Carlo method to automatically improve the manually tuned multiple feedback control loops. Damage Equivalent Loads are quantified for minimising the cost function and automatically update the control parameters. The preliminary results presented here show the potential of this novel optimisation methodology to improve the mechanical fatigue loads of the desired components whereas maintaining the overall performance of the wind turbine system. This methodology provides a good balance between the computational complexity and result effectiveness. The study is carried out with the fully coupled non-linear NREL 5-MW wind turbine model mounted on the ITI Energy's barge and the FASTv8 code.

Keywords: floating offshore wind turbine; aerodynamic platform stabiliser; wave rejection; feedback loop; control; optimisation

---

1. Introduction

Wind energy is becoming a real green energy solution to reduce the exhaust gases produced by the fossil fuel of conventional energy production plants. Actually, wind energy can be harvested in offshore water areas thanks to the bottom fixed and Floating Offshore Wind Turbine (FOWT) technologies (see Figure 1), providing many advantages in comparison to its onshore counterpart [1]. The available amount of offshore wind energy is enough to supply most of the worldwide energetic needs (https://carnegiescience.edu/node/2248, last accessed: 12 June 2019). The bottom fixed technology presents the limitation of being installed in water depths less than 30–40 m [2]. This limitation significantly reduces the available offshore areas worldwide for the installation of bottom-fixed wind turbines. Thus, the alternative solution to overcome this limitation is to mount the wind turbines on floating platforms, i.e., the FOWTs.

Several FOWT prototypes have been launched since the first Blue H in 2007, such as Hywind, WindFloat, SWAY, VolturnUS, Kabashima, Mitsui, Shimpu, Hamakaze, Floatgen and Hibiki [3,4], among others. Furthermore, after the Hywind Scotland Pilot Park [5], several floating wind farms are now currently in the developing stage: Kincardine in Scotland [6], Windfloat Atlantic in Portugal [7], and four more in France (Groix Belle Ile, Golfe du Lion, Eolmed, and Provence Grand Large). The growing trend in these sort of projects confirms the industry interest in the FOWT technology.

The FOWT technology is actually in the development stage and not yet mature; this means that there are many different floating platform designs, each one holding its strengths and weaknesses [3]. All these platform designs can be classified into three main groups, i.e., Spar-buoys (Spar), Tension Leg Platforms (TLP) and Semi-submersible platforms (Semi-sub), as can be seen in Figure 1. These different designs considerably differ in terms of building processes as well as in performance characteristics. In general, platforms with larger hydrostatic restoring stiffness and hydrodynamic damping cost more to build, mount and deploy, whereas less stable platforms can affect to the FOWT's performance [8]. In this sense, the control strategy implemented for above rated wind speed (Region III) in FOWTs is somehow conditioned by the platform design and performance. On the one hand, controllers implemented in platforms with high hydrodynamic stiffness can focus their duties on improving and optimising the overall system performance and reducing the mechanical loads of the components. On the other hand, controllers implemented in platforms with low hydrodynamic stiffness should mainly ensure the stability of the system due to the negative platform damping effect [9], and then, improve the performance of the system as much as possible without destabilising the system.

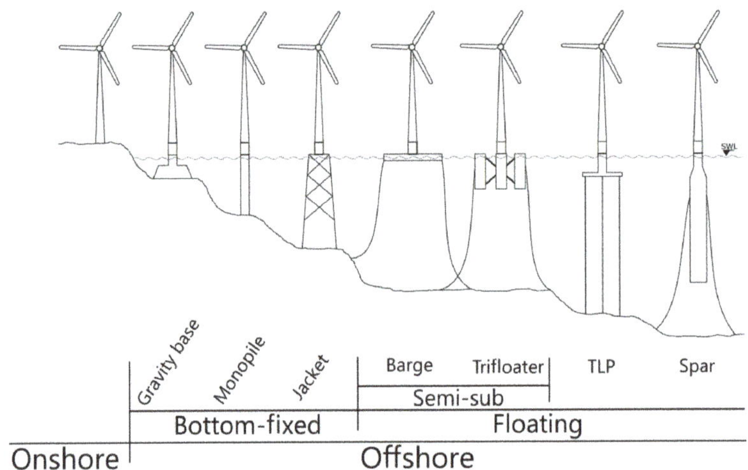

Figure 1. Illustrative representation of the main offshore wind turbine concepts.

Several optimal controllers have been used to achieve improved behaviour in FOWT systems according to the platform technology. The Model Predictive Control (MPC) is mainly used for Spar-type platform-mounted wind turbines as found in [10,11], among others, or for hybrid Spar platform such as the concrete torus [12], although one study for the Barge type platform is also found [13]. Similarly, the optimal Linear Quadratic Regulator (LQR) controller is widely used in Spar-type FOWT models as found in [14,15] or [16], among others, but some researches for TLP [17,18] and Barge [19,20] floating technologies are also found. Finally, the H_∞ is used for the Spar and TLP type FOWT systems, but not for Semi-sub ones, as found in [21,22] and in [23], respectively.

In practice, wind turbine manufacturers often favor simple controllers based on the Proportional and Integral (PI) feedback loops over more complex ones [24]. In previous studies, in combination with the conventional PI controller, the Aerodynamic Platform Stabiliser (APS) [25,26] and Wave Rejection (WR) [27] feedback control loops have been designed for above rated wind speed working region to reduce the platform-pitch motions, while improving the generator speed regulation and reducing the tower-base and blade-root loads of the NREL 5-MW wind turbine mounted on the ITI Energy barge. The APS control loop is designed to reduce the platform-pitch dynamics produced by the interactions of the conventional generator speed PI controller and platform dynamics while regulating the generator speed. Additionally, the WR control loop is designed to improve the APS control loop performance in rough sea states due to the excessive platform-pitch loads registered in some working conditions. The process followed to manually tune these control loops is: (1) the FOWT system is designed and modelled; (2) the FOWT model is linearised around the working points of interest; (3) the control loops are manually tuned through the loop shaping of the linearised transfer functions; and (4), if the designed control loops fulfil the time- and frequency-domains as well as the load performance analysis specifications, then (5) the manually tuned control loops can be optimised. This process is depicted by the flowchart in Figure 2.

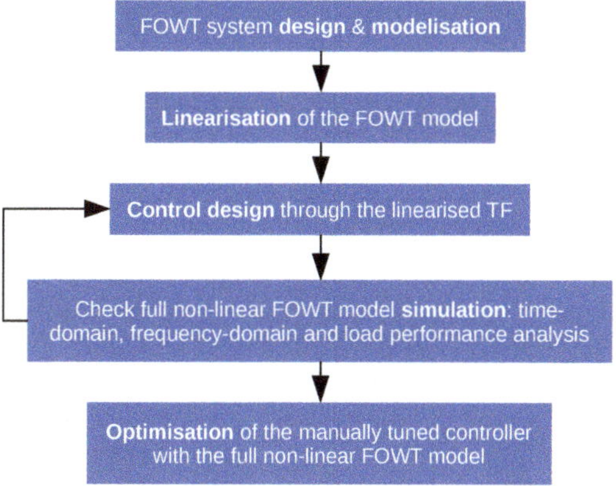

Figure 2. Process carried out for FOWT control design.

Both designed APS and WR control loops include several control parameters, so it is tedious to improve the performance of the FOWT system by manual tuning. Such a fact motivates us to apply an optimisation methodology to automatically find the control parameters which will improve the system performance. The proposed optimisation methodology in this manuscript is based on the Monte Carlo method [28]. This method is especially appropriate for systems with significant uncertainties in inputs and large numbers of coupled degrees of freedom, such as the FOWTs. The novelty of this method ensures the minimisation of the cost function through the computation of the wind turbine Damage Equivalent Loads (DEL). The randomised algorithm is useful for numerically solving the optimisation problem through the simple random search of the control parameters around the best obtained solution. The random search is based on the arbitrary selection of the parameters within a constrained region. Although other optimisation methodologies can be found in the literature, e.g., gradient descent [29] or genetic [30] algorithms, the Monte Carlo method is numerically simpler to implement taking into account the complexity of the FOWT non-linear system and cost function characteristics. Furthermore, the random behaviour of this method prevents from getting stuck in a locally optimal point while providing a more global optimal solution.

In regards to the conclusions collected in [27], after the implementation of the WR control loop, the need to improve the tower-base DEL was summarised. Therefore, the optimisation methodology proposed in this study quantifies the DEL of the FOWT simulation results to improve the tower-base DELs automatically tuning the APS and WR control loops . All the simulations have been carried out with FASTv8 [31].

This paper is organised as follows. Section 2 provides an overview of the manually tuned APS and WR control loops processes. Section 3 describes the optimisation methodology used for automatically improving the manually tuned controllers. Section 4 presents the results obtained with the optimised APS and WR controllers in comparison to the manually tuned ones and, to the conventional generator speed Detuned PI controller performance taken from [32] as a reference. Finally, Section 5 summarises the conclusions achieved during the development of this work as well as recommendations for future work.

2. Manually Tuned Controllers

The APS and WR feedback control loops have been independently tuned: first, the APS is tuned to reduce the platform-pitch dynamics due to the coupling between the thrust in the rotor of the wind turbine and the low hydrodynamic stiffness of the platform while regulating the generator speed; and second, the WR is tuned to improve the APS control loop performance due to the excessive platform-pitch loads registered in some working conditions, concretely, in rough sea states. The APS control loop measures the nacelle-pitch angular velocity, e.g., through an Inertial Measurement Unit (IMU), and the WR measures the blade-flapwise moments, e.g., through strain gauges or optical fibre, both control loops contributing to the blade-pitch angle regulation as can be seen in Figure 3.

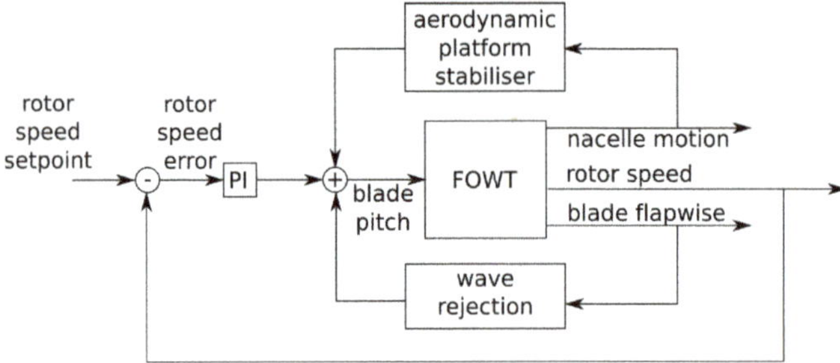

Figure 3. Conventional PI, Aerodynamic Platform Stabiliser and Wave Rejection feedback control loops.

The APS and WR feedback control loops have been designed following the procedure of Figure 4. First, the FOWT model is linearised around certain above rated wind speeds, such as in this study, every 2 m/s from 13 to 25 m/s. Since the ITI Energy barge model presents very low hydrodynamic stiffness the linearisation process has been carried out with the alternative linearisation strategy of trimming the generator torque explained in [33]. Next, the Single Input Single Output (SISO) transfer functions for the tuning process are selected: for the APS control loop, the transfer function from the blade-pitch to the nacelle-pitch angular velocity; and for the WR, from the blade-pitch to the blade-flapwise bending moment. Then, the loop shaping of each control loop is done in open-loop configuration as explained in [26] for the APS, and in [27] for the WR. The last step includes the verification of the tuned control loops in a closed-loop configuration from wind to platform-pitch and to generator speed, as shown in [26] and [27], respectively.

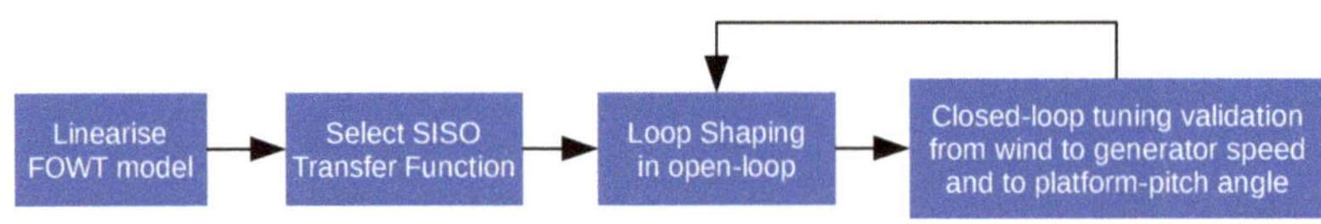

Figure 4. Process carried out for the APS and WR control loops manual tuning.

Once each closed-loop transfer function response shows an acceptable behaviour, the sensitivity of the system and the Power Spectral Density (PSD) of the platform-pitch angle as well as the generator speed are analysed. The sensitivity of the system is to be reduced as much as possible while improving the PSDs results. In this study, the results of the reference Detuned PI controller [32] must be improved. If the requirements of lowering system sensitivity and improving the PSDs are not fulfilled, then the feedback control loop process is restarted, taking into the account the aspects that must be improved.

An additional strategy for switching both feedback control loops has been implemented to increase the effectiveness of each control loop duty as explained in [27]. The control loop switching strategy must deal with the on-line adjustment of each control loop weighting factor depending on the sea state to maximise its effectiveness.

After several retuning iterations, the manually tuned feedback control loops fulfil the FOWT performance requirements: the platform-pitch dynamics have been improved while regulating the generator speed, and the quality of the harvested electric energy is better than that of the reference Detuned PI controller; and, the blade-root-flapwise and -edgewise fatigue loads show a reduction below the reference in all sea state conditions. However, the tower-base-pitch loads are still above the reference in some rough sea state conditions. Since there are two feedback control loops with several parameters to be modified, an optimisation methodology has been implemented to automatically obtain the control parameters. The objective of this optimisation is to reduce the tower-base-pitch DELs, while maintaining the blade-root ones below the reference and keeping the performance of the overall system.

3. Optimisation Methodology

Optimal control deals with the problem of finding a control law to achieve the optimum system performance for a given criterion. The proposed optimisation methodology optimises the manually tuned feedback APS and WR controllers analysing the simulation results and automatically finding the optimal control parameters according to the defined cost function. Concretely, the cost function is based on the quantification of the DELs of the tower-base-pitch and blade-root results. The automatic finding of the control parameters is based on the random search strategy, which arbitrarily selects the control parameters of the APS and WR loops. The random search starts from the manually tuned parameters and the search range is limited within a set of constrains. The constrains are defined during the manually tuning process, where one can learn within which values the behaviour of the designed controller does not work as desired. The controller is improved every time the cost function is minimised, starting from the initial value obtained with the manually tuned controller results.

The proposed optimisation methodology takes the pole, zero and static gains of the manually tuned feedback control loops and automatically changes their control parameter values minimising the predefined cost function. This method will be referred to as Pole and Zero Optimiser (PZO). The process for the PZO methodology is illustrated by the block diagram of Figure 5. The process is divided in two main parts: the initialisation, where the initial cost function value is computed with the manually tuned control parameters; and the iterative loop, where the control parameters are automatically changed to minimise the cost function. The optimisation process step by step is explained in detail next.

In the first step of the PZO process, the control parameters of the manually tuned controller are inserted. The parameters of the APS control loop are: the static gain (K_{APS}) and the real pole frequency (Real Pole$_{APS}$). The parameters of the WR control loop are: the static gain (K_{WR}), the complex pole frequency (Complex Pole$_{WR}$) and damping coefficient (ζ_{WR}), and the real zero frequency (Real Zero$_{WR}$). The static gain (K_c) of the control loop switching strategy is also included. In the second step, the simulations of the FOWT system are run, in this case with FASTv8 [31], for the following working conditions: wind speeds from 13 to 25 m/s, every 2 m/s; and significant wave heights from 0 to 6 m, every 1 m. After the simulations, the DEL of each working condition is computed. Then the partial terms of the cost function are defined: the mean and the maximum values of all the computed DELs (Equations (1) and (2), respectively). Since the main objective is to reduce the tower-base loads while maintaining the blade-root ones below the reference, the partial terms are repeated for all the measurement of blade-root-flapwise (M_{yB}), -edgewise (M_{xB}), and tower-base-pitch (M_{yT}) bending moments, given as:

$$J_{mean} = [mean(DEL_{M_{yB}}), mean(DEL_{M_{xB}}), mean(DEL_{M_{yT}})] \tag{1}$$

$$J_{max} = [max(DEL_{M_{yB}}), max(DEL_{M_{xB}}), max(DEL_{M_{yT}})] \tag{2}$$

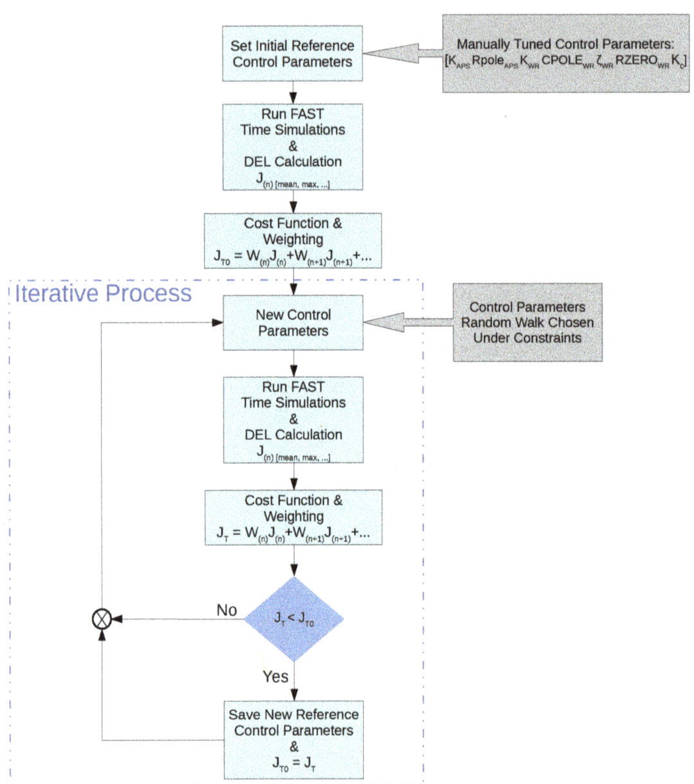

Figure 5. Process carried out for the manually tuned control parameters optimisation.

In the third step, the initial cost function is computed weighting (W) each partial term, as shown in Equation (3). The weighting of each partial term has been assigned according to the improvement interest, which has been heuristically chosen analysing the first PZO trials.

$$J = \sum_{i=1}^{3} \left[W_{mean(i)} \times J_{mean(i)} + W_{max(i)} \times J_{max(i)} \right] \tag{3}$$

Once the initial cost function with the manually tuned control parameters is computed, the iterative loop of the PZO process starts. In this step, the new control parameters are selected through a random search. In the first optimisation loop the randomly chosen parameters are around the manually tuned control solution and, for the next loops, the parameters selected by the random search are around the last optimised control solution. A random value α^i is added to the actual control parameters P^i (for $i = 1$ to the number of control parameters) to obtain the new ones P_r^i, as can be seen in Equation (4). All the parameters are within the set of constraints Λ^i that is defined with the knowledge acquired during the manually tuning process. During that process, one can learn between what range of control values the operation of the controller and FOWT system performance are acceptable.

$$P_r^i = P^i + \alpha^i, P^i \wedge P_r^i \in \Lambda^i \tag{4}$$

In the next steps, simulations are run with FASTv8 [31] for the same working conditions as during the initialisation part. The DELs and cost function are computed for these new selected control parameters. If the cost function value obtained with the new control parameters is smaller than the values of previous iterations then these control parameters as well as the current value of the cost function are saved. After that, as well as if the cost function has not been minimised, the process is repeated going back to the previous step where the control parameters are selected through the random search around the last saved control parameters. Then, the PZO iterative process continues as explained. The PZO process can be concluded when the cost function is not minimised after a number of interactions.

4. Result Discussion

The proposed PZO process has been implemented with Matlab code whereas FASTv8 and the feedback control loops have been implemented in Simulink [34]. In the first part of the Matlab code the manually tuned control parameters and the initial cost function are computed. In the second part, the iterative optimisation process of the FOWT system simulations and the selection of the randomly chosen control parameters are carried out. The FOWT simulations are run in Simulink with FASTv8 and the NREL 5-MW wind turbine mounted on the ITI Energy barge model, where the FASTv8's code has not been modified at all. The randomly chosen control parameters of the APS and WR control loops have been obtained with the rand command of Matlab. After simulations, the new cost function value and the comparison to the previous one are computed in Matlab.

The PZO process has been repeated a number of times, first to adjust the weighting of the cost function partial terms, and second, to verify the convergence of the optimisation process. The selected weights for the mean and max (W_{mean} and W_{max}) of the tower-base-pitch, blade-root-flapwise and -edgewise partial terms are 2, 1.2 and 1, respectively. The weight of each partial term have been keep constant throughout the optimisation process. The obtained cost function values during the iterative optimisation process are shown in Figure 6. One can see the stochastic cost function values obtained during the process due to the completely random input control values. The initial cost function value obtained with the manually tuned control parameters is minimised until the iteration 31. In the next iterations, until the optimisation process is interrupted, the cost function value has no longer been minimised. Considering the reduction of the minimisation rate in the last three minimisations and the lack of new minimised value in the last 34 iterations, it can be said that the chances for a new minimisation is unlikely. Otherwise, the minimisation rate will be so small that the impact on the system performance will be negligible considering the impact of the control parameters obtained in the last minimised iterations.

Note that since the control parameter values have several decimals, the number of possible solutions can be huge. However, from the FOWT performance point of view the variation of the least significant digits of the control parameters does not impact in the behaviour of the controller. Due to the stochastic nature of PZO process and the huge number of possible solutions, it may be a more optimal solution, even if the final FOWT operation shows hardly any improvements. Therefore, it can be said that the proposed PZO methodology is suboptimum. The manually tuned feedback control loop parameters, the optimised ones and the established constrains for the PZO process are summarised in Table 1.

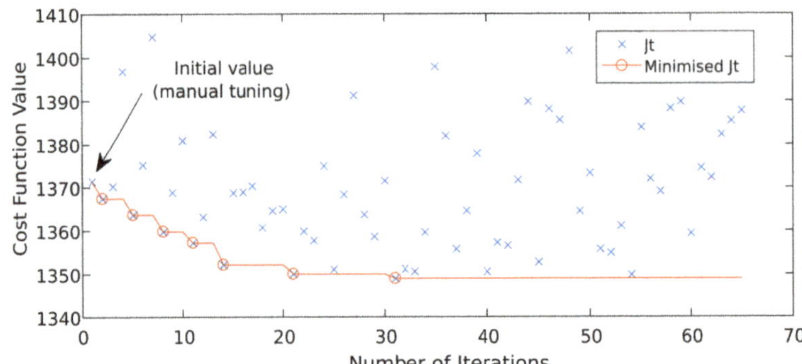

Figure 6. Cost function value results during the iterative optimisation process.

Table 1. Manually tuned and optimised control parameters values after applying the PZO method.

Control Parameter	Manually	Optimised	Constrains
K_{APS} (–)	7	3.6945	$2-8$
Real Pole$_{APS}$ (Hz)	0.00278	0.00380	$0.001-0.007$
K_{WR} (–)	7.6×10^{-7}	4.19×10^{-7}	$1\times10^{-7}-1\times10^{-6}$
Complex Pole$_{WR}$ (Hz)	0.0884	0.120	$0.05-0.3$
ζ_{WR} (–)	1.00	0.19	$0.1-1.0$
Real Zero$_{WR}$ (Hz)	0.0035	0.00731	$0.001-0.1$
K_c (–)	2	1.75	$0.5-3$

All the simulations have been carried out with FASTv8 [31] and the fully non-linear FOWT model, with all Degrees Of Freedom (DOF) enabled [34]. The stochastic wind fields have been generated with the Turbsim v1.03.00 module. The most turbulent wind case scenarios have been selected (wind turbine class—A, according to the IEC61400-1 design standard [35]), to demonstrate the vigorous blade-pitch regulation with the designed controllers in harsh wind conditions. The irregular sea state conditions have been generated with the Hydrodyn v2.03 module [36] where the sea state is derived from the Joint North Sea Wave Project (JONSWAP) wave spectrum, as prescribed by IEC61400-3 offshore standard [37].

Figure 7 contains the DEL results of the simulations carried out at all working conditions with the control parameters shown in Table 1. The results of the manually and optimised controllers are shown normalised according to the reference Detuned PI results. One can see how the objective of reducing the tower-base-pitch DELs has been achieved with the results obtained by using the optimised controller below the reference Detuned PI controller ones in the most of the FOWT working conditions. From 17 m/s wind speed to above all the DEL, results are below the reference with the optimised controller. Some results below this wind speed are above the reference; in contrast, such conditions are hardly expected to happen according to [32]. This region is also highly influenced by the switching between the below and above rated wind speed controllers. Therefore, it is believed that these results can be improved with a smooth controller switching strategy. Furthermore, the second objective of keeping the blade-root-flapwise and -edgewise DELs below the reference has been achieved as well. The tower-base-roll DELs have also been reduced, although the reductions are not significant due to the magnitude difference in comparison to tower-base-pitch loads, which are between 13.6 and 80.5% smaller depending on the working condition.

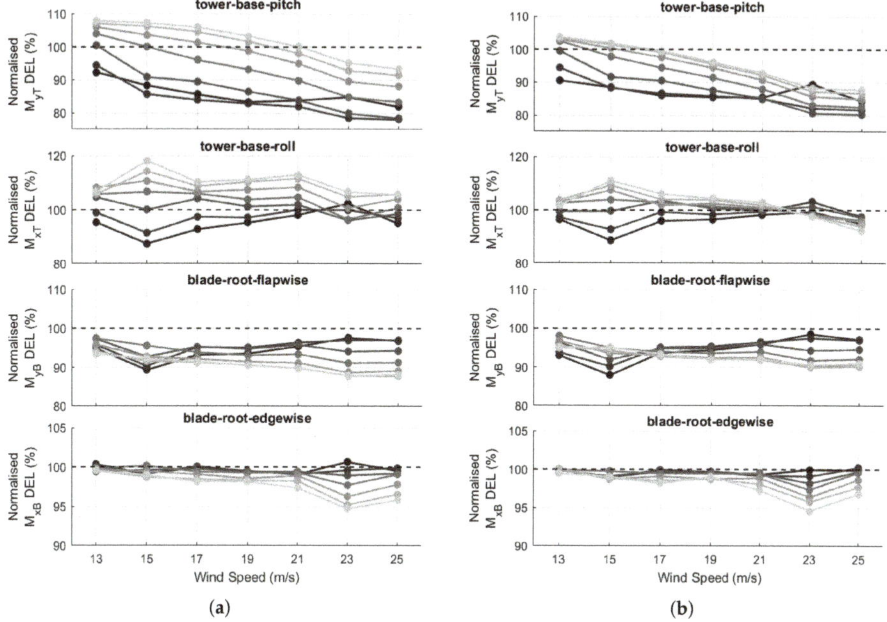

Figure 7. Damage Equivalent Load relatives to the baseline Detuned PI results (dashed lines) for $H_s = 0...6$ m (decreasing darkness for increasing H_s). (**a**) Manually tuned APS & WR. (**b**) Optimised APS & WR.

The improvements in the DELs results do not negatively affect the overall performance of the FOWT. The performance of the optimised controller is quite similar to the manually tuned one as can be seen in calm (a) and rough (b) sea states in Figure 8. This is due to the controller parameters being slightly modified with respect to those of the manually tuned controller (as shown in Table 1) to achieve the objectives defined by the cost function in the previous section, but without altering the essence of the original controller. However, the generated electric power (P_g) quality has been improved and not only in calm sea states avoiding the gaps when extreme waves occur, but also in rough sea states where the improvement is much more accused. The generator speed (Ω_g) excursions have been also reduced in both sea state conditions as well as the generator torque (M_g), which shows smother regulation. This provides more conservative operation reducing the strokes in the gear-box. The platform-roll (θ_{xP}) and -pitch (θ_{yP}) oscillations have been reduced successfully, mainly in rough sea states which is the main contributor to the positive tower-base DEL results shown in Figure 7. In addition, the blade-pitch (θ) activity shows less excursions than the reference Detuned PI and the manually tuned controller contributing to a minor deterioration of the blade-pitch actuator.

The Standard Deviation (STD) quantifies the amount of dispersion of the set of variables shown in the time-domain results of Figure 8. One can see the platform-pitch and -roll STD reductions in rough sea states with the optimised controller in Figure 9b, not increasing the blade-root-flapwise and -edgewise ones. The platform-pitch and -roll STD have been reduce up to 10 and 14%, respectively. The generator speed and generator torque regulation have been slightly worsen with the optimised controller compared to the results obtained with the manually tuned controller, as it can be seen in Figure 9b. However, still showing better results than the reference Detuned PI controller. These results are consistent with those of Figure 8.

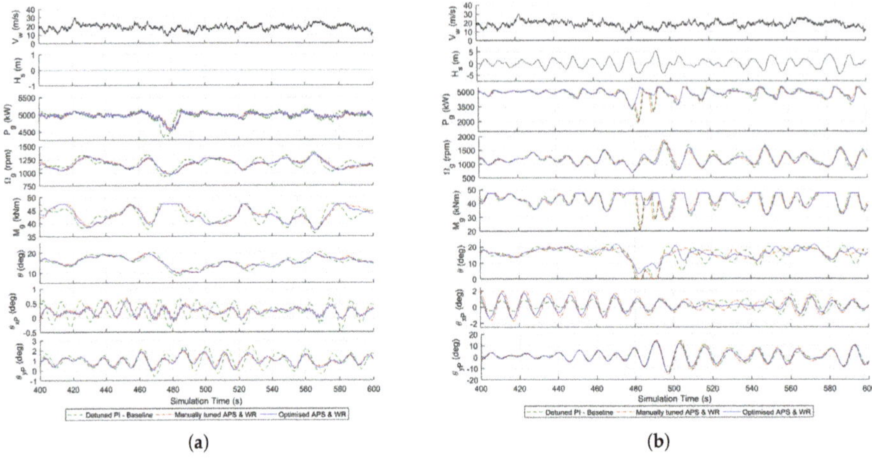

Figure 8. Time-domain simulation results. (**a**) $V_{hub} = 19$ m/s and $H_s = 0$ m. (**b**) $V_{hub} = 19$ m/s and $H_s = 6$ m.

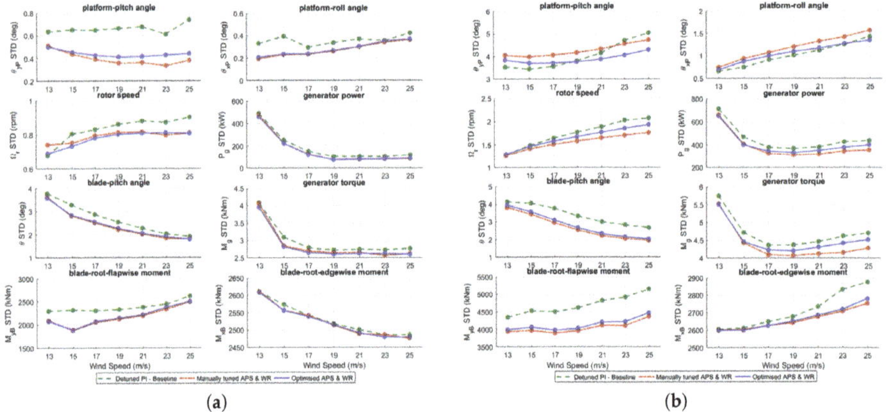

Figure 9. Time-domain simulation Standard Deviations results. (**a**) $H_s = 0$ m. (**b**) $H_s = 6$ m.

5. Conclusions and Recommendations

In this article, a novel optimisation methodology for the manually tuned FOWT feedback controllers for an above rated wind speed region is presented. This method automatically updates the manually tuned feedback control loop parameters to improve the system performance minimising the described cost function. In this study, the previously presented APS [25,26] and WR [27] control loops for the NREL 5-MW ITI Energy barge FOWT models are used for the optimisation process. These two control loops present several parameters to be manually optimised, therefore, a novel optimisation methodology has been presented in this article based on Monte Carlo, showing the potential to improve the manually tuned controller performance through the objectives of: (1) reduce the tower-base DELs in all the contemplated working conditions below the reference from 17 m/s wind speed to above; (2) keep the blade-root DELs below the reference; and (3), preserve the overall FOWT performance hardly affected in comparison to the reference Detuned PI controller results. Efforts are currently being carried out at IK4-IKERLAN to further improve this preliminary optimisation method and analyse different variants with the aim of comparing the results obtained in this study as well as investigating other optimisation methods, e.g., genetic algorithms, in order to examine the trade off between code complexity, computational time and performance results.

Author Contributions: This paper is part of J.O.'s PhD thesis conducted at the University of the Basque Country UPV/EHU. J.J. contributed in the design and implementation of this work. I.E. contributed in the theoretical and result analysis of this work. S.A.-Q. contributed to the final version of the manuscript. A.P.-A. supervised the project.

Funding: This work has been partially funded by the Spanish Ministry of Economy and Competitiveness through the research project DPI2017-82930-C2-2-R.

Conflicts of Interest: The authors declare no conflict of interest.

References

1. Bilgili, M.; Yasar, A.; Simsek, E. Offshore wind power development in Europe and its comparison with onshore counterpart. *Renew. Sustain. Energy Rev.* **2011**, *15*, 905–915. [CrossRef]
2. Musial, W.; Butterfield, S.; Ram, B. Energy from offshore wind. In Proceedings of the Offshore Technology Conference, Houston, TX, USA, 1–4 May 2006.
3. Rhodri, J.; Ros, M.C. *Floating Offshore Wind: Market and Technology Review*; Carbon Trust Report Prepared for the Scottish Government; Carbon Trust: London, UK, 2015.
4. Shyam Kularathna, A.H.T.; Suda, S.; Takagi, K.; Tabeta, S. Evaluation of co-existence options of marine renewable energy projects in Japan. *Sustainability* **2019**, *11*, 2840. [CrossRef]
5. Jiang, Z.; Zhu, X.; Hu, W. Modeling and analysis of offshore floating wind turbines. In *Advanced Wind Turbine Technology*; Springer International Publishing: Berlin/Heidelberg, Germany, 2018.
6. Gonzalez, S.F.; Diaz-Casas, V. Present and future of floating offshore wind. In *Floating Offshore Wind Farms*; Springer International Publishing: Berlin/Heidelberg, Germany, 2016.
7. Roddier, D.; Cermelli, C.; Weinstein, J.; Byklum, E.; Atcheston, M.; Utsunomiya T.; Jorde, J.; Borgen, E. State-of-the-Art. In *Floating Offshore Wind Energy: The Next Generation of Wind Energy*; Springer International Publishing: Berlin/Heidelberg, Germany, 2016.
8. Wayman, E.N.; Sclavounos, P.D.; Butterfiel, S.; Jonkman, J.; Musial, W. *Coupled Dynamic Modeling of Floating Wind Turbine Systems*; NREL/CP-500-39481; National Renewable Energy Laboratory (NREL): Golden, CO, USA, 2006.
9. Jose, A.; Falzarano, J.; Wang, H. A study of negative damping in floating wind turbines using coupled program FAST-SIMDYN. In Proceedings of the 1st International Offshore Wind Technical Conference (IOWTC), San Francisco, CA, USA, 4–7 November 2018.
10. Schlipf, D.; Sandner, F.; Raach, S.; Matha, D.; Cheng, P.W. *Nonlinear Model Predictive Control of Floating Wind Turbines*; International Society of Offshore and Polar Engineers (ISOPE): Mountain View, CA, USA, 2013.
11. Chaaban, R.; Fritzen, C.P. Reducing blade fatigue and damping platform motions of floating wind turbines using model predictive control. In Proceedings of the 9th International Conference on Structural Dynamics (EURODYN), Porto, Portugal, 30 June–2 July 2014.
12. Lemmer, F.; Raach, S.; Schlipf, D.; Cheng, P.W. Prospects of linear model predictive control on a 10 MW floating wind turbine. In Proceedings of the International Conference on Ocean, Offshore and Arctic Engineering, St. John's, NL, Canada, 31 May–5 June 2015.
13. Yang, F.; Song, Q.W.; Wang, L.; Zuo, S.; Li, S.S. Wind and wave disturbances compensation to floating offshore wind turbine using improved individual pitch control based on fuzzy control strategy. *Abstr. Appl. Anal.* **2014**, *2014*, 968384. [CrossRef]
14. Christiansen, S.; Bak, T.; Knudsen, T. Minimum thrust load control for floating wind turbine. In Proceedings of the IEEE International Conference on Control Applications (CCA), Dubrovnik, Croatia, 3–5 October 2012.
15. Namik, H.; Stol, K. Individual blade pitch control of a spar-buoy floating wind turbine. *IEEE Trans. Control. Syst. Technol.* **2014**, *22*, 214–223. [CrossRef]
16. Ramos, R.L. Linear quadratic optimal control of a spar-type floating offshore wind turbine in the presence of turbulent wind and different sea states. *J. Mar. Sci. Eng.* **2018**, *6*, 151. [CrossRef]
17. Namik, H.; Stol, K. Individual blade pitch control of a floating offshore wind turbine on a tension leg platform. In Proceedings of the 48th AIAA Aerospace Sciences Meeting Including the New Horizons Forum and Aerospace Exposition, Orlando, FL, USA, 4–7 January 2010.
18. Ma, Y.; Sclavounos, D.; Cross-Whiter, J.; Arora, D. Wave Forecast and its application to the optimal control of offshore floating wind turbine for loads mitigation. *Renew. Energy* **2018**, *128*, 163–176. [CrossRef]

19. Zuo, S.; Song, Y.D.; Wang, L.; Song, Q.W. Computationally inexpensive approach for pitch control of offshore wind turbine on barge floating platform. *Sci. World J.* **2013**, *2013*, 357849. [CrossRef] [PubMed]

20. Bagherieh, O.; Nagamune, R. Gain-scheduling control of a floating offshore wind turbine above rated wind speed. *Control Theory Technol.* **2015**, *13* 160–172. [CrossRef]

21. Bakka, T.; Karimi, H.R. Robust H∞ dynamic output feedback control synthesis with pole placement constraints for offshore wind turbine systems. *Math. Probl. Eng.* **2012**, *2012*, 616507. [CrossRef]

22. Hara, N.; Nihei, Y.; Iijima, K.; Konishi, K. Blade pitch control for floating wind turbines: Design and experiments using scale model. In Proceedings of the IEEE Conference on Control Technology and Applications (CCTA), Kohala Coast, HI, USA, 27–30 August 2017.

23. Betti, G.; Farina, M.; Guagliardi, G.A.; Marzorati, A.; Scattolini, R. Development of a control-oriented model of floating wind turbines. *IEEE Trans. Control. Syst. Technol.* **2014**, *22*, 69–82. [CrossRef]

24. Lackner, M. Controlling platform motions and reducing blade loads for floating wind turbines. *Wind Eng.* **2009**, *33*, 541–553. [CrossRef]

25. Olondriz, J; Elorza, I.; Trojaola, I. Pujana, A.; Landaluze, J. On the effects of basic platform design characteristics on floating offshore wind turbine control and their mitigation. *J. Phys. Conf. Ser.* **2016**, *753*, 052008. [CrossRef]

26. Olondriz, J.; Elorza, I.; Jugo, J.;Alonso-Quesada, S.; Pujana-Arrese, A. An advanced control technique for floating offshore wind turbines based on more compact barge platforms. *Energies* **2018**, *11*, 1187. [CrossRef]

27. Olondriz, J.; Jugo, J.; Elorza, I.; Alonso-Quesada, S.; Pujana-Arrese, A. A blade feedback control for floating offshore wind turbines. *J. Phys. Conf. Ser.* **2019**, *1222*, 012014. [CrossRef]

28. Kroese, D.P.; Rubinstein, R.Y. Monte Carlo Methods. *Wiley Interdiscip. Rev. Comput. Stat.* **2012**, *4*, 48–58. [CrossRef]

29. Tyatyushkin, A.I. Numerical optimization methods for controlled systems with parameters. *Comput. Math. Math. Phys.* **2017**, *57*, 1592–1606. [CrossRef]

30. Whitley, D. A genetic algorithm tutorial. *Stat. Comput.* **1994**, *4*, 65–85. [CrossRef]

31. Jonkman, B.; Jonkman, J. *FAST v8.16.00a-bjj*; National Renewable Energy Laboratory: Golden, CO, USA, 2016.

32. Jonkman, J. *Dynamics Modeling and Loads Analysis of an Offshore Floating Wind Turbines*; NREL/TP-500-41958; National Renewable Energy Laboratory (NREL): Golden, CO, USA, 2007.

33. Olondriz, J.; Jugo, J.; Elorza, I.; Alonso-Quesada, S.; Pujana-Arrese, A. Alternative linearisation methodology for aero-elastic Floating Offshore Wind Turbine non-linear models. *J. Phys. Conf. Ser.* **2018**, *1037*, 062019. [CrossRef]

34. Jonkman, J.M.; Buhl, M.L., Jr. *FAST User's Guide*; NREL/EL-500-38230; National Renewable Energy Laboratory (NREL): Golden, CO, USA, 2005.

35. *Wind Turbines—Part 1: Design Requirements*, 3rd ed.; Technical Report IEC61400-1; International Electrotechnical Commission (IEC): Geneva, Switzerland, 2005.

36. Jonkman, J.M.; Robertson, A.N; Hayman, G.J. *HydroDyn User'sGuide and Theory Manual*; National Renewable Energy Laboratory (NREL): Golden, CO, USA, 2014.

37. *Wind Turbines—Part 3: Design Requirements for Offshore Wind Turbines*; Technical Report IEC61400-3; International Electrotechnical Commission (IEC): Geneva, Switzerland, 2009.

Article

A Comparison of the Dynamic Performance of Conventional and Ternary Pumped Storage Hydro

Soumyadeep Nag [1,*], Kwang Y. Lee [1] and D. Suchitra [2]

[1] Department of Electrical and Computer Engineering, Baylor University, Waco, TX 76706, USA; kwang_y_lee@baylor.edu

[2] Department of Electrical Engineering, SRM University, Kattankulathur 603203, India; suchitra.d@ktr.srmuniv.ac.in

* Correspondence: Soumyadeep_Nag@baylor.edu; Tel.: +1 2544207420

Received: 31 July 2019; Accepted: 10 September 2019; Published: 12 September 2019

Abstract: With decreasing costs of renewable energy harvesting devices, penetration of solar panels and wind turbines have increased manifold. Under such high levels of penetration, coping with increased intermittency and unpredictability and maintaining power system resiliency under reduced inertia conditions has become a critical issue. Pumped storage hydro (PSH) is the most matured and economic form of storage that can serve the purpose of capacity for over 4 to 8 h. However, to increase network inertia and add required flexibility to low inertia power systems, significant paradigm shifting modifications have been engineered to result in the development of Ternary PSH (TPSH). In this paper a test system to consider governor interaction is constructed. The dynamic models of conventional PSH (CPSH) and TPSH are constructed and integrated to the test system to examine the effect of CPSH and TPSH in the hydraulic short circuit (TPSH-HSC). The ability and the effect of mode change (from pump to turbine) without the loss synchronism of TPSH has also been examined. Results display the superior capability and effect of TPSH with its HSC capability to contribute to frequency regulation during pumping mode and the effect of rapid mode change, as compared to its primitive alternative, CPSH.

Keywords: energy storage; pumped storage hydro; ternary pumped storage hydro; dynamic simulation; dynamic modeling; inertia; renewable energy; power system

---

1. Introduction

1.1. Problem Statement

With the introduction of large-scale renewable energy sources (RES), power system inertia is dropping [1]. The key underlying reason behind this is the fact that the inclusion of renewables in the schedule removes steam-driven synchronous generators out of the schedule. The removal of synchronous generators from the schedule simultaneously removes the services that are being provided to the grid by the thermal generators. These services include the inertia response and reserve, primary control and reserve, and secondary control and reserve. To exacerbate the situation, many thermal power plants that are now aging, are being decommissioned. From press releases, these aging thermals are planned to be replaced mostly by combined cycle gas turbine plants [2] but some are planned and being converted into solar farms [3–5]. Apart from aging, the other cause of retirement is the cost of operation. Especially in regions with high renewable penetration the cost of operating an outdated coal-based or gas-based plant can be comparatively high. Among other reasons for early retirement is that the plant has become unreliable or it does not meet the emissions standard of the region [6].

In such a low inertia system, a significant frequency deviation (that can be caused by the loss of a generator) may cause highly configured inverters to provide virtual support through virtual synchronous generation, but most others (especially small roof-top photovoltaic (PV) units) are programmed to disconnect from the grid and cause further deviation in frequency [7]. It is obvious that gas turbine based peaking plants can be deployed but, adding gas turbines diminishes the purpose of renewable energy completely as they are a key source of greenhouse gases and NO_x and SO_x pollutants.

Energy storage is a critical component of future power systems. If managed efficiently, energy storage can provide several beneficial services like cost savings through avoided capacity investments, curtailment mitigation and increased and firmed renewable penetration, ancillary and capacity reserves, black start capabilities, load shifting, transmission support, and congestion relief among a few others [8]. As such, modern large-scale green initiatives and investments are inclined more towards solar + storage or wind + storage rather than having the RES only [9]. A study conducted in [10] states that the size of required storage increases almost linearly as renewable energy penetration increases. However, there is an optimal size of energy storage beyond which penetration levels off. This was observed in two different systems examined and the authors arrived at the conclusion that it was 70% to 20% of the daily average demand.

Pumped storage is by far the most matured form of grid scale energy storage with discharge durations ranging from 1 to 6 h. PSH can provide synchronous inertia and primary and secondary frequency regulation services. Pumped storage hydro (PSH) relies on gravity as the driving force. When the price of electricity is less than a carefully selected optimum value, the plant pumps water to an elevated reservoir. When there is a shortage of power on the grid, it unleashes the water stored in the reservoir down to the turbines to produce energy and delivers it to the grid. This describes the basic operation of the PSH plants that has been prevalent for the past several decades. However, this basic use of price differentials or arbitrage is no more than enough to meet today's power system's needs. The need for power system resiliency and flexibility has caused power system operators to look for storage devices that can be deployed for long duration and serve system regulatory needs.

Around the world, there is 170 GW of installed PSH plants [11] that includes 3211 MW as ternary PSH (TPSH) and 9545 MW as adjustable speed PSH (ASPSH) but the 157 GW is still conventional PSH (CPSH). The main drawbacks of CPSH include the pump mode inflexibility (which is because of the use of a synchronous machine) apart from which the slow mode changing time and low part-load generation efficiencies are important short-comings of CPSH.

To overcome these drawbacks, PSH technology has seen a paradigm shift during the past 10 years with technological advancements in the fields of electromechanical design and construction and power electronics. Two such remarkable modifications are the TPSH and the ASPSH. Given a majority of PSH plants are at the end of life or above their half-life [12], many existing PSH plants are considering the upgrade during renovations and new PSH plants are looking forward to these modified units.

Moreover, markets around the world are creating opportunities for energy storage to participate in capacity and regulation markets. Realizing the benefits and obstacles of widescale use of energy storage, Federal Energy Regulatory Commission (FERC) order-841 issued on 15 February 2018 in USA directs regional operators and independent system operators (ISOs) e.g., California ISO, New York ISO, Pennsylvania-New Jersey-Maryland Interconnection (PJM), etc. to remove barriers and lay down legal foundations for the participation of energy storage in the concerned region's wholesale markets, i.e., the capacity market, energy market, and ancillary service markets [13]. On the face of increased penetration from renewable energy, regional markets have realized the potential of energy storage and have responded to FERC order-841 through their compliance plans [14]. Most of these ISOs are now in the process of developing participation models for energy storage systems while trying to comply with the directives of FERC order-841. As of now, except CAISO, most other ISOs are either partially or completely non-compliant with FERC order-841. Most of these regional operators need to figure out a market structure or model in which the energy storage devices can participate [15].

1.2. A General Classification of PSH

Pumped storage facilities can be classified as a closed or open circuit. Open-circuit PSH refers to those plants that have their lower reservoir connected to a river or a sea. Closed-circuit PSH refers to those plants that have no other source of water except that which is already stored in the upper and lower reservoirs. Pumped storage facilities can be further classified into sub-surface and surface reservoir types. While the surface reservoir is the most common type of PSH, the sub-surface reservoir type, one or more reservoirs, and the powerhouse will be underground. This concept allows the usage of abandoned mines, wells, shelters, and shafts to be utilized as a reservoir [16].

1.3. Other Notable Modifications of PSH

Although PSH is a solution to the ever-fluctuating renewable energy resources, many regions in the world do not have the geographic ability required to build a PSH. To overcome this problem, the following solutions have been devised:

1. STENSEA (stored energy in the sea) where a concrete tank, equipped with a pump-turbine-generator system, is placed at the bottom of the ocean bed. This structure generates electricity when filled and uses electricity to pump water out of the system [17]. The concept is now at a technology readiness level between 5 and 6. It has been successfully tested in a lab, which was followed by a similar environment testing in a lake.
2. Wind turbines with water storage are a recent invention and are the only source where the two forms of energy are physically integrated. Here wind turbines pump water up their tower and into tall tanks beside them when excess energy is available. When energy is required these tanks are emptied through a pump-turbine to generate electricity [18]. The pilot project is already completed for this concept. Four such turbines are in operation in Germany's Swabian–Franconian forest with turbines supplied by General Electric (GE) renewable energy.
3. Gravity power modules implement the principle of a hydraulic jack and gravity to generate electricity. A very heavy mass slides down a hole in the ground pushing water below the mass into the pump-turbine (through a narrow hole) to generate electricity. The same pump turbine is used to pump the mass up to its initial level to store excess energy. [19].

This work contributes to the construction of dynamic models of a governor-turbine system for TPSH and examines the benefits of TPSH over CPSH. The dynamic model is constructed to display the generating, pumping, and hydraulic short circuit mode. The authors have considered an IEEE 9 bus test system and evaluated the effect of significant PV penetration by simulating an increase in the load event. Simultaneously, the authors have examined the benefit of TPSH over CPSH in enhancing system resiliency in terms of system frequency profile. This paper is divided into five sections. Section 2 describes ASPSH and TPSH and details the modes of operation of the TPSH. Section 2 also details some differences between ASPSH and TPSH. Section 3 presents the dynamic modeling of TPSH. Section 4 presents the test system for the study along with certain simulation details and Section 5 presents the results of the study followed by a brief description of the test case and results.

2. Advanced Pumped Storage Hydro Plants

2.1. Adjustable Speed PSH plants (ASPSH)

There are two variants of adjustable speed PSH (ASPSH) as displayed in Figure 1.

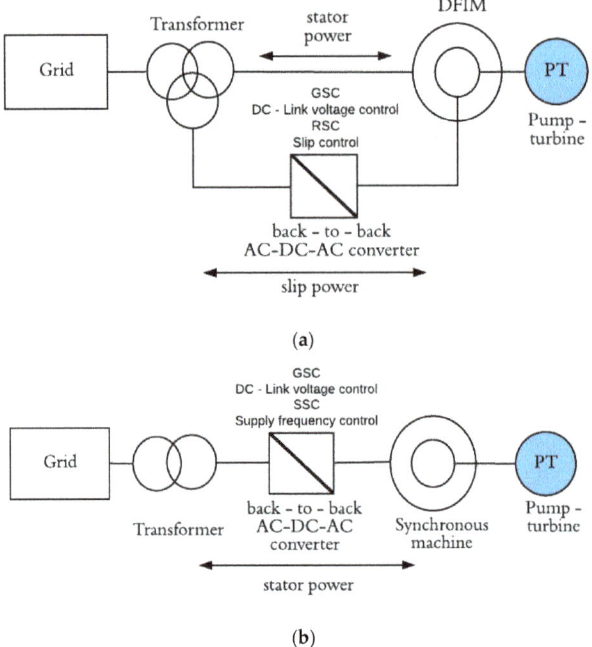

Figure 1. Adjustable speed pumped storage hydro: (**a**) Employing doubly fed induction machine (DFIM) with a rotor side converter; and (**b**) employing synchronous machine. GSC—Grid side converter, SSC—Stator side converter, RSC—rotor side converter.

Type-1: Here, the plant uses a doubly fed induction machine (DFIM) with a reversible pump-turbine, instead of the conventional fixed-speed synchronous machine coupled with a reversible pump-turbine [20]. This version is inspired from type-3 wind turbines. The back-to-back converter has two separate parts namely, the grid-side converter (GSC) and the rotor-side converter (RSC). The RSC controls the torque (real power) and the reactive power exchanged with the grid through the stator. The GSC regulates the DC-link voltage and attempts to keep it constant. By utilizing this configuration, the speed of the machine can be controlled, which will in turn control the power consumed during the motoring phase or generating phase.

Type-2: Here, a fully fed converter is coupled with a synchronous machine, which is coupled with a reversible pump-turbine. This is inspired from the type-4 wind turbines. This plant implements a synchronous machine with the stator connected to the grid via a back-to-back voltage source converter and the required filter. The rotor is extended to provide excitation. The stator-connected converter helps to control the speed and hence the power transfer to or from the machine. The excitation is usually a controller to maintain a fixed stator voltage. Here, the size of the converter required is equal to that of the machine and hence the floor space requirements are more than that of type-1, which translates into increased capital investment cost. Like type-1, generation and pumping speeds can be varied by the converter system. This configuration allows the operator to bypass the converter generation operation while including it during the pumping operation, which adds to network regulation.

2.2. Ternary Speed PSH Plants (Synchronous Speed)

A TPSH consists of a turbine, pump, and a synchronous machine on the same shaft. The turbine can be a Francis turbine or a Pelton wheel depending upon the available head. The pump is usually a single or multi-stage centrifugal pump. The three systems are separated by two clutches on each side of the salient-pole synchronous machine. The turbine and the pump share the same penstock, which also implies that the TPSH system can operate only in the pumping mode or the generating mode at any instant of time.

2.2.1. Generator Mode of TPSH

Figure 2a illustrates the generator mode. During the generation mode, the TPSH plant operates like a conventional hydro plant, where the pump-side clutch c2 separates (status 0) the pump from the shaft while c1 connects the turbine to the shaft (status 1). A governor is used to regulate the speed and power of the generator. The turbine valve (Vt) remains open and the pump valve (Vp) remains closed.

2.2.2. Pump Mode

Figure 2b displays the pump mode operation. During the pumping mode, the turbine remains disconnected from the shaft using clutch c1 (status 0) and the pump is connected to the shaft via c2 (status 1). The turbine valve Vt remains closed for pure pumping mode.

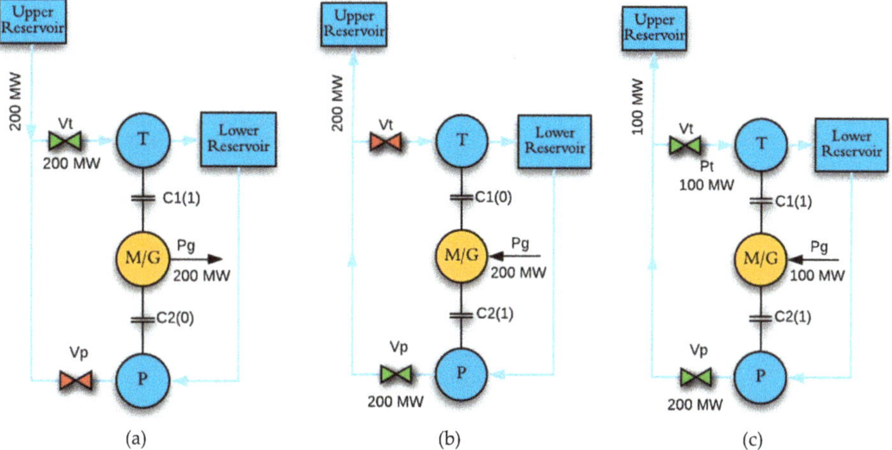

Figure 2. Modes of operation of ternary pumped storage hydro (TPSH) with the status of valves and clutches: (**a**) Generator mode; (**b**) pure pump mode; and (**c**) hydraulic short circuit mode (HSC). Clutches c1 and c2 are either in open mode (status 0) or closed mode (status 1). The turbine valve Vt and pump valve Vp are either in the open mode (status green) or in the closed mode (status red).

2.2.3. Hydraulic Short-Circuit Mode

Figure 2a displays the hydraulic short-circuit mode (HSC) mode. During the HSC mode, a part of (or all) the water pumped by the pump is transferred to the turbine by opening Vt while Vp remains open. Both the clutches c1 and c2 also connect the turbine and pump to the shaft. This in turn causes the turbine to produce torque that is supplied to the shaft. This mechanical torque from the turbine will reduce the current drawn from the grid while still rotating at the synchronous speed. The remaining part of the water is pumped up to the upper reservoir.

The hydraulic short circuit mode takes advantage of the fact that the head generated by the pump is either equal or greater than that of the upper reservoir at any instant of its operation as long as the pump rotates at synchronous speed. The turbine being connected right next to the pump experiences the pump generated head, which is in fact the rated head of the turbine, which enables it to generate power.

In Figure 2a 100 MW of electrical power is consumed from the grid. Since the pump rotates at synchronous speed it generates rated head and part of the discharge at rated head (worth 100 MW) is supplied to the turbine, which generates equivalent mechanical power. Hence the net work done, which is found through the amount of water raised to the upper reservoir is only 100 MW while 100 MW is supplied to the shaft.

The use of HSC allows the plant to follow power commands from the automatic generation control (AGC) and provide a secondary frequency. It also allows the plant to respond to network frequency disturbances and provide primary frequency control. Table 1 highlights the pros and cons of the advanced pumped storage units in comparison to CPSH.

Table 1. Comparison of adjustable speed PSH (ASPSH) and TPSH with respect to conventional PSH (CPSH).

Attribute	ASPSH	TPSH
Converter requirement	Type 1 requires converters rated 10% of the machine, Type 2 requires fully rated converters	Does not require converters for grid interfacing.
Efficiency	Peak efficiency will be reduced compared to the CPSH. Part-load efficiencies can be improved with speed variation.	Peak efficiency same as that of CPSH. Part-load efficiencies are better than CPSH as TPSH uses a pure turbine instead of pump-turbine.
Mode-change time	Time required to change from the pumping to generating mode is in the order of a few minutes.	Time required to change from the pumping to generating mode is significantly reduced. Seamless mode change can be performed as shaft rotated in a single direction.
Shaft inertia	Same as CPSH.	The shaft consists of the electrical machine and the separate pump and a separate turbine. Thus, in the hydraulic short circuit mode the shaft has comparatively more inertia.
Design optimization	The pump-turbine is optimized as a pump and then as a turbine. Thus, during off-design operating conditions as a turbine, the efficiency decreases drastically.	The pump and the turbine can be separately optimized. This leads to higher cycle efficiency.
Frequency regulation	Possible but only with auxiliary control.	Fully capable.
Natural inertia and synchronous frequency response	The converter's natural tendency is to act against grid frequency. The direction of change of speed and change of torque (and hence the current) are opposite.	Its natural tendency is to support the grid frequency, which allows the machine to slow down and release energy into the network.
Range of flexibility	Limited range of flexibility is available. Range is limited by cavitation limits and converter size.	Higher range of flexibility due to the use of a separately optimized turbine and pump.
Smooth start capability for pumping mode.	Possible but only with auxiliary control.	Fully capable as the pump can be started with the turbine.
Possibility of failure	Increases as the number of components increases.	Low because the number of components is comparatively lower.

3. Dynamic Modeling of TPSH

To realize the above-mentioned functionalities of the TPSH plant, the following model has been adapted from [21] and [22]. Compared to [22], [21] gives a better idea of pump performance but provides a model that simulates the turbine and pump modes separately. While [22] does unite the pump and turbine models, however, there are certain model corrections that need to be noticed. First, there is no need for a power distribution coefficient as the pump does not participate in reference tracking. Secondly, the integration of Δh to generate the discharge should be negative to realize the effect of the pump. Finally, the pump and turbine mechanical powers are added together, which should be a subtraction operation instead. Apart from that, the fact that the system has variable inertia has not been stated or accommodated by the model in [21] or [22]. This work is not focused on correcting these flaws but tries to incorporate the required corrections to regenerate the dynamics of the plant.

Figure 3 displays the schematic of the governor turbine system. The penstock dynamics for the pump and the turbine and their interaction can be modeled by [23]:

$$\frac{dq_t}{dt} = \frac{H_t - h - h_l}{Tw},\tag{1a}$$

$$\frac{dq_p}{dt} = \frac{-[H_p - h - h_l]}{Tw},\tag{1b}$$

where H_t, H_p, h, and h_l define the available head for the turbine, generated head by the pump, head at turbine entrance q_{base}, and head loss due to friction as $h_l = f_p Q^2$, and Tw is the water starting time constants to accommodate the interaction between the turbine and pump or vice versa.

Equations (1a) and (1b) relates head and discharge without accommodating any interactions and can be generalized as:

$$\frac{dq}{dt} = \frac{\Delta h}{Tw},\tag{2}$$

where, the water starting time constant is given as:

$$Tw = \frac{Lq_{base}}{Agh_{base}}.\tag{3}$$

Here, L, q_{base}, A, g, and h_{base} define the length of the penstock, base flow (= 1), cross section of the penstock, the gravitational constant and the base head (= 1).

The above model can also be represented in matrix form as:

$$\begin{bmatrix} Tw_{tt} & Tw_{tp} \\ Tw_{pt} & Tw_{pp} \end{bmatrix} \begin{bmatrix} \frac{dq_t}{dt} \\ \frac{dq_p}{dt} \end{bmatrix} = \begin{bmatrix} \Delta h_t \\ \Delta h_p \end{bmatrix}.\tag{4}$$

Here Tw_{tt}, Tw_{tp}, Tw_{pp}, and Tw_{pt} are the water starting time constants to accommodate the interaction between the turbine and pump or vice versa.

Equation (5) represents the interactions between the turbine and pump. To accommodate multiple units, this can be expanded in a similar manner [21]. The non-linear relation between gate, head and discharge is given by:

$$q_t = G\sqrt{h},\tag{5}$$

where G is the turbine or pump gate.

The turbine is equipped with a PIDGOV type governor turbine model that controls the main inlet valves of the turbine and that of the pump. The non-linearity between the gate and power required has been ignored. The characteristics of the turbine and pump are simulated by:

$$P_m = A_t h((q - q_{nl})) - DG\Delta\omega,\tag{6}$$

where $\Delta\omega$, and q_{nl} define the speed deviation from synchronous speed and the no load discharge required to maintain 0 power output while meeting all losses and yet maintain synchronism; D and A_t define the damping and power coefficients of the turbine to account for turbine characteristics; and P_t and P_p define the power output of the turbine and pump.

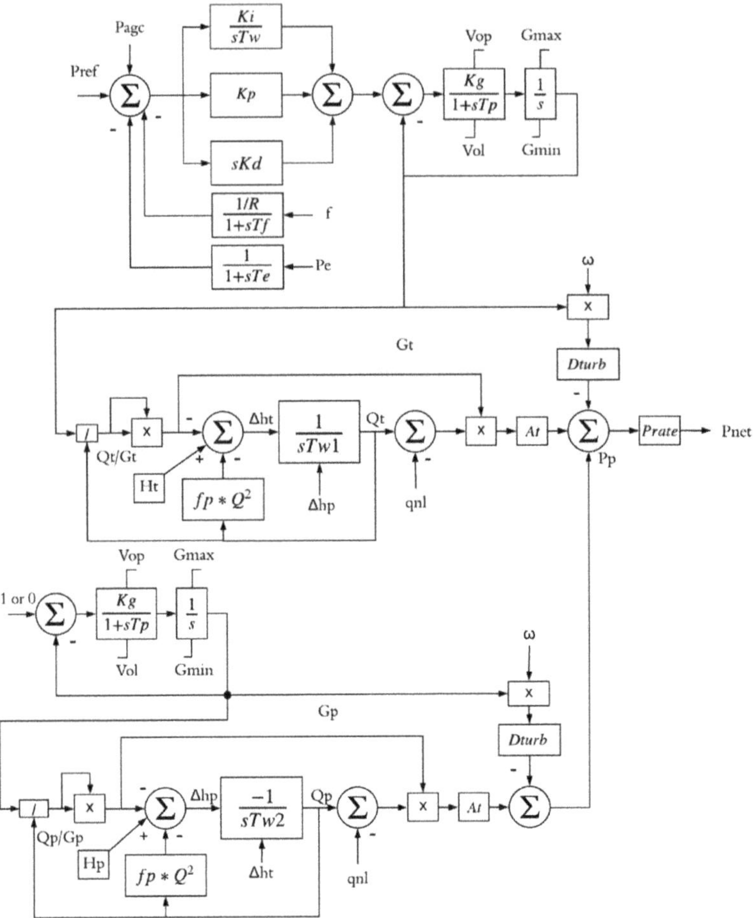

Figure 3. Dynamic model of TPSH with a PID governor for the turbine.

This model accounts for the no-load losses of water and the frictional losses. The model uses the coefficient A_t to relate discharge to the mechanical power output,

$$A_t = \frac{Turbine\,MW\,rating}{(Generator\,MVA\,rating)hr(q - q_{nl})}.$$ (7)

Equations (5)–(7) can be repeated for the pump in a similar way. The net power from the system is the sum of the pump power and turbine power, which is also the shaft power that is given to the electrical machine,

$$P_{shaft} = P_t + P_p.$$ (8)

While the turbine accepts continuous gate commands, the pump accepts open and close commands. This governor turbine system is then connected to a salient pole machine model. The pump and turbine heads are modeled, respectively, by:

$$H_p = [H_0 + H_1(q) + H_2(q)^2](1 + \Delta\omega)^2,$$ (9)

$$H_t = 1(H_p < 1) + H_p(H_p > 1). \tag{10}$$

Equation (9) can be obtained by identifying the characteristics of the pump given by the manufacturer. Equation (10) states the possibility of head at turbine entrance being greater than 1 due to system frequency fluctuations and pump speed variations. However, since the pump acts against a rated head produced by the upper reservoir, the head at the turbine entrance cannot be less than 1.

4. The Test System, Simulation, and Solver

Figure 4 demonstrates the test system used to compare the dynamic behavior of TPSH and CPSH plants. The IEEE 9-bus system was modeled with governors to represent system dynamics. The system consists of three thermal plants equipped with IEEE-T1 governor systems with similar parameters but varying capacities [24]. The exciter system and voltage dynamics were assumed to be unaffected. The system parameters can be found in Appendix A.

To realize the effect of the loss of inertia, the thermal plant at bus 1 rated 500 MW was replaced by a PV plant of equal capacity and a CPSH plant of 500 MW was placed at bus 7 to form the baseline cases. For the purpose of comparison, the CPSH was replaced by a TPSH of 500 MW with the governor and turbine structure as indicated in Figure 3. CPSH was simulated using the governor turbine structure as indicated in Figure 3 with the turbine set-point as 0 and the clutch c1 disengaged. Simulating the CPSH the water starting time matrix was also modified to represent a single value Tw_{pp}.

MATLAB code has been used to reconstruct the differential equations relevant to each block. The equations are then initialized and simulated. To simulate the system, a runge-kutta 4 solver was implemented. The use of this solver helps simulate all differential equations and accommodate the time varying inertia of the shaft, which would not be convenient in Simulink.

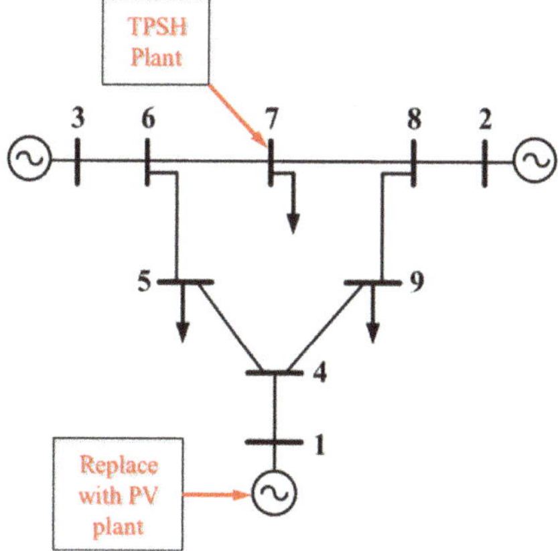

Figure 4. IEEE 9-bus test system indicating PV and PSH plants.

5. Results and Discussion

The test cases analyzed here focused mainly on the pump mode operation of the plant. In the generating mode the TPSH and the CPSH were basically the same as they both had governor control of the turbine. The difference, however, was that the TPSH employs a turbine rather than a pump-turbine, which enables it to operate over a wider range of power output.

5.1. TPSH-HSC Reference Tracking

Reference tracking was examined to understand the regulatory capability of TPSH in the pump mode using HSC and observe the expected non-linear characteristics of the hydro plant. Power references were provided to the plant and the plant tracks the references. The tracking performance of the plant in the pump mode has been displayed in Figure 5. At t = 40 s a step of 0.5 was applied, followed by a step of 0.9 at t = 100 s, and 0.75 at t = 175 s.

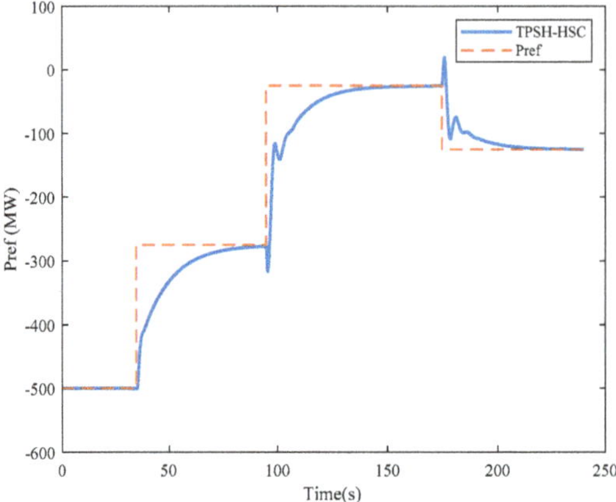

Figure 5. Reference tracking performance during the HSC mode. Negative power indicates pumping mode.

The reverse response behavior of the turbine was visible at t = 100 s and at t = 175 s. The extent or degree of reverse response (non-linearity) increased as the loading level increased, which was mainly because of the change in effective water time constant. More details about such behavior can be found in [25].

5.2. Baseline Analysis of Test System in Pump Mode of CPSH

This baseline analysis answers the question as to what happens when large capacity PV farms were added to the system in the presence of a CPSH operating as a pump. The addition of the PV farm caused the retirement of a similar capacity thermal plant or exclusion of the thermal plant from scheduling. However, this resulted in the loss of inertia and to make matters worse, removed the primary and secondary frequency response services that were provided by the thermal plant. This situation was simulated by removing the 500 MW thermal governor and plant from the system and adding a 500 MW PV farm into the system whose dispatch was adjusted to match that of the thermal plant that it replaced. The system inertia was reduced by 3 s. The steady-state dispatch value was obtained through simulation without the PV farm. The size of the test system did not allow the simulation of a plant trip so load change from 33% to 66% was simulated to the system at 50 s, which caused the system frequency to drop significantly as to initiate governor action.

Before the addition of RES the frequency nadir was 59.72 Hz. Figure 6a reveals that when the system inertia was reduced and primary frequency services previously provided by plant 3 (Appendix A) existed no more, the frequency nadir now decreased to 59.68 Hz. The usual frequency limit set by FERC was ±0.5 Hz hence the system was not in critical condition as to trigger load shedding or the tripping of PV inverters but the possibility of such events was valid in the face of further reduction of inertia with further addition of RES. Figure 6b illustrates that under the given test conditions, the thermal plants experience more governor action. This will result in higher wear and tear or thermal governor. Hence, there is an ardent need for flexibility of network resources, which helps increase network resiliency to cope with such events.

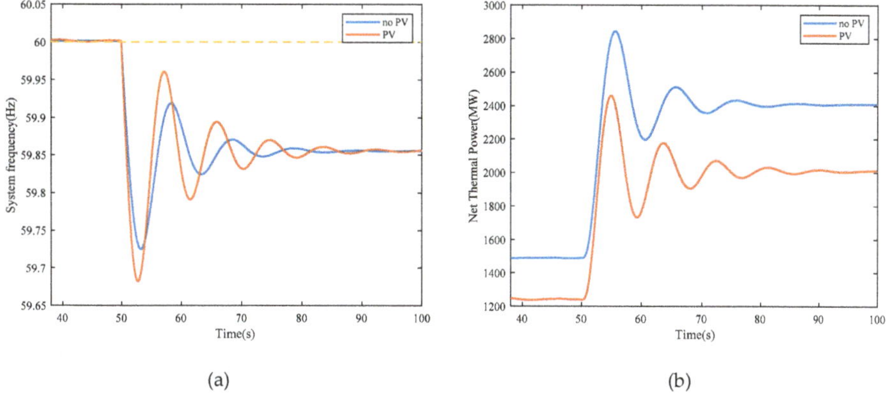

Figure 6. Baseline analysis of the test system in pump mode of CPSH.

5.3. Comparison of Pump Mode Response from CPSH, TPSH in HSC, and Mode Change Capability of TPSH without PV

This case described the benefits of the ability of the TPSH system to change its mode rapidly from pumping mode to generation mode in case of severe disturbances. A comparison of the governor response of CPSH and TPSH and the unique capability of the TPSH to change its mode rapidly was studied here. This case did not consider PV.

System frequency dropped rapidly as the load was applied at $t = 50$ s. The effect of the governor action of the thermal and hydro generators on system frequency was visible in Figure 7a. Thermal governors respond with instantaneous droop-based control through their IEEE-T1 governor system. The PIDGOV-based hydro action was visible when the TPSH-HSC was observed after $t = 60$ s, where the system frequency climbed up to be higher than that with CPSH. The HSC mode helped the plant to respond to the AGC systems command and to the change in system frequency.

The TPSH-MC describes the effect of a mode change (MC). The plant trip detection is done with wide area monitoring systems and the command is sent to the TPSH plant. After receiving the command, the plant engages the turbine and HSC to change the mode. The algorithm then inspects if the turbine has saturated. If the turbine saturates at its maximum, the pump outlet gate is closed, and it is mechanically disconnected from the shaft with the help of a clutch.

The figures display the regulatory capability of the TPSH and its ability to change its mode. Figure 7b shows the TPSH unit could move from −500 MW to 500 MW, thereby adding 1000 MW of regulatory capability to the existing network. Figure 7c shows the response of the turbine. The turbine equipped with the PID governor ramped up to a rated capacity of 500MW. Figure 7d shows that after the turbine saturated (in Figure 7c) at its rated output, the pump gate closed. As a result, pump power decreased to its minimum, after which the pump was disconnected from the shaft.

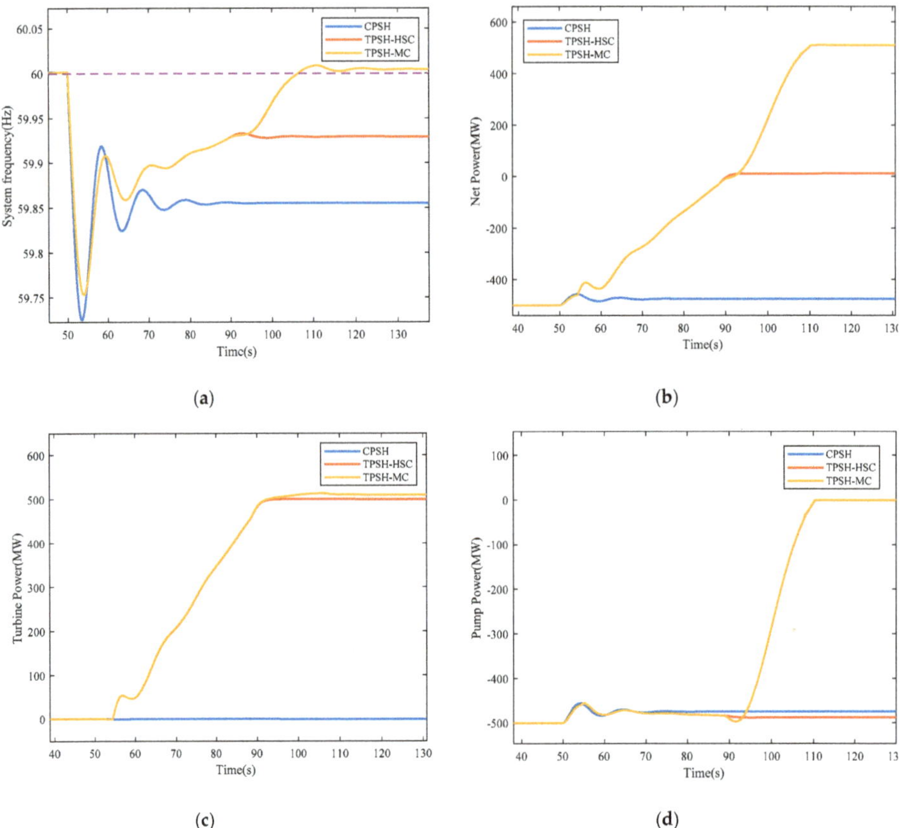

Figure 7. Comparison of CPSH, TPSH in HSC-mode (TPSH-HSC), and TPSH with HSC and mode change capability (TPSH-MC): (**a**) System frequency response; (**b**) net power production or consumption; (**c**) turbine power; and (**d**) pump power without PV.

5.4. Comparison of the Pump Mode Response from CPSH, TPSH in HSC-mode, and Mode Change Capability of TPSH in the Presence of PV

This case illustrates the contribution of TPSH over CPSH in a low inertia power system. Here, bus 1 was replaced with an equal capacity PV plant, which removed 3 s of inertia and the primary frequency services provided by the thermal plant.

With the HSC mode and the MC capability the frequency nadir was 59.75 Hz as compared to the 59.68 Hz of the case with CPSH. Until $t = 90$ s, the effect of the PIDGOV hydraulic governor was evident in the frequency response after which the turbine governor saturated, and the pump gates were closed. Frequency oscillations were evidently reduced by the HSC mode and the MC capability added to the regulation as stated before and aided in system frequency recovery as shown in Figure 8.

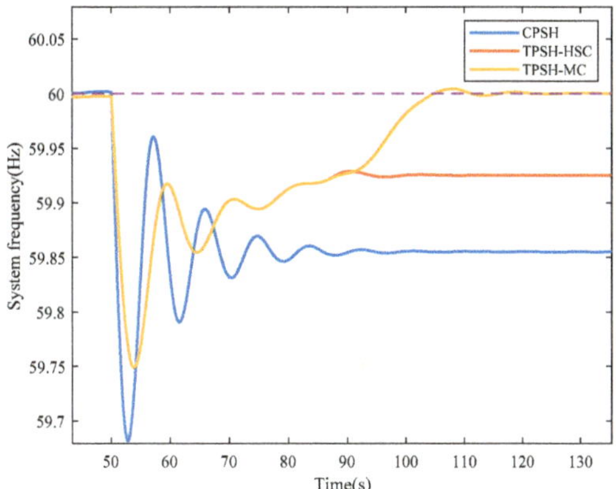

Figure 8. Comparison of CPSH, TPSH in HSC mode (TPSH-HSC), and TPSH with HSC and mode change capability (TPSH-MC) with PV.

6. Conclusions

The reduction in system synchronous inertia is a rising issue among power system engineers. In case of a disturbance in power balance, the low inertia power system can experience cascaded failures. To host more renewable energy and retain the resiliency of the system, pumped storage is a great alternative to electrochemical storage.

To understand the benefits of ternary pumped storage in a reduced inertia scenario, a dynamic model of ternary pumped storage plant was constructed and compared with that of a conventional pumped storage plant by integrating them to an IEEE 9-bus test-system simulated in MATLAB. Certain short-comings of similar works were accommodated. An increase in the load event was simulated to create a disturbance. Baseline analysis with and without PV indicated the fall in frequency nadir as the system inertia reduced when the disturbance was triggered. Results show that the hydraulic short circuit allowed the ternary pumped storage plant to track power references and hence provided regulation for the power system even in the pump mode. This capability to provide pump mode frequency regulation will be extremely helpful for modern power systems as PSH will also provide inertia to the system compared to electrochemical alternatives. Lastly, the HSC allows the TPSH plant to rapidly travel from the pumping mode to the generating mode, which provides twice the regulation capability in the case of severe outages in low inertia power systems.

Author Contributions: The following statements should be used, Conceptualization, S.N.; K.Y.L. and D.S.; Methodology, S.N. and D.S.; Software, S.N. and D.S; Validation, K.Y.L. and D.S.

Funding: This research received no external funding.

Conflicts of Interest: The authors declare no conflict of interest. The funders had no role in the design of the study; in the collection, analyses, or interpretation of data; in the writing of the manuscript, or in the decision to publish the results.

Appendix A

The following Table A1 describes the parameters of the hydraulic governor system for the TPSH. Identical parameters have been used for the CPSH system.

Table A1. Parameters for the hydraulic governor turbine system.

Hydro Penstock, Turbine, and Pump Parameters	Value (unit)
Turbine damping (D)	0.5 pu/pu
Water starting time for turbine (Tw_{tt})	2.13 s
Water starting time for pump (Tw_{tp})	2.13 s
Water starting time for turbine-pump (Tw_{pp})	0.13 s
Water starting time for pump-turbine (Tw_{pt})	0.13 s
Pump head discharge coefficients (H_0, H_1, H_2)	700, −0.2, −0.001
Turbine and pump no-load losses (q_{nl})	4.3 m^3/s
Turbine and pump rating (P_{rate})	500 MW
Generator rating	500 MVA
Inertia for pump and turbine	3 s, 3 s

Hydro Governor Parameters	Value (unit)
Servo motor gain (K_g)	0.98
Servo motor Time constant (T_p)	0.2 s
Permanent droop (R)	0.05 pu
Electrical power filter time constant (T_e)	5 s
Frequency filter time constant (T_f)	5 s
PID gains (K_p, K_i, K_d)	0.875, 0.00025, 0
Maximum opening (V_{op}) and closing rate (V_{ol})	0.1 pu/s and −0.1 pu/s

References

1. Du, P.; Matevosyan, J. Forecast System Inertia Condition and Its Impact to Integrate More Renewables. *IEEE Trans. Smart Grid* **2018**, *9*, 1531–1533. [CrossRef]
2. Malley, E. Coal Power Plant Post-Retirement Options. *Power Mag.* **2016**, *160*, 60. Available online: https://www.powermag.com/coal-power-plant-post-retirement-options/?pagenum=6 (accessed on 6 July 2016).
3. Delhi to Replace Retired Thermal Power Plant with Solar Project. *CleanTechnica* **2019**. Available online: https://cleantechnica.com/2019/07/22/delhi-to-replace-retired-thermal-power-plant-with-solar-project/ (accessed on 23 July 2019).
4. Brown, F. Ontario utility OPG makes room for solar at demolished coal power plant. *Pv Mag. Int.* **2018**. Available online: https://www.pv-magazine.com/2018/03/01/ontarios-utility-opg-makes-room-for-solar-at-demolished-coal-power-plant/ (accessed on 06 December 2016).
5. Canadian Coal-Fired Power Plant Transformed into Solar Farm Yale E360. 2019. Available online: https://e360.yale.edu/digest/canadian-nanticoke-coal-fired-power-plant-transformed-in-solar-farm (accessed on 23 July 2019).
6. Scott, A. General Electric to Scrap California Power Plant 20 Years Early. 2019. Available online: https://www.reuters.com/article/us-ge-power/general-electric-to-scrap-california-power-plant-20-years-early-idUSKCN1TM2MV (accessed on 23 July 2019).
7. Aemo.com.au. 2019. Available online: https://aemo.com.au/-/media/Files/PDF/Response-of-Existing-PV-Inverters-to-Frequency-Disturbances-V20.pdf (accessed on 06 December 2016).
8. Energy Storage Exchange. 2019. Available online: https://energystorageexchange.org/projects (accessed on 23 July 2019).
9. Antal, B.A. Pumped Storage Hydropower: A Technical Review. Master's Thesis, University of Colorado–Denve, Denver, CO, USA, 2014.
10. Applications of Energy Storage Technology. Energy Storage Association, 2019. Available online: http://energystorage.org/energy-storage/applications-energy-storage-technology (accessed on 23 July 2019).
11. Nrel.gov. 2019. Available online: https://www.nrel.gov/docs/fy10osti/47187.pdf (accessed on 23 July 2019).
12. Solomon, A.A.; Child, M.; Caldera, U.; Breyer, C. How much energy storage is needed to incorporate very large intermittent renewables? *Energy Procedia* **2017**, *135*, 283–293. [CrossRef]
13. Energy Storage Association. 2019. Available online: http://energystorage.org/ (accessed on 24 July 2019).

14. Ferc.gov. 2019. Available online: https://ferc.gov/whats-new/comm-meet/2018/021518/E-1.pdf (accessed on 24 July 2019).
15. E. 841, Energy Storage Association Unveils Initial Assessment of Regional Grid Operator Compliance with Federal Energy Regulatory Commission Order 841. Energy Storage Association, 2019. Available online: http://energystorage.org/news/esa-news/energy-storage-association-unveils-initial-assessment-regional-grid-operator (accessed on 24 July 2019).
16. Pumped Hydro-Power. Energy Storage Association, 2019. Available online: http://energystorage.org/energy-storage/energy-storage-technologies/pumped-hydro-power (accessed on 25 July 2019).
17. STENSEA—Storing Energy at Sea, Fraunhofer-Institut für Energiewirtschaft und Energiesystemtechnik. 2019. Available online: https://www.iee.fraunhofer.de/de/projekte/suche/laufende/stensea-storing-energy-at-sea.html (accessed on 23 July 2019).
18. Pilot Project—Max Bögl Wind, AG. *Max Bögl Wind AG* **2019**. Available online: https://www.mbrenewables.com/en/pilot-project?cn-reloaded=1 (accessed on 25 July 2019).
19. Gravity Power Module. Energy Storage. Grid-Scale Energy Storage. 2019. Available online: http://www.gravitypower.net/ (accessed on 25 July 2019).
20. Nag, S.; Lee, K. DFIM-Based Variable Speed Operation of Pump-Turbines for Efficiency Improvement. *IFAC PapersOnLine* **2018**, *51*, 708–713. [CrossRef]
21. Kratirov, V.; Guzwoski, L. *Modelling Ternary Pumped Storage Units*; Argonne National Laboratory: Oak Ridge, TN, USA, 2013.
22. Dong, Z.; Nelms, R.; Muljadi, E.; Tan, J.; Gevorgian, V.; Jacobson, M. Development of dynamic model of a ternary pumped storage hydropower plant. In Proceedings of the 2018 13th IEEE Conference on Industrial Electronics and Applications (ICIEA), Wuhan, China, 31 May–2 June 2018; pp. 656–661. [CrossRef]
23. Demello, F.P.; Koessler, R.J.; Agee, J.; Anderson, P.M.; Doudna, J.H.; Fish, J.H.; Hamm, P.A.; Kundur, P.; Lee, D.C.; Rogers, G.J.; et al. Hydraulic turbine and turbine control models for system dynamic studies. *IEEE Trans. Power Syst.* **1992**, *7*, 167–179. [CrossRef]
24. Report, I. Dynamic Models for Steam and Hydro Turbines in Power System Studies. *IEEE Trans. Power Appar. Syst.* **1973**, *92*, 1904–1915. [CrossRef]
25. Pradhan, S.K.; Tripathy, S.K. Dynamic stability improvement of a hydraulic turbine generating unit using an optimal PID controller. In Proceedings of the 2015 International Conference on Industrial Instrumentation and Control (ICIC), Pune, India, 28–30 May 2015; pp. 115–120. [CrossRef]

Article

Economy Mode Setting Device for Wind-Diesel Power Plants

Andrey Dar'enkov, Elena Sosnina *, Andrey Shalukho and Ivan Lipuzhin

Department of Electric Power Engineering, Power Supply and Power Electronics, Nizhny Novgorod State Technical University n.a. R.E. Alekseev, 603950 Nizhny Novgorod, Russia; darenkov@nntu.ru (A.D.); shaluho@nntu.ru (A.S.); lipuzhin@nntu.ru (I.L.)
* Correspondence: sosnyna@yandex.ru

Received: 20 December 2019; Accepted: 5 March 2020; Published: 10 March 2020

Abstract: The article is devoted to the problem of reducing fuel consumption in a diesel generator set (DGS) as a part of a wind-diesel power plant (WDPP). The object of the research is a variable speed DGS. The goal is to develop the WDPP intelligent control system, providing an optimal shaft speed of an internal combustion engine (ICE). The basis of the intelligent control system is an economy mode setting device (EMSD), which controls the fuel supply to the ICE. The functional chart of EMSD has been presented. The main EMSD blocks contain a controller and an associative memory block. The associative memory block is a software model of an artificial neural network that determines the optimal shaft speed of the ICE. An algorithm for the WDPP intelligent control system has been developed and tested using the WDPP Simulink model. The EMSD prototype has been created, and its research has been conducted. Dependences of the change in specific and absolute fuel consumption on the load power have been obtained for two 4 kW DGS: with constant rotation speed and variable rotation speed DGS with EMSD. It has been established that the use of EMSD in the mode of low loads allow one to reduce fuel consumption by almost 30%. The error in determining the optimal engine speed using EMSD prototype is not more than 15%.

Keywords: wind-diesel power plant; intelligent control system; diesel-generator set; internal combustion engine; artificial neural network; fuel economy

---

1. Introduction

Wind-diesel power plants (WDPP) are widely used for power autonomous consumers all round the world [1]. The main components of the WDPP are a wind turbine and a diesel generator set (DGS). DGS provides guaranteed power supply to consumers, and the wind turbine can reduce fuel consumption and harmful emissions into the environment [2]. However, commercially available WDPP have certain disadvantages like an excessive diesel fuel consumption during DGS operation, low efficiency, an incomplete use of a wind turbine at low wind speeds [3].

The studies aimed at solving these problems can be divided into two main groups:

- Optimization of the WDPP structure
- Controlling WDPP operating modes and ensuring its stability

The first group includes studies devoted to determining an optimal quantity and capacity of the power units in the system of the WDPP. The article [4] proposes a specific methodology for substantiating the parameters, placing and evaluating the WDPP effectiveness based on a multi-level system assessment of the resource potential and climate information. The article [5] considers the influence of the share of the wind turbine generator capacity in the installed capacity in the WDPP and the composition of consumers on the amount of replacement of the DGS energy output. The article [6] presents an optimal day-ahead scheduling of the WDPP and considers the role of battery storage system in frequency response. The optimal sizing of the PV/Wind/Diesel/Battery storage stand-alone hybrid system has been presented in the article [7] to minimize the cost of energy supplied by the system while increasing the reliability and efficiency of the system presented by the loss of power supply probability. The article [8] proposes a two-stage stochastic optimization framework to determine an optimal size of energy storage devices in a WDPP. The article [9] introduces an optimization method that takes economic and technical indexes into account. The method optimizes the capacity of the battery and DGS under the premise of satisfying the technical indexes, and it also increases the capacity of the wind turbine and photovoltaic array by considering the economic index.

The second group includes studies on improving the WDPP efficiency. The articles [10,11] proposed an additional adaptive voltage compensation controller for static var compensator by using a double sliding-mode optimal strategy to maintain the system bus voltage stability. In the article [12], a composite control strategy has been investigated for a photovoltaic, wind turbine, and DGS-based off-grid power generation system to achieve high performance while supplying unbalanced nonlinear loads. The article [13] introduces a frequency regulation scheme to share the active power, not supplied by the wind power, among DGS with the aim of reducing the total fuel consumption of the DGS. The article [14] presents a wind energy conversion system with a diesel generator. The proposed system regulates the loading of diesel generator to achieve low-specific fuel consumption.

The use of DGS with the variable shaft speed as a part of the WDPP is the most effective solution [15]. As a rule, a change in the shaft speed depending on the load power is stepwise. This allows one to reduce specific fuel consumption by up to 20%. However, fuel economy can be increased by a smooth regulation of the speed based on the multi-parameter characteristic of the internal combustion engine (ICE) as a part of the DGS.

The multi-parameter characteristic is individual for each ICE; it has a nonlinear character and shows the dependence of specific fuel consumption on the shaft speed and pressure on the ICE piston. The multi-parameter characteristic varies depending on the external and internal ICE operating conditions. External working conditions for ICE are temperature, pressure, and humidity of ambient air, etc. Internal conditions are the brand and quality of fuel and the wear of ICE internal parts, etc. WDPP manufacturers do not provide information on such characteristics, as a rule. It is impossible to calculate a multi-parameter characteristic; it can only be obtained experimentally. There are no developments allowing an optimal selection of the ICE speed in the absence of a multi-parameter characteristic.

The aim of the research is the development of an intelligent control system for WDPP, which provides effective operating modes under changing external and internal working conditions.

The economy mode setting device (EMSD) has been proposed to be used as the basis of the intelligent control system. EMSD automatically calculates the value of the optimal speed of the ICE for the current value of the load power under changing external and internal conditions.

Section 2 describes EMSD. Section 3 presents an algorithm of the intelligent control system. Section 4 is devoted to simulation. Section 5 describes the EMSD prototype. Section 6 is devoted to the experimental research and discussion.

2. Economy Mode Setting Device (EMCD)

A functional chart of EMSD is presented in Figure 1.

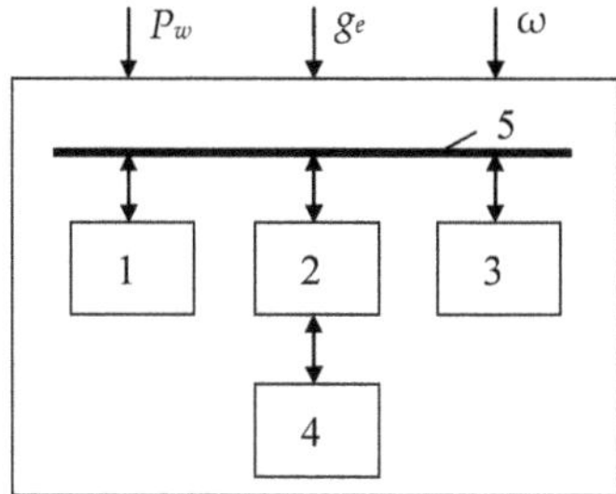

Figure 1. Functional chart of economy mode setting device (EMSD): (1) learning controller, (2) management controller, (3) associative memory block, (4) data memory block, (5) information bus, (P_ω) DGS power, (g_e) ICE fuel consumption, and (ω) ICE shaft speed.

EMSD is an adaptive controller that uses the principle of artificial neural networks. EMSD consists of four main elements: a management controller, a learning controller, an associative memory block, and nonvolatile memory. A common information bus, providing bi-directional exchange of information and expansion of the system in the case of controlling multiple ICEs, combines the elements.

EMSD optimizes the speed of the engine shaft by controlling the ICE fuel supply depending on the internal and external working conditions. The control signal is generated using the associative memory block, which is a software model of an artificial neural network. The optimal ICE shaft rotation speed is determined in real time. The principle of EMSD operation is as follows.

The sensors record the following parameters: DGS power (P), ICE fuel consumption (g_e), and ICE shaft speed (ω). These parameters are transferred to the learning controller, where they are compared with similar parameters available in the data memory block.

If the parameters of the current mode match to the parameters of one of the modes stored in the data memory block, the learning controller reads information about the position of the fuel pump rail (value h) and transmits it to the management controller. The management controller forms a control signal in the ICE fuel supply system. As a result, the fuel pump rail moves in the right direction by the required number of steps.

If the input data of the learning controller does not match to any mode stored in the data memory block, then the learning controller performs several learning cycles of the memory block. After training, a new optimal position of the fuel pump rail is formed, which is transmitted to the management controller. Additionally, parameters of a new mode are formed from the input (P, g_e, and ω) and output (h) parameters, which are stored in the data memory block.

A block scheme of the neural network of the EMSD associative memory block is presented in Figure 2.

The neural network of the associative memory block consists of three layers: input, hidden, and output. The input layer distributes the input data. The hidden layer consists of two neurons whose activation function is hyperbolic tangent. The signals of the neurons of the hidden layer are displayed linearly in the output layer. The output layer sums up the signals of the hidden layer and normalizes them. Neural network learning is implemented according to the algorithm of back error propagation, and it is realized by the learning controller.

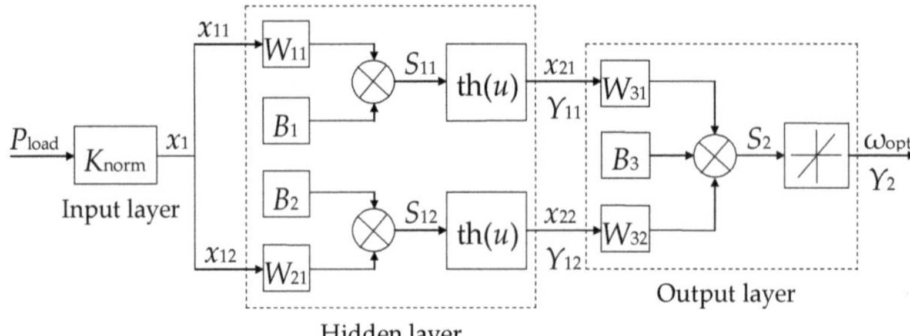

Figure 2. The block scheme of the associative memory block: (P_{load}) load power vector, (K_{norm}) normalization coefficient required to bring the load power value to the unit basis of the neural network, (x_{ij}) input *i* of the neuron (*j*) of the layer *n*, (W_{ij}) weight ratio of the corresponding neuron input, (B_i) weight of additional input for the *i*-th neuron initialization, (Y_{ij}) output *i* of the neuron (*j*-1) of the layer *n*, (S_{ij}) calculated function on the *j*-th layer of neurons, and (ω_{opt}) optimal value of the rotation speed at the output of the learning controller.

At the beginning of learning process, the load power vector P_{load} is applied to the inputs of the first layer of the network. At the outputs of the first layer, the following values are calculated:

$$x_{i1} = P_{\text{load}} \cdot K_{\text{norm}}; \tag{1}$$

$$x_{i2} = \text{th}(x_{i1} \cdot W_{i1} + B_i); \tag{2}$$

$$s_j^{(n)} = \sum_{i=0}^{M} y_i^{(n-1)} \cdot W_{ij}^{(n)}; \tag{3}$$

$$y_j^{(n)} = \text{th}\left(s_j^{(n)}\right), \ \omega_q \equiv P_q, \tag{4}$$

where

x_{ij}^n is the input *i* of the neuron (*j*) of the layer *n*;

$W_{ij}^{(n)}$ is the weight ratio of the corresponding neuron input;

K_{norm} is the normalization coefficient required to bring the load power value to the unit basis of the neural network;

B_i is the weight of additional input for the *i*-th neuron initialization;

M is the number of neurons in the layer *n*;

$y_{i(j-1)}^n$ is the output *i* of the neuron (*j* − 1) of the layer *n*;

P_q is the component *q* of the input vector *P*; and

ω_q is the component *q* of the output speed array ω.

For the output layer and the hidden layer *n*, the error value (δ) and the weight change (ΔW_{ij}) are calculated:

$$\delta_j^{(n)} = \left(y_j^{(n)} - d_j\right)\frac{\partial y_j}{\partial s_j}; \tag{5}$$

$$\Delta W_{ij}^{(n)} = -\beta \cdot \delta_j^{(n)} y_i^{(n-1)}, \tag{6}$$

where d_j is an ideal (desired) state of the neuron and β is the learning coefficient.

Next, the values of the network link weights are corrected:

$$W_{ij}^{(n)}(t) = W_{ij}^{(n-1)}(t-1) + \Delta W_{ij}^{(n)}(t), \tag{7}$$

where t is the number of the learning era of the associative memory block.

The value of the optimal rotation speed (ω_{opt}) is determined in the output:

$$\omega_{opt} = \left(\sum_{i=1}^{n} W_{i2} x_{i2} + B_i \right) \cdot 1/K_{norm},$$ (8)

where x_{i1} and x_{i2} are the inputs of the i-th neurons of layers 1 and 2 of the neural network, respectively, and K_{norm} is the normalization coefficient needed to reduce the load power value P to the unit basis of the neural network.

The calculation continues until the network operation error (E_ω) becomes less than permissible. Network operation error is determined by:

$$E_\omega = \frac{1}{2} \sum_{t=1}^{p} \left(\omega_t - \omega_{opt} \right)^2,$$ (9)

where

ω_t is an actual value of the rotation speed at the output of the neural network at the era t;

ω_{opt} is an optimal value of the rotation speed at the output of the learning controller; and

p is the number of the learning era corresponding to the permissible error.

A necessary condition for the correct learning of the neural network is the presence of a learning set; that is, a set of logical pairs $\omega_{opt} = f(P)$. A table array of speed values $\omega_{opt} = f(P)$ for discrete values of power P is stored in a data memory block.

3. Algorithm of the WDPP Intelligent Control System

The WDPP intelligent control system consists of four modules:

- A network learning management module
- A module for determining the minimum fuel consumption of ICE
- A control module for the optimal position of the ICE fuel pump rail
- A module for managing associative memory and data memory of the neural network

A block-chart of the algorithm of the WDPP intelligent control system is shown in Figure 3.

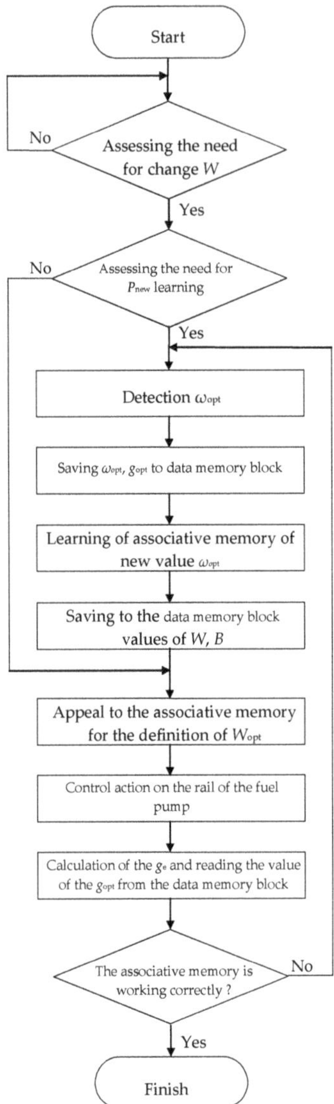

Figure 3. The algorithm block-chart of the wind-diesel power plant (WDPP) intelligent control system.

The input data of the algorithm are an actual value of the DGS power (P_{DGS}), an ICE fuel consumption (g_e), a rotational speed of the ICE shaft (ω_{ICE}), and an actual value of the output power of the wind generator (P_w).

The management controller monitors change in the value of the load power during the WDPP operation. If the load power changes by more than a predetermined value P_{step}, then a new value P_{new} is fixed. Further, the management controller determines that the new power value belongs to the range of values stored in the associative memory block.

Based on the artificial neural network of the associative memory block, the optimal value of the ICE speed is determined. In accordance with the optimal value of the ICE speed, a control signal for the fuel pump rail h is generated. This provides an efficient mode of WDPP operation.

The developed algorithm significantly expands the limits of the efficient operation of the WDPP and makes the intelligent control system adaptive towards to the external and internal operating conditions of the ICE.

4. Simulation

A Simulink model of the WDPP has been created to test the developed algorithm. The Simulink model allows one to study the static and dynamic modes of WDPP taking into account the mutual influence of WDPP units on each other.

The model is built on the basis of differential equations describing the functioning of the main units of the WDPP:

- A wind turbine, consisting of a turbine and a synchronous generator
- DGS, consisting of a diesel engine and a synchronous generator
- Energy storage systems
- Three-phase active-inductive load

The block scheme of the Simulink model is in Figure 4.

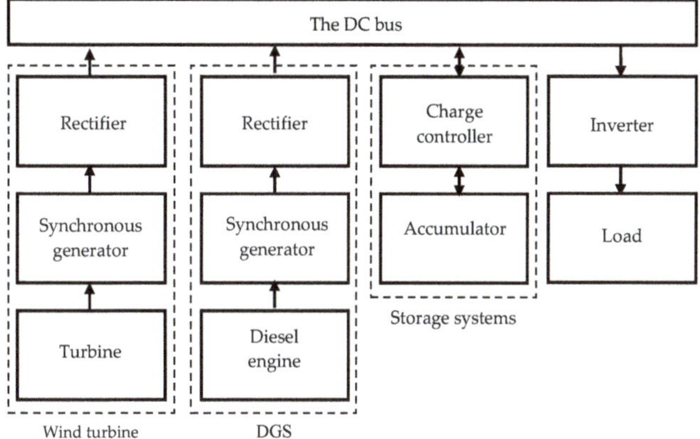

Figure 4. The block-scheme of the WDPP Simulink model. DGS: diesel generator set.

Mathematical models have been developed to simulate the operation of each WDPP unit. The models are differential equations that describe the processes of change of physical quantities such as current, voltage, frequency, torque, etc.

So, the operation of a synchronous generator as part of a wind turbine and DGS is described by a system of equations based on the law of electromagnetic induction and Ohm's law [16]:

$$U_{abc} = \frac{d}{dt}\Psi_{abc} + R_{abc}I_{abc} + H_s U_x,$$

$$\frac{d}{dt}\psi_f = e_f - r_f i_f,$$

(10)

where

U_{abc} is the stator voltage vector;

Ψ_{abc} is the stator flux distribution vector;

R_{abc} is the stator windings resistance vector;

I_{abc} is the stator current vector;

H_s is the matrix of independent circuits [17];

U_x is the vector of the voltage across stator clamps;

ψ_f is the excitation winding flux distribution;
e_f is the electromotive force of the excitation winding;
r_f is the resistance of the excitation winding; and
i_f is the current of the excitation winding.

Differential equations allow one to determine the phase currents and voltages for static and dynamic modes of operation based on the initial parameters for a specific model of a synchronous generator. The detailed descriptions of the mathematical models of WDPP units and their study results are given in [16].

A system of matrix differential equations has been developed to study WDPP operation modes, taking into account the mutual influence of WDPP units on each other:

$$\begin{cases} |\frac{d}{dt}Y_D = V_D(Y_D, Y_G), \\ |\frac{d}{dt}Y_G = V_G(Y_D, Y_G, Y_W, Y_E), \\ |\frac{d}{dt}Y_W = V_W(Y_G, Y_W, Y_E), \\ |\frac{d}{dt}Y_E = V_E(Y_G, Y_W, Y_E). \end{cases} \tag{11}$$

Vectors Y with corresponding indices represent variable values describing the operation of the diesel engine (D), the synchronous generator (G), the wind turbine (W), and the load (E). Vectors Y are determined for each WDPP unit with appropriate mathematical models. So, the variables for the synchronous generator are the stator voltage and current vectors, which are determined by Equations (10). The variables for the diesel engine are the engine speed and the position of the fuel pump rail. The variables for the wind turbine are wind turbine power and torque.

All equations have been compiled in relative units that facilitate their joint integration. The basic values correspond to the nominal passport data of the elements [16].

The algorithm of the WDPP intelligent control system for WDPP consisting of a 4 kW wind turbine and a 4 kW diesel generator has been tested on the Simulink model. Modeling was carried out for relative values of the load power in the range from 0.00 to 1.00 in increments of 0.05. Figure 5 shows the obtained dependence of the ICE rotation speed on the load power P.

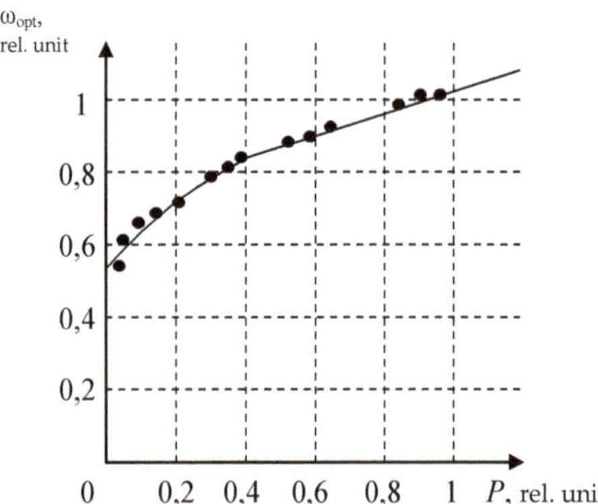

Figure 5. The dependence of the internal combustion engine (ICE) rotation speed on the load power obtained with the Simulink model: (P) load power, (ω_{opt}) optimal value of the ICE speed.

Dots indicate discrete values calculated by using the learning controller. The solid line shows a continuous dependence $\omega_{opt} = f(P)$ obtained with the associative memory block. The difference between the values ω_{opt}, calculated with the associative memory block and the discrete values determined by the learning controller for the corresponding P values, does not exceed 0.03% of the nominal value of the ICE speed.

The research has shown that the developed algorithm of the WDPP intelligent control system provides a minimum error in determining the optimal ICE shaft speed.

5. Economy Mode Setting Device (EMSD) Prototype

An EMSD prototype has been developed. In addition to the learning and management controller, associative and data memory blocks (for generating a control signal), the prototype includes a block of parameters from measuring sensors (for obtaining initial data) and a stepper motor control unit (for changing the position of the fuel pump rail).

The appearance of the prototype is shown in Figure 6.

Figure 6. EMSD prototype: (1) programmable control unit, (2) linear stepper motor, and (3) learning and management controller.

Learning and management controllers are combined in one device (3), which is based on the TM4C123GH6 microcontroller. Its memory capacity is 256 kB, which makes it possible to store the control algorithm, process sensor signals, and also store intermediate data during the functioning of the neural network. To control the position of the fuel pump rail, a linear stepper motor (2) is used, which is connected to a programmable control unit (1). The control unit generates control pulses with a frequency of up to 10 kHz and an accuracy of 0.2%.

6. Experimental Studies and Discussion

The appearance of the test bench is shown in Figure 7. The test bench consists of the following parts: (1 and 2) DGS and wind turbine simulators based on an asynchronous variable frequency drive and an alternator, (3) a DGS based on a gasoline ICE and synchronous generator, (4) connecting terminals for connecting a wind turbine, (5) an accumulator battery, (6) a programmable logic controller, (7) an autonomous inverter, (8) measuring equipment, (9) learning and management controller, and (10) programmable control unit.

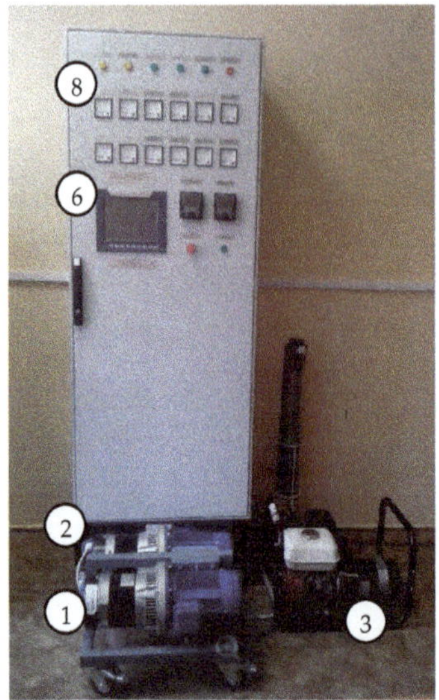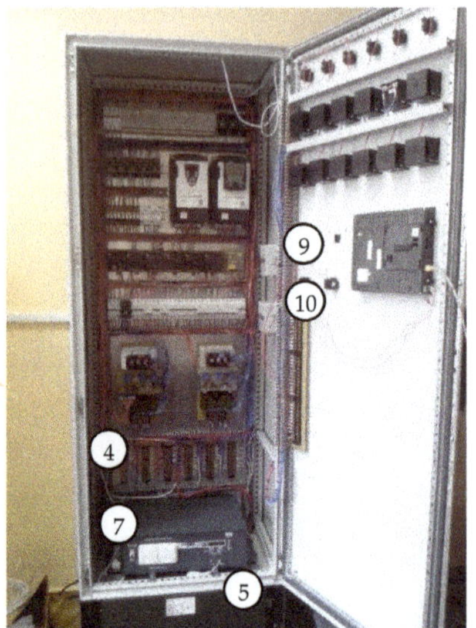

Figure 7. The test bench. (1 and 2) DGS and wind turbine simulators based on an asynchronous variable frequency drive and an alternator, (3) a DGS based on a gasoline ICE and synchronous generator, (4) connecting terminals for connecting a wind turbine, (5) an accumulator battery, (6) a programmable logic controller, (7) an autonomous inverter, (8) measuring equipment, (9) learning and management controller, and (10) programmable control unit.

Technical parameters of the test bench are shown in Table 1.

Table 1. Technical parameters of the test bench. DGS: diesel generator set.

Parameter	Value	Units
Wind turbine rated power	4	kW
DGS-rated power	4	kW
Wind speed	changes from 0 to 10	m/s
Active load	changes from 0 to 8 with step 0,1	kW

The programmable logic controller (6) provides an overall control of the test bench. EMSD prototype is designed to control the operation of DGS simulator (1) and DGS with the gasoline ICE (3).

The EMSD prototype is connected to the frequency converter as part of the DGS simulator. Therefore, it is able to adjust the DGS speed. DGS simulator (1) and wind turbine simulator (2) are used for testing and tuning the software of the WDPP intelligent control system.

The main purpose of the EMSD prototype is to control the DGS based on a gasoline ICE and synchronous generator (3). The learning and management controller (9) as part of EMSD prototype receives information from the sensors about the DGS power, the ICE fuel consumption, and the rotation speed of the ICE shaft and determines the optimal value of the ICE speed according to the algorithm (Figure 3). Information is transmitted to a programmable control unit (10), which forms a control signal for a linear stepper motor. The linear stepper motor changes the position of the ICE fuel pump rail according to the optimal engine speed. The placement of the learning and management controller, the programmable control unit, and the linear stepper motor as the part of the EMSD prototype is shown in Figure 8.

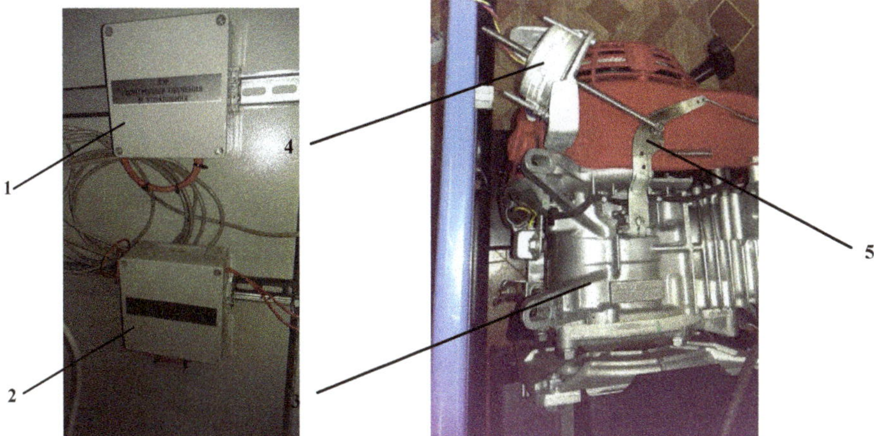

Figure 8. Placement of the EMSD prototype blocks on the test bench: (1) learning and management controller, (2) programmable control unit of the stepper motor, (3) ICE, (4) linear stepper motor, and (5) rail of the ICE fuel pump.

Figure 8 shows: (1) the learning and management controller, (2) programmable control unit of the stepper motor, (3) ICE, (4) linear stepper motor, and (5) rail of the ICE fuel pump.

A test bench has been used to study the parameters of the WDPP operation modes with EMSD. Figure 9 shows the dependences of the specific g_e (Figure 9a) and absolute g (Figure 9b) fuel consumption of 4 kW DGS with variable rotation speed. The results were obtained by calculation (using the multi-parameter characteristics of the ICE) and experimentally (using the EMSD prototype).

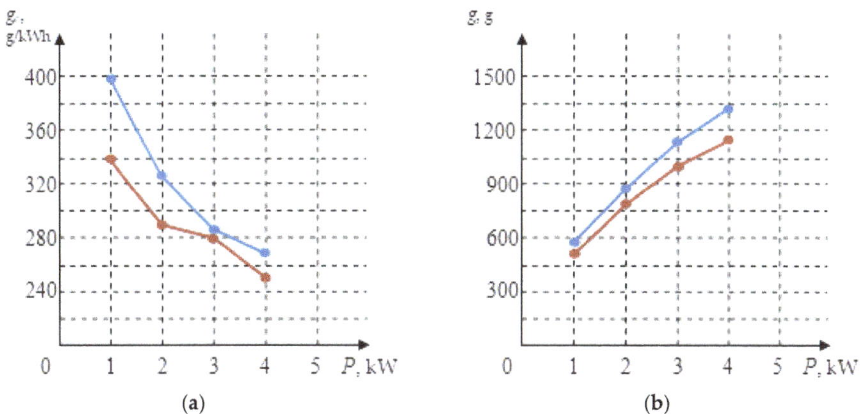

Figure 9. Dependence of specific (**a**) and absolute (**b**) fuel consumption for DGS with variable rotation speed: (P) load power, (g_e) specific fuel consumption, (g) absolute fuel consumption.

In Figure 9, the blue lines show the calculated characteristics, and the red lines show the experimental characteristics. The obtained results show that the error in determining the optimal engine speed using EMSD prototype is not more than 15%.

Figure 10 shows that the dependences of changes in specific (Figure 10a) and absolute (Figure 10b) fuel consumptions on the load power have been studied for two types of 4 kW DGS: with a constant (blue lines) and variable speed when using EMSD (red lines).

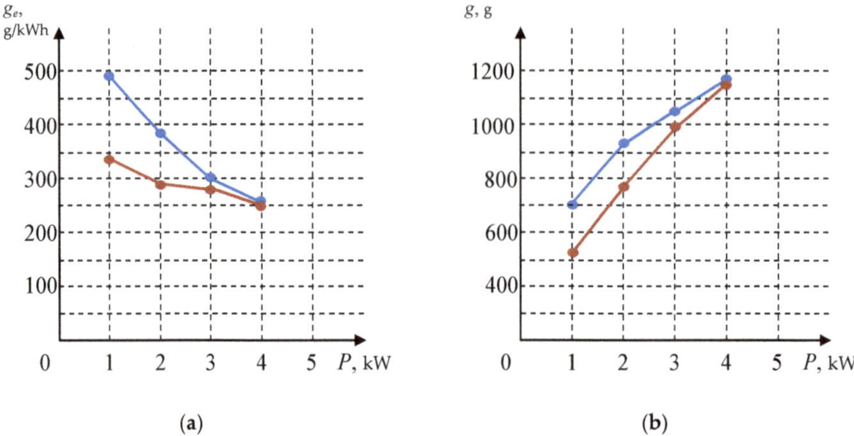

Figure 10. Dependence of specific (**a**) and absolute (**b**) fuel consumption for DGS with constant rotation speed (blue lines) and DGS with variable rotation speed with EMSD (red lines): (P) load power, (g_e) specific fuel consumption, (g) absolute fuel consumption.

In the figures, one can see that the greatest difference in the fuel consumption corresponds to the mode of low loads. The difference in the fuel consumption is decreasing with an increase in the load power (Figure 10). The fuel economy takes place due to the regulation of the engine rotation using EMSD and is the greatest with a load power of 1 kW and amounts to almost 30%.

Also, the research results have shown that the intelligent management system based on EMSD provides:

- A stabilization of the current value of the output voltage when changing wind speed and load power
- A total harmonic distortion of the output voltage that does not exceed 5%

7. Conclusions

The article is devoted to solving the problem of the low efficiency of WDPPs.

An WDPP intelligent control system has been developed, which provides effective operating modes when changing external and internal conditions. It has been proposed to use the EMSD as the basis of the intelligent control system. The key block of EMSD is an associative memory block, which is a software model of an artificial neural network. Using this unit, the optimal ICE speed for the current load power value is automatically determined. The proposed approach allows one to expand the limits of WDPP effective operation.

The algorithm of the intelligent control system has been tested on the WDPP Simulink model. The simulation results have shown that the error in determining the values of the ICE shaft rotation speed does not exceed 0.03%. This confirms the high accuracy of the algorithm.

An EMSD prototype has been created. Research of the prototype has been conducted on a test bench consisting of simulators of a 4 kW DGS and 4 kW wind turbine. The research results have shown that the use of EMSD can significantly improve the efficiency of the WDPP. The error in determining the optimal engine speed using an EMSD prototype is not more than 15%. Diesel fuel economy in the low-load area was almost 30% for variable-speed ICE with EMSD compared with the constant-speed ICE.

Author Contributions: Conceptualization, A.D. and E.S.; data curation, I.L.; formal analysis, I.L.; funding acquisition, E.S.; investigation, A.D., A.S., and I.L.; methodology, A.D.; project administration, E.S., A.S., and I.L.; resources, A.D.; software, A.D.; supervision, E.S.; validation, A.D., A.S., and I.L.; visualization, A.S.;

writing—original draft, A.S. and I.L.; and writing—review and editing, A.D. and E.S. All authors have read and agree to the published version of the manuscript.

Funding: This research was funded by the Ministry of Science and Higher Education of the Russian Federation (state task № 13.2078.2017/4.6 on 31 May 2017).

Conflicts of Interest: The authors declare no conflicts of interest. The funders had no role in the design of the study; in the collection, analyses, or interpretation of data; in the writing of the manuscript; or in the decision to publish the results.

Abbreviations

DGS	Diesel generator set
EMSD	Economy mode setting device
ICE	Internal combustion engine
WDPP	Wind-diesel power plant

References

1. Ochoa, D.; Martinez, S. Modeling an isolated hybrid wind-diesel power system for performing frequency control studies. A case of study: San Cristobal Island, Galapagos-Ecuador. *IEEE LATAM* **2019**, *17*, 775–787. [CrossRef]
2. Saad, Y.; Nohra, C.; Younes, R.; Abboudi, S.; Ilinca, A.; Ibrahim, H.; Feger, Z. Study of an optimized wind-diesel hybrid system for canadian remote sites. In Proceedings of the 2017 IEEE Electrical Power and Energy Conference (EPEC), Saskatoon, SK, Canada, 22–25 October 2017; pp. 1–6.
3. Khvatov, O.S.; Dar'enkov, A.B. Power plant based on a variable-speed diesel generator. *Russ. Electr. Engin.* **2014**, *85*, 145–149. [CrossRef]
4. Elistratov, V.V.; Bogun, I.V.; Kasina, V.I. Optimization of wind-diesel power plants parameters and placement for power supply of Russia's Northern Regions consumers. In Proceedings of the 2019 16th Conference on Electrical Machines, Drives and Power Systems (ELMA), Varna, Bulgaria, 6–8 June 2019; pp. 1–5.
5. Chernov, D.A.; Karpov, N.D.; Tyagunov, M.G. Features of design and operation of wind-diesel complexes. In Proceedings of the 2018 International Ural Conference on Green Energy (UralCon), Chelyabinsk, Russia, 4–6 October 2018; pp. 34–38.
6. Nguyen-Hong, N.; Nakanishi, Y. Optimal scheduling of an isolated wind-diesel-battery system considering forecast error and frequency response. In Proceedings of the 2018 7th International Conference on Renewable Energy Research and Applications (ICRERA), Paris, France, 14–17 October 2018; pp. 464–468.
7. Zaki Diab, A.A.; Sultan, H.M.; Mohamed, I.S.; Kuznetsov, O.N.; Do, T.D. Application of different optimization algorithms for optimal sizing of PV/wind/diesel/battery storage stand-alone hybrid microgrid. *IEEE Access* **2019**, *7*, 119223–119245. [CrossRef]
8. Nguyen-Hong, N.; Nguyen-Duc, H.; Nakanishi, Y. Optimal sizing of energy storage devices in isolated wind-diesel systems considering load growth uncertainty. *IEEE Trans. Ind. Appl.* **2018**, *54*, 1983–1991. [CrossRef]
9. Zhu, C.; Liu, F.; Hu, S.; Liu, S. Research on capacity optimization of PV-wind-diesel-battery hybrid generation system. In Proceedings of the 2018 International Power Electronics Conference (IPEC-Niigata 2018-ECCE Asia), Niigata, Japan, 20–24 May 2018; pp. 3052–3057.
10. Mi, Y.; Ma, C.; Fu, Y.; Wang, C.; Wang, P.; Loh, P.C. The SVC additional adaptive voltage controller of isolated wind-diesel power system based on double sliding-mode optimal strategy. *IEEE Trans. Sustain. Energy* **2018**, *9*, 24–34. [CrossRef]
11. Zargar, M.Y.; Mufti, M.; Lone, S.A. Voltage and frequency control of a hybrid Wind-Diesel system using SVC and predictively controlled SMES. In Proceedings of the 2017 6th International Conference on Computer Applications in Electrical Engineering-Recent Advances (CERA), Roorkee, India, 5–7 October 2017; pp. 25–30.
12. Rezkallah, M.; Chandra, A.; Saad, M.; Tremblay, M.; Singh, B.; Singh, S.; Ibrahim, H. Composite control strategy for a PV-wind-diesel based off-grid power generation system supplying unbalanced non-linear loads. In Proceedings of the 2018 IEEE Industry Applications Society Annual Meeting (IAS), Portland, OR, USA, 23–27 September 2018; pp. 1–6.

13. Esmaeilian, H.R.; Fadaeinedjad, R. A remedy for mitigating voltage fluctuations in small remote wind-diesel systems using network theory concepts. *IEEE Trans. Smart Grid* **2018**, *9*, 4162–4171. [CrossRef]

14. Tiwari, S.K.; Singh, B.; Goel, P.K. Control of wind–diesel hybrid system with BESS for optimal operation. *IEEE Trans. Ind. Appl.* **2019**, *55*, 1863–1872. [CrossRef]

15. Dar'enkov, A.B.; Guzev, S.A.; Fedorov, O.V. Autonomous power plant with variable speed based on multi-windings generator. In Proceedings of the 2017 International Conference on Industrial Engineering, Applications and Manufacturing (ICIEAM), St. Petersburg, Russia, 16–19 May 2017; pp. 1–5.

16. Baikov, A.I.; Dar'enkov, A.B.; Sosnina, E.N. Simulation modeling of a wind-diesel power plant. *Russ. Electr. Engin.* **2018**, *18*, 161–167. [CrossRef]

17. Baikov, A. *Modelirovanie Elektroprivodov s Ventil'nymi Preobrazovatelyami*; LAP LAMBERT Academic Publishing: Saarbrucken, Germany, 2014.

Article

A Firefly Algorithm Optimization-Based Equivalent Consumption Minimization Strategy for Fuel Cell Hybrid Light Rail Vehicle

Han Zhang [1,2], Jibin Yang [3,4], Jiye Zhang [1,2,*], Pengyun Song [1,5] and Xiaohui Xu [3,4]

[1] State Key Laboratory of Traction Power, Southwest Jiaotong University, Chengdu 610031, China
[2] School of Information Science and Technology, Southwest Jiaotong University, Chengdu 611756, China
[3] Key Laboratory of Fluid and Power Machinery, Ministry of Education, Xihua University, Chengdu 610039, China
[4] Key Laboratory of Automobile Measurement and Control & Safety, School of Automobile & Transportation, Xihua University, Chengdu 610039, China
[5] College of Electrical and Information Engineering, Southwest Minzu University, Chengdu 610041, China
* Correspondence: jyzhang@home.swjtu.edu.cn; Tel.: +86-139-8073-3229

Received: 3 June 2019; Accepted: 9 July 2019; Published: 11 July 2019

Abstract: To coordinate multiple power sources properly, this paper presents an optimal control strategy for a fuel cell/battery/supercapacitor light rail vehicle. The proposed strategy, which uses the firefly algorithm to optimize the equivalent consumption minimization strategy, improves the drawback that the conventional equivalent consumption minimization strategy takes insufficient account of the global performance for the vehicle. Moreover, the strategy considers the difference between the two sets of optimized variables. The optimization objective is to minimize the daily operating cost of the vehicle, which includes the total fuel consumption, initial investment, and cycling costs of power sources. The selected case study is a 100% low-floor light rail vehicle. The advantages of the proposed strategy are investigated by comparison with the operating mode control, firefly algorithm-based operating mode control, and equivalent consumption minimization strategy. In contrast to other methods, the proposed strategy shows cost reductions of up to 39.62% (from operating mode control), 18.28% (from firefly algorithm-based operating mode control), and 13.81% (from equivalent consumption minimization strategy). In addition, the proposed strategy can reduce fuel consumption and increase the efficiency of the fuel cell system.

Keywords: fuel cell; power control; multi-objective optimization; equivalent consumption minimization strategy; firefly algorithm; hybrid light rail vehicle

---

1. Introduction

With the environmental deterioration and energy crisis, developing green energy and new energy vehicles has become the consensus of government and public [1–3]. Compared with the small-power new energy vehicles like the plug-in hybrid vehicles and hybrid electric vehicles, the hybrid light rail vehicles are more suitable for urban public transportation, due to its high-power level and large-capacity.

As typical green energy, proton exchange membrane (PEM) fuel cell (FC) is suitable for the vehicle [4–6]. Proton exchange membrane fuel cell (PEMFC) has many advantages such as high reliability and efficiency, however, its dynamic response is slow. Designers often add an energy storage system (ESS) to assist the PEMFC system. The ESS can provide the extra energy, store the regenerative energy, and extend the cruising range [7].

The ESS often contains a battery (BAT) pack and/or a supercapacitor (SC) pack [8]. According to the different ESS, the PEMFC hybrid system can be divided into three types: (1) "PEMFC + BAT" [9],

(2) "PEMFC + SC" [10], (3) "PEMFC + BAT + SC" [11]. Compared with the other two combinations, the optimized "PEMFC + BAT + SC" hybrid vehicle has the best fuel economy [12]. A 100% low-floor light rail vehicle (LF-LRV) is powered by "PEMFC + BAT + SC" hybrid system [13]. As the primary power source of the LF-LRV, the PEMFC provides the average power demand. Furthermore, the SC supplement the power during acceleration, because of its high specific power. As the second power compensation, the battery improves the dynamic performance of the whole hybrid system. Besides, the battery and SC can absorb braking energy [14].

Due to the introduction of ESS, there are multiple power sources in the hybrid system. The power control strategy is used to split the power demands among the power sources and meet the load power. The strategy is mainly divided into two types: rule-based and optimization-based [15]. The rules of the rule-based strategy can be made by power source characteristics. Ahmadi et al. [16] proposed an operating mode control (OMC) for the FC hybrid vehicle, and subsequently used the genetic algorithm (GA) to optimize the OMC. Caux et al. [17] implemented GA to optimize the fuzzy logic strategy and developed an online fuzzy logic strategy for the FC hybrid vehicle. Other examples of the rule-based strategy include: power follower strategy [18], model predictive control [19], sliding mode control [20], and so on. The rule-based strategy has strong applicability; however, its control performance depends on the experience of designers.

Different from the rule-based strategy, the optimization-based strategy needs to define an optimization objective and use optimization methods to obtain the control parameters or control law. By the length of time horizon, the optimization-based strategy consists of the global optimization strategy (GO) and instantaneous optimization strategy (IO) [21]. The GO can obtain the global optimum solution for the driving cycle. Hu et al. [22] and Xu et al. [23] applied the dynamic programming to a FC hybrid vehicle, respectively, and achieved good results. However, the main insufficiency of dynamic programming is the curse of dimensionality, and this limits its application in complex systems. Another example of GO is based on the meta-heuristic algorithms, such as GA and particle swarm optimization (PSO). Li et al. [24] used the GA to optimize the control strategy for the hybrid tramway, and fuel economy is improved. Olivier et al. [25] utilized PSO to optimize five driving parameters for the FC vehicle, in order to maximize vehicle performance. Zhang et al. [26] used the multi-population GA and the artificial fish swarm algorithm to reduce the operating costs of the high-power hybrid vehicle. GO can obtain good global performance, however, these methods are hard to real-time apply due to the heavy computation load.

Instantaneous optimization methods can obtain the control command in real time. As one of the typical IO, equivalent consumption minimization strategy (ECMS) is widely used in the small and medium power vehicles [27–29] because of its excellent instantaneous performance, but less used in the high-power light rail vehicle. Torreglosa et al. [30] developed a control method based on ECMS and improved the hydrogen consumption for an FC/battery light rail vehicle. Hong et al. [3] presented a method to introduce a dynamic power factor to ECMS for an FC/battery locomotive, in order to achieve less fuel consumption and high efficiency of the system. Zhang et al. [31] improved the fixed balance coefficient of conventional ECMS and presented a control strategy based on the hybrid dynamic degree for an FC/battery light rail vehicle.

The ECMS has been verified to effectively reduce fuel consumption for the "PEMFC + BAT" vehicle. However, the strategies cannot be used directly in the "PEMFC + BAT + SC" vehicle, as its solving process only considers FC and battery. To solve this problem, Zhang et al. [21] and Garcia et al. [32] used the SC to compensate the gap of load power that the FC and battery cannot meet, and developed the ECMS for FC/battery/SC light rail vehicle respectively. Yan et al. [33] proposed an equivalent energy consumption strategy for FC/SC light rail vehicle, and aims to minimize energy consumption in multiple modes. Nevertheless, the ECMS mentioned before cannot obtain the global optimal solution because of its aim to minimize instantaneous fuel/energy consumption. In fact, the global optimal solution is useful for the further decrease of the consumption of the light rail vehicle. Moreover, the optimization objective of ECMS only considers fuel/energy consumption. Actually, some other factors such as the total fuel consumption, sizing, and lifecycle of power sources, are also important for the overall evaluation of the operating cost of the vehicle.

In order to mitigate the problems mentioned above, this paper presents an optimal control strategy for the PEMFC/battery/SC light rail vehicle. The strategy uses the firefly algorithm (FA) to optimize the ECMS, and the aim is to obtain the global optimal solution. The optimization objective includes the total fuel consumption cost, initial investment, and replacement cost of power sources. The selected case study is a 100% low-floor light rail vehicle. The paper is structured as follows. In Section 2, the model of the vehicle is introduced. In Section 3, an improved ECMS combined conventional ECMS with OMC is designed. In Section 4, the FA is introduced. Subsequently, the optimization objective is defined. Based on this objective, an optimal control strategy integrates FA and ECMS is developed at last. The results are discussed in Section 5. In Section 6, the conclusions are summarized.

2. Vehicle Modeling

A 100% LF-LRV is assembled by the CRRC Tangshan Co. Ltd. and Southwest Jiaotong University [34], as shown in Figure 1a. The vehicle includes two motor units, powered by the PEMFC system, SC pack, and battery pack [13]. The PEMFC system is the primary power source that provides the average traction power. Besides, the SC pack and battery pack are utilized for the ESS to help the PEMFC system during the vehicle cruise and acceleration and can absorb energy during braking. The diagram of the hybrid system's power flow is shown in Figure 1b.

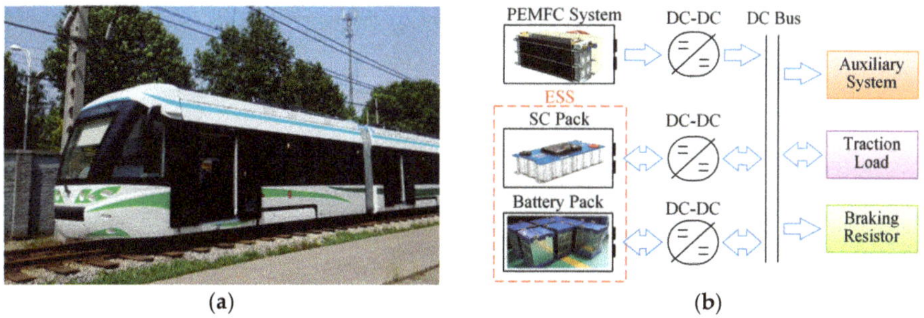

(a) (b)

Figure 1. The low-floor light rail vehicle: (**a**) realistic images; (**b**) structure of the hybrid system.

In order to test the performance of the proposed strategy, a backward/forward LF-LRV model is established. The model consists of eight blocks: the driving cycle, train longitudinal dynamics, wheel dynamics, transmission system, motor, DC bus/auxiliary load, control strategy, and hybrid system. The description of the whole model has been proposed in our previous research [35,36], and a brief introduction of the hybrid system model is given below.

2.1. PEMFC

PEMFC is a kind of electrochemical device, which can transform the chemical energy of the fuel into electrical energy. The anode and the cathode electrode of the PEMFC are separated by a proton exchange membrane [37]. Based on the Ballard HD6 Module [38], the PEMFC model is built. The stack output voltage can be expressed as [39]:

$$U_f = U_{foc} - U_a - U_{fo} \tag{1}$$

With

$$\begin{cases} U_{foc} = k_{fc}(U_n - U_c) \\ U_a = \frac{1}{t_d s + 1} N_f A \ln\left(\frac{i_f}{i_{fo}}\right) \\ U_{fo} = R_{fo} i_f \end{cases} \tag{2}$$

where U_{foc} is the open-circuit voltage, U_a is the activation voltage, U_{fo} is the ohmic voltage, k_{fc} is a fit coefficient, U_n is the Nernst voltage, U_c is the voltage drop caused by a decrease in the gas concentration of the reactants, t_d is the dynamic response time constant, N_f is the number of cells, i_f is the cell output current, and R_{fo} is inner resistance of a stack. The efficiency of the PEMFC system is described in Figure 2a.

2.2. Battery

In LF-LRV, the lithium-ion battery is selected as the ESS. The lithium-ion battery is a green and rechargeable power device. It can transform the chemical energy of the active materials into the electrical energy and can be recharged by the converse process. The battery is utilized, capturing the braking energy coordinate with SC, and supplying a portion of load power coordinate with PEMFC. In the study, a Li-ion battery model was applied. The output voltage of the battery is defined as [40]:

$$\begin{cases} U_b = U_{bo} - R_b i_b - k_b \frac{Q_b}{Q_b - i_b t}\left(i_b t + i_b^*\right) + \alpha_b e^{(-\beta_b i_b t)} & (i_b \geq 0) \\ U_b = U_{bo} - R_b i_b - k_b \frac{Q_b}{i_b t - 0.1 Q_b} i_b^* - k_b \frac{Q_b}{Q_b - i_b t} i_b t + \alpha_b e^{(-\beta_b i_b t)} & (i_b < 0) \end{cases} \tag{3}$$

where U_{bo}, i_b, Q_b, and i_b^* are the constant voltage, output current, maximum capacity, and filtered current of the battery, respectively, k_b, α_b, and β_b are the parameters of polarization constant, exponential zone amplitude, and exponential zone time constant inverse, respectively.

The state of charge (SOC) is used to evaluate the current status of the power source. The relation between SOC and open circuit voltage (OCV) of the battery measured in the experiment is shown in Figure 2b, and the SOC is calculated as:

$$SOC_b = 1 - \frac{1}{Q_b} \int i_b dt \tag{4}$$

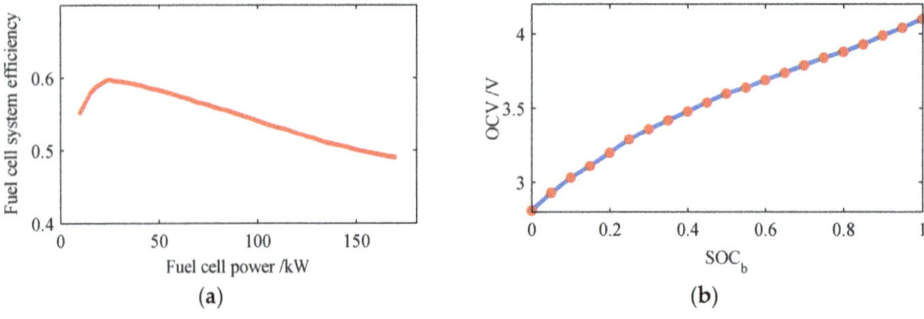

Figure 2. The data of power sources: (**a**) efficiency of proton exchange membrane fuel cell system; (**b**) open circuit voltage-state of charge curve of battery.

2.3. Supercapacitor

A supercapacitor is a storage unit, which stores electrical energy on high-surface-area conducting materials [41]. Compared with battery, SC is suitable to restrain the power fluctuation in short-time for the high-power hybrid vehicle, due to its structure of an electric double layer capacitor. In this work, an SC equivalent electrical model is selected. The model consists of an ideal capacitance (the voltage is U_{sc}), an equivalent series resistance (R_{ss}), and an equivalent parallel resistance (R_{sl}). They simulate the SC ideal performance, the resistance, and the self-discharging losses, respectively. The SC output voltage is expressed as [42]:

$$\begin{cases} U_s = U_{sc} - i_s R_{ss} \\ i_s = i_{sc} + i_{sl} \\ i_{sc} = -C_{sc} \cdot \dfrac{dU_{sc}}{dt} \\ i_{sl} = \dfrac{U_{sc}}{R_{sl}} \end{cases} \tag{5}$$

where i_s, i_{sl}, and i_{sc} are the currents flowing through R_{ss}, R_{sl}, and the ideal capacitance respectively, and C_{sc} is the capacitance. The SOC of the SC is calculated as:

$$SOC_s = \frac{U_s^2}{U_{smax}^2} \tag{6}$$

where U_{smax} is the maximum voltage of SC.

As the battery and SC can release and absorb the energy, their power and current can get positive or negative values. Moreover, the validity of PEMFC, battery, and SC models mentioned above have been verified by the contrast of experimental data in our previous research [39]. Besides, the efficiency model of DC-DC converters has been considered as a constant value (90%), due to the backward/forward LF-LRV model mainly simulate the power flow throughout the power sources [43,44].

3. Equivalent Consumption Minimization Strategy

The current power control strategy of the LF-LRV is based on the OMC. However, the strategy has some deficiencies, such as the large fuel consumption of the fuel cell, the low utilization rate of the battery, and the less efficiency of the vehicle [24]. So, the strategy needs to be improved, and the IO strategy is considered as an improvement scheme because of its good fuel economy.

The IO strategy controls the power split of power sources according to an instantaneous optimization objective. In order to consider fuel and electricity consumption in a single objective, the equivalent fuel consumption was defined by Paganelli et al. [45]. Based on that, Xu et al. [46] proposed an ECMS to optimize the fuel economy for an FC/battery city bus. Due to the conventional ECMS not considering the factor of SC, the strategy needs to be improved for the PEMFC/battery/SC light rail vehicle. First, the conventional ECMS is detailed, and finally, an improved ECMS which considers the factor of SC is presented.

3.1. Conventional ECMS

Conventional ECMS aims to convert the battery's electrical energy into the equivalent fuel consumption. For the vehicle, the strategy calculates the optimal power of the FC system ($P_{f,opt}$), which minimizes hydrogen fuel consumption (C_1) and consists of FC hydrogen consumption (C_f) and battery equivalent hydrogen consumption (C_b). The optimization problem can be formulated as [30,31]:

$$P_{f,opt} = \overset{argminC_1}{P_f} = \overset{argmin(C_f + k_c C_b)}{P_f} \tag{7}$$

where P_f is the output power of the FC system, k_c is the penalty co-efficient.

Otherwise, C_f can be expressed as a function of P_f:

$$C_f = a_f P_f + b_f \tag{8}$$

where a_f and b_f are fit coefficients.

C_b is calculated based on the battery power P_b:

$$C_b = P_b \cdot \sigma \cdot \frac{C_{f,avg}}{P_{f,avg}} \tag{9}$$

with:

$$\sigma = \begin{cases} \dfrac{1}{\eta_{bc,avg} \cdot \eta_{bd}} & P_b \geq 0 \\ \eta_{bc} \cdot \eta_{bd,avg} & P_b < 0 \end{cases} \tag{10}$$

where $C_{f,avg}$ and $P_{f,avg}$ are the average hydrogen consumption and average power of the FC system respectively, η_{bd} and η_{bc} are the discharging efficiencies and charging efficiencies of the battery respectively, $\eta_{bd,avg}$ and $\eta_{bc,avg}$ are the mean efficiencies.

η_{bd} and η_{bc} are defined by:

$$\begin{cases} \eta_{bd} = \dfrac{1}{2}\left(1 + \sqrt{1 - \dfrac{4R_{bd}P_b}{U_b^2}}\right) \\ \eta_{bc} = 2\Big/\left(1 + \sqrt{1 - \dfrac{4R_{bc}P_b}{U_b^2}}\right) \end{cases} \tag{11}$$

where R_{bd} and R_{bc} are the discharging resistances and charging resistances of the battery respectively.

The penalty coefficient k_c is expressed by:

$$k_c = 1 - 2\mu_b \cdot \frac{SOC_b - 0.5 \cdot (SOC_{bh} + SOC_{bl})}{SOC_{bh} + SOC_{bl}} \tag{12}$$

where μ_b is the balance factor, and it is 0.6 in [21,30,46], SOC_{bh} and SOC_{bl} are the upper and lower limit of SOC_b respectively.

Now, some variables can define as follows:

$$\begin{cases} \rho_1' = k_c / \eta_{bc,avg} \\ x_{min} = \sqrt{1 + 4U_{b,min} \cdot (U_{b,min} - U_b)/U_b^2} \\ x_{max} = \sqrt{1 + 4U_{b,max} \cdot (U_{b,max} - U_b)/U_b^2} \end{cases} \tag{13}$$

where, $U_{b,min}$ and $U_{b,max}$ are the minimum and maximum voltage of battery respectively. Based on Equations (8–13), the solutions of the optimization problem raised by Equation (7) can be expressed:

$$\begin{cases} P_{b,opt} = \begin{cases} (U_b - U_{b,min})U_{b,min}/R_{bd} & \rho_1' \leq x_{min} \\ \left(1 - (\rho_1')^2\right)U_b^2/4R_{bd} & x_{min} < \rho_1' \leq 1 \\ 0 & 1 < \rho_1' \leq 1/(\eta_{bc,avg}\eta_{bd,avg}) \\ \left[1 - (\rho_1'\eta_{bc,avg}\eta_{bd,avg})^2\right]U_b^2/4R_{bc} & 1/(\eta_{bc,avg}\eta_{bd,avg}) < \rho_1' \leq x_{max}/(\eta_{bc,avg}\eta_{bd,avg}) \\ -(U_{b,max} - U_b)U_{b,max}/R_{bc} & \rho_1' \geq x_{max}/(\eta_{bc,avg}\eta_{bd,avg}) \end{cases} \\ P_{f,opt} = P_m + P_{aux} - P_{b,opt} \end{cases} \tag{14}$$

where $P_{b,opt}$ is the optimal power of the battery, P_m is the output power of the electric motor, P_{aux} is the power consumed by the auxiliary components of the hybrid vehicle.

ECMS has the advantages of good optimal effect and high real-time. However, the strategy can only obtain the optimized power of FC and battery. For the PEMFC/battery/SC LF-LRV scenario, the strategy should be improved in the next section.

3.2. Improved ECMS

Operating mode control is a typical rule-based strategy. Due to its good adaptation and low-cost computing, it is widely used for FC hybrid vehicles [16,47]. However, the fuel costs of OMC are often more expensive compared with ECMS, due to its control effect relying on expertise. To use the ECMS to control the power distribution for PEMFC/battery/SC light rail vehicle, an improved ECMS combined with conventional ECMS and OMC has been tried.

The improved ECMS is based on an assumption: compared to FC and battery, SC aims to generate the peak powers in short periods, and the equivalent consumption of SC can be ignored [21,32]. The strategy contains two processes: the ECMS process and the OMC process. In the ECMS process, $P_{b,opt}$ and $P_{f,opt}$ can be calculated according to Section 3.1, and they are used as the input variables of the OMC process.

In the OMC process, the slope limitation and maximum/minimum limitation is considered. FC actual output power P_f can be determined at first. Then, six modes are defined. According to correlated variables and conditions, battery actual power output P_b and SC actual power output P_s can then finally be determined. The structure of the improved ECMS is depicted in Figure 3, where P_{fmax} and P_{fmin} are the maximum power and minimum power limitations of the FC, respectively, P_d is the power demand of DC bus, C_1, C_2, C_3, and C_4 represent the judgment conditions, M_1, M_2, M_3, M_4, M_5, and M_6 represent the operating modes of the vehicle.

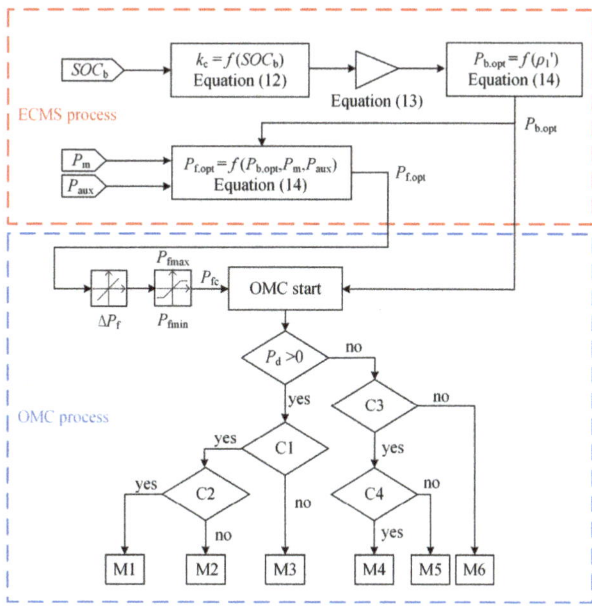

Figure 3. The structure of improved equivalent consumption minimization strategy (ECMS).

Because the energy storage system of the LF-LRV contains two power sources: battery and supercapacitor, the improved ECMS can be divided into two types by the priority of energy supply in ESS: battery-prior ECMS (BP-ECMS) and supercapacitor-prior ECMS (SP-ECMS). Their judgment conditions and operating modes are summarized in Table 1, where P_{bmax} and P_{bmin} are the maximum power and minimum power limitations of the battery, respectively, P_{smax} and P_{smin} are the maximum power and minimum power limitations of SC, respectively, SOC_{bh} and SOC_{sh} are the maximum SOC limitations of battery and SC, respectively, P_{br} is the output power of braking resistor. The comparison of BP-ECMS with SP-ECMS will be shown in Section 5.

So far, the ECMS has been applied to the PEMFC/battery/SC light rail vehicle. ECMS aims to solve the optimization problem in Equation (7), which is to minimize the instantaneous fuel consumption. However, to evaluate the performance of the vehicle comprehensively, many more factors need to be considered such as the total fuel consumption, replacement and investment costs of power sources. These factors can be considered as global performance. The next section attempts to enhance the global performance of ECMS for the hybrid vehicle.

Table 1. The conditions and modes of operating mode control (OMC) process for ECMS.

Condition/Mode	BP-ECMS	SP-ECMS
C1	$P_{b,opt} < P_{bmax}$	$P_d - P_f - P_{b,opt} < P_{smax}$
C2	$P_{b,opt} > P_{bmin}$	$P_d - P_f - P_{b,opt} > P_{smin}$
C3	$\left(P_{b,opt} > 0\right)\&\left(SOC_s > SOC_{sh}\right)$	$\left(P_{b,opt} > 0\right)\&\left(SOC_s > SOC_{sh}\right)$
C4	$SOC_b < SOC_{bh}$	$SOC_b < SOC_{bh}$
M1	$P_b = P_{b,opt},\, P_s = P_d - P_f - P_b$	$P_s = P_d - P_f - P_{b,opt},\, P_b = P_d - P_f - P_s$
M2	$P_b = P_{bmin},\, P_s = P_d - P_f - P_b$	$P_s = P_{smin},\, P_b = P_d - P_f - P_s$
M3	$P_b = P_{bmax},\, P_s = P_d - P_f - P_b$	$P_s = P_{smax},\, P_b = P_d - P_f - P_s$
M4	$P_s = P_{smax},\, P_b = P_d - P_f - P_s$	$P_s = P_{smax},\, P_b = P_d - P_f - P_s$
M5	$P_b = P_{b,opt},\, P_s = P_{smax},$ $P_{br} = P_d - P_f - P_b - P_s$	$P_b = P_{b,opt},\, P_s = P_{smax},$ $P_{br} = P_d - P_f - P_b - P_s$
M6	$P_b = P_{b,opt},\, P_s = P_d - P_f - P_b$	$P_s = P_d - P_f - P_{b,opt},\, P_b = P_d - P_f - P_s$

4. Optimal Control Strategy

Researchers often use the meta-heuristic algorithms to improve the global performance of the control strategy for the hybrid vehicle [24,25,44]. With this in mind, this section attempts to optimize the ECMS by the meta-heuristic algorithm. First, a meta-heuristic algorithm - the firefly algorithm, is detailed. Then, an optimization objective is introduced to evaluate the global performance of the vehicle. Finally, the optimizing ECMS with the firefly algorithm is proposed.

4.1. Firefly Algorithm

Yang [48] developed a meta-heuristic optimizer named the firefly algorithm, which is inspired by the flashing light of fireflies. Compared with other classical meta-heuristic algorithms, such as GA and PSO, the FA showed a better optimization efficiency [48]. Therefore, the algorithm is widely used in many engineering fields [49–51] and is adopted in this paper.

In FA, the higher the brightness of the firefly is the more attractive it is, and the attraction of each firefly can be calculated:

$$\beta = \beta_0 \cdot e^{-\gamma r^2} \tag{15}$$

where β_0 is the attraction at $r = 0$. The distance between any two fireflies i and j at x_i and x_j respectively, is $r_{ij} = x_i - x_j$. For any given two fireflies, x_i and x_j, the movement of firefly i is attracted to another firefly j, which is more attractive:

$$x_i = x_i + \beta_0 \cdot e^{-\gamma r_{ij}^2}\left(x_j - x_i\right) + \alpha \varepsilon_i \tag{16}$$

where the third term represents random movement, α and ε_i are the randomization parameter and a vector of random numbers respectively.

To calculate the attraction of the firefly by Equation (15), an optimization objective will be defined in the next section.

4.2. Optimization Objective

For the PEMFC/battery/SC light rail vehicle, reducing the fuel consumption of the FC system might lead to an increase in the use of the ESS and hence accelerate the degradation of it, and vice versa. This can be regarded as a multi-objective problem.

To solve the multi-objective problem, one needs to define an optimization objective, which determines the optimized direction. Considering the scenario of this paper, the objective is defined as follows:

$$\min_{X \in \Omega} C_{tram}(X) = [C_{fc}(X), C_{sc}(X), C_{bat}(X)] \tag{17}$$

With

$$\begin{cases} C_{fc} = (FC_{fuel} + FC_{ca} + FC_{re} + FC_{m})/365 \\ C_{sc} = (SC_{ca} + SC_{re} + SC_{m} + SC_{chg})/365 \\ C_{bat} = (BAT_{ca} + BAT_{re} + BAT_{m} + BAT_{chg})/365 \end{cases} \tag{18}$$

where C_{tram} is the total operating cost of the vehicle, C_{fc}, C_{sc} and C_{bat} are the cost models of the PEMFC system, SC pack, and battery pack, respectively.

FC_{fuel} is the annualized total fuel cost due to the PEMFC system operation:

$$FC_{fuel} = m_{fuel} \cdot C_{fuel} = C_{fuel} \cdot \int \frac{P_f}{LHV \cdot \eta_{fc}} dt \tag{19}$$

where m_{fuel} is the fuel consumption of the PEMFC, C_{fuel} is the cost of the fuel, LHV is the low heat value of hydrogen, and η_{fc} is the PEMFC efficiency.

FC_{ca} is the annualized capital cost related to the initial investment:

$$FC_{ca} = (P_{dc\_fc}C_{dc} + Ca_{fc}C_{fc}) \cdot CRF \tag{20}$$

where P_{dc_fc} is the DC-DC rated power of the PEMFC system, C_{dc} is the cost of the DC-DC, Ca_{fc} is the rated power of the PEMFC system, C_{fc} is the cost of the PEMFC, and CRF is the capital recovery factor.

FC_{re} is the annualized replacement cost:

$$FC_{re} = \sum_{i=1}^{N_{r\_fc}} (1+I)^{-i \cdot Life_{fc}} \cdot Ca_{fc} \cdot C_{fc} \cdot CRF \tag{21}$$

With

$$N_{r\_fc} = ceil(T/Life_{fc} - 1) \tag{22}$$

where ceil(x) is the function that returns the smallest integer that is not less than x, I is the interest rate, T is the lifetime of the system, $Life_{fc}$ is the life expectancy of the PEMFC and FC_m is the average maintenance cost.

SC_{ca}, SC_{re}, SC_m, and SC_{chg} are the annualized capital cost, the annualized replacement cost, the annualized maintenance cost, and the annualized charged cost of the SC pack, respectively. BAT_{ca}, BAT_{re}, BAT_m, and BAT_{chg} are the annualized capital cost, the annualized replacement cost, the annualized maintenance cost, and the annualized charged cost of the battery pack, respectively. They are defined as follows:

$$\begin{cases} ESS_{ca} = (P_{dc\_ess}C_{dc} + Ca_{ess}C_{ess}) \cdot CRF \\ ESS_{re} = \sum_{i=1}^{N_{r\_ess}} (1+I)^{-i \cdot Life_{ess}} \cdot Ca_{ess} \cdot C_{ess} \cdot CRF \\ ESS_{chg} = \frac{E_{ess}(SOC_{init\_ess} - SOC_{end\_ess}) \cdot C_{grid}}{\xi_{dc}} \end{cases} \tag{23}$$

where the ESS/ess can represent the BAT or SC scenario. E_{ess} is a function that calculates the energy of the battery or supercapacitor from the termination state SOC_{init_ess} to the initial state SOC_{end_ess}, C_{grid} is the cost of the energy from the grid, and ξ_{dc} is the efficiency of the DC-DC. The definitions of other symbols in Equation (23) are similar to the PEMFC scenario. FC_m, SC_m, and BAT_m are considered the fixed value.

In Equation (18), the FC_{ca}, SC_{ca}, and BAT_{ca} are related to the sizing of the corresponding power sources. Besides, the FC_{re}, SC_{re}, and BAT_{re} are related to the lifecycle of the power sources. Moreover, the SC_{chg} and BAT_{chg} are used to calculate the electricity costs which the power sources recharge to the initial state. The cost models in Equation (18) are expressed in economic terms, and have been detailed by our previous research in [26].

The feasible solution space Ω is subject to the following constraints:

$$
\begin{cases}
30\% \leq SOC_b \leq 90\% \\
30\% \leq SOC_s \leq 100\% \\
10\ \text{kW} \leq P_f \leq 170\ \text{kW} \\
-200\ \text{kW} \leq P_b \leq 200\ \text{kW} \\
-300\ \text{kW} \leq P_s \leq 300\ \text{kW} \\
-20\ \text{kW/s} \leq \frac{dP_f}{dt} \leq 20\ \text{kW/s}
\end{cases}
\tag{24}
$$

These constraints aim to ensure that the power sources do not overcharge or over-discharge, while also taking the limit of device operating conditions into account. It is worth to mention that the optimization objective Equation (18) is used to evaluate the global performance of the hybrid vehicle.

4.3. Control Strategy

So far, a new control framework can be proposed, and its aim is to get better global performance than the ECMS. The framework, named the firefly algorithm-based ECMS (F-ECMS), is applied to FA in order to optimize the variables of ECMS proposed in Section 3.2, and the objective is to minimize the total operating cost for PEMFC/battery/SC light rail vehicle.

The choice of optimized variables affects the results of the optimization problem. As shown in Figure 3, the ECMS contains two processes. In the ECMS process, the calculation of $P_{b,opt}$ depends on the balance factor μ_b and SOC_b, as seen in Equations (12)–(14). The relationships between $P_{b,opt}$ and SOC_b, with different balance factor (BF), are illustrated in Figure 4a. It means that the balance factor can affect $P_{b,opt}$, and so affect $P_{f,opt}$. For this reason, the balance factor μ_b is selected as the optimized variables.

Moreover, in the OMC process, the maximum power and minimum power limitations of power sources (e.g., P_{fmax}, P_{bmax}, P_{smax} etc.) have a large impact on the choice of operating mode, and thus affect the vehicle performance, as seen in Figure 3 and Table 1. So, these power limitations of power sources are considered as the candidates of optimized variables.

After the optimized variables are selected, the process of F-ECMS can be determined:

Step 1: Determine the number of iterations X and fireflies N.

Step 2: According to the selected optimal variables, initial the swarm of firefly within the confined region.

Step 3: Take out the nth firefly, it represents a solution. Put it to the ECMS (Section 3.2).

Step 4: Apply the above-mentioned ECMS to the vehicle model (Section 2). Then the model is used for simulation.

Step 5: According to the simulation results, calculate the global performance index C_{tram} (Section 4.2).

Step 6: Repeat steps 3 to 5 until the corresponding performance index of all fireflies are acquired.

Step 7: Calculate the attraction of each firefly with Equation (15) and update the locations of all fireflies with Equation (16).

Step 8: Repeat steps 6 and 7 until X iterations are complete.

Step 9: The location of the most attractive firefly is the best solution for the optimization problem.

The flowchart of the F-ECMS process is summarized in Figure 5.

In order to reveal the impact of the optimized variables selection on optimization effect, a comparison between two different sets of optimized variables is made. The one set is to optimize the signal variable: balance factor μ_b. Another set is to optimize multiple variables: balance factor and power limitations. The optimized process and results of two sets are shown in Figure 4b, where 'S-OV' means the situation of signal optimized variable, 'M-OV' means the situation of multiple optimized variables, and 'B-data' means the situation of the benchmark data produced by ECMS. The higher attraction means a better effect. It can be seen that:

1. Only optimizing the balance factor of F-ECMS can obtain a better result (2.875) than ECMS (2.826).
2. Optimizing the balance factor and power limitations can obtain the best result (3.279).

So, both balance factor and power limitations are selected to the optimized variables for F-ECMS in this paper.

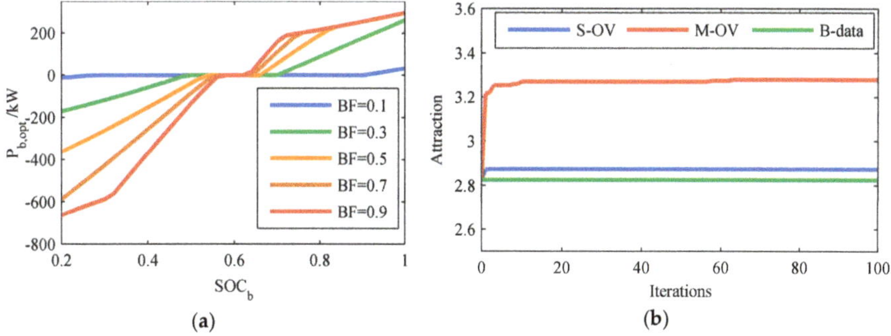

Figure 4. The choice of the optimized variables: (a) the relationship between $P_{b,opt}$ and SOC_b; (b) optimized process of the firefly algorithm-based equivalent consumption minimization strategy (F-ECMS).

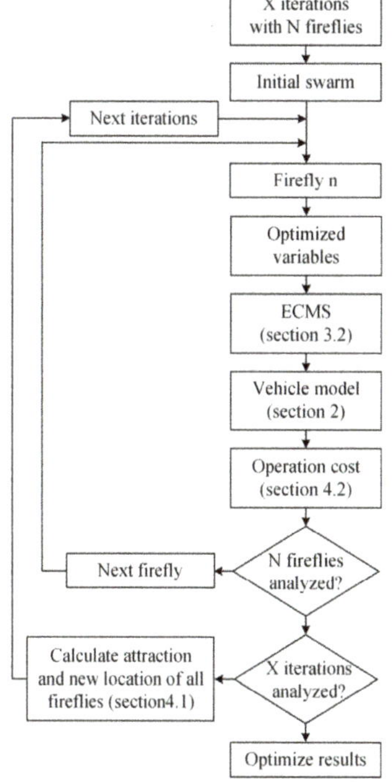

Figure 5. The process of F-ECMS.

5. Results and Analysis

This section reports the validation of the proposed strategy in the LF-LRV scenario. In order to verify the control strategy, the model of the vehicle is constructed using MATLAB/Simulink, as shown in Figure 6. The model according to the target speed profile, calculates the speed demand, torque demand, or power demand of each part for the hybrid vehicle. The comparison between actual data and model simulation of the vehicle has been proposed in our previous research [52], and the error is less than 1%.

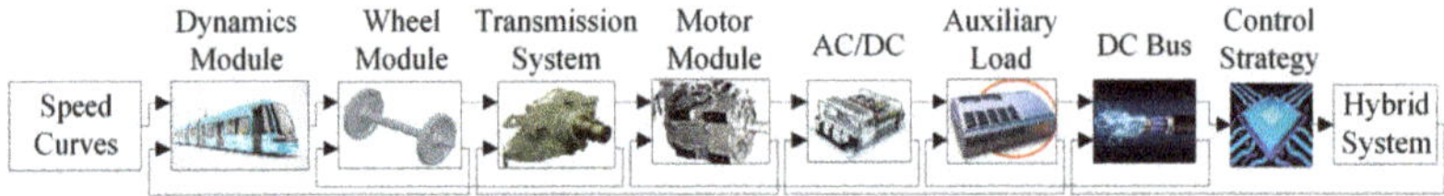

Figure 6. Vehicle simulation model.

The driving cycle is based on the Guanggu T1 line in Wuhan, China, which is 15 km in length and has 23 stations. The corresponding target speed profile of the vehicle as shown in Figure 7.

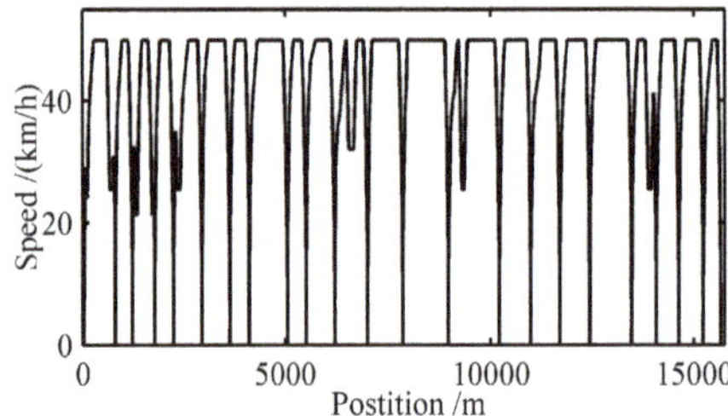

Figure 7. Target speed profile.

To accomplish the driving cycle, the PEMFC system's power of the vehicle should be above the cycle's average power, which is 200.8 kW. So, two PEMFC systems are selected, and its rated power is 150 kW. The peak traction power and braking power of the driving cycle are 696.7 kW and 636.7 kW. So, the hybrid system, in a motor unit, consists of an FC system (150 kW), two battery packs (40 Ah) and three SC packs (45 F). The basic parameters of the power sources are summarised in Table 2, and the parameters of the objective function can be found in [26].

Table 2. Basic parameters of power sources.

Power source	Parameter	Value
PEMFC	Maximum power (kW)	170
	Rated voltage (V)	540
	Maximum current (A)	320
	Cell number	735
Battery cell	Rated voltage (V)	3.2
	Rated capacity (Ah)	40
	Operating voltage (V)	2.5–3.8
	Operating temperature (K)	253–318
SC cell	Rated voltage (V)	2.7
	Rated capacity (F)	3000
	Maximum current (A)	210
	Operating temperature (K)	233–335

To show the performance of the strategy mentioned in this paper clearly, two strategies are chosen as benchmarks: one is the OMC strategy [26], and another is FA-based OMC (F-OMC), which uses the

FA to optimize the power limitations of OMC. Their corresponding operation cost of the vehicle is as shown in Figure 8a. Because of the FA optimizing, the F-OMC ($373.20/day) can get lower costs than OMC ($505.13/day). It is worth to mention that in Figure 8a, the 'Fuel' cost represents the fuel cost consumed by the FC system, and the 'FC' cost includes the cost of initial investment, replacement, and maintenance for the FC system.

First, two types of ECMS: BP-ECMS and SP-ECMS, have been evaluated. The corresponding operating costs are shown in Figure 8a. Compared with the OMC ($505.13/day), the costs based on the BP-ECMS ($374.45/day) and SP-ECMS ($353.84/day) can be decreased to varying degrees. It is because the ECMS can optimize fuel consumption, which is the major part of the operating cost. Besides, the cost of SP-ECMS is lower than BP-ECMS. The reason is as follows; when compared with BP-ECMS, the usage of battery in SP-ECMS is higher, which results in a reduction of fuel cost. It can also be seen that the result of SP-ECMS is better than F-OMC ($373.20/day), due to the good instantaneous performance of the ECMS. As the operating costs of SP-ECMS precede BP-ECMS as mentioned above, the SP-ECMS is used to represent the ECMS in a followed comparison. Moreover, in this paper, the F-ECMS combines the FA and SP-ECMS.

Then, the F-ECMS has been evaluated. The result is shown in Figure 8a. Compared with SP-ECMS ($353.84/day), the operating costs of F-ECMS ($304.99/day) could be further reduced, due to the introduction of FA. It also proves that F-ECMS has the best global performance in this paper.

To observe the fuel economy of the strategies, the total fuel consumption resulting for the driving cycle are shown in Figure 8b. The fuel consumption of the F-ECMS is 1.68 kg, the OMC is 3.43 kg, the F-OMC is 2.47 kg, and the SP-ECMS is 2.12 kg. It means that, compared with OMC, the introduction of the ECMS (i.e. SP-ECMS) or FA (i.e. F-OMC) can improve the fuel economy for the hybrid vehicle, and combine both of them (i.e. F-ECMS) to achieve the best results.

In order to demonstrate the performance advantage of the F-ECMS compared with SP-ECMS more clearly, the operating point distributions of the FC system are shown in Figure 9a. The FC system output power is mainly distributed in two regions: low-power region and high-power region. The low-power region of SP-ECMS and F-ECMS is around 18 kW, which means that the vehicle operates in the regenerative braking phase. Because of the optimization of FC power limitations, the high-power region of F-ECMS around 115 kW. In contrast, the high-power region of SP-ECMS around 170 kW. This should result in the average efficiency of F-ECMS (55.85%) is high than SP-ECMS (55.35%).

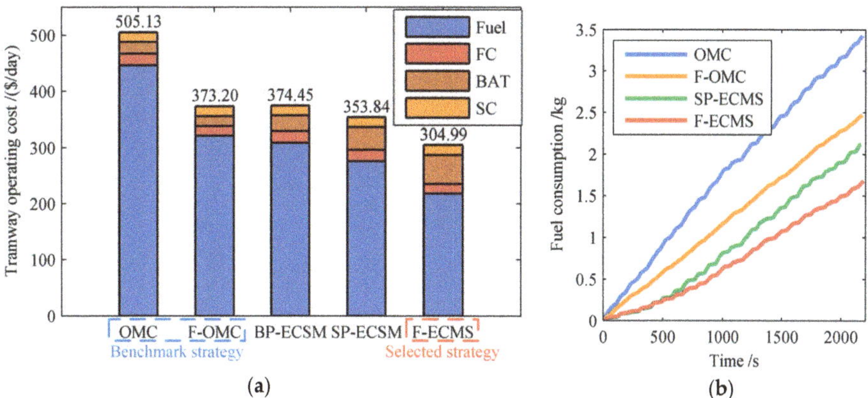

Figure 8. The results of different strategies: (**a**) optimal solutions and operating cost; (**b**) fuel consumption.

The fluctuation distribution comparison of FC power between F-ECMS and SP-ECMS is shown in Figure 9b, and the lower standard deviation means the better life expectancy of the FC system. By using FA to optimize the SP-ECMS, the limiting effect of F-ECMS (7.6176) is better than SP-ECMS (9.3744).

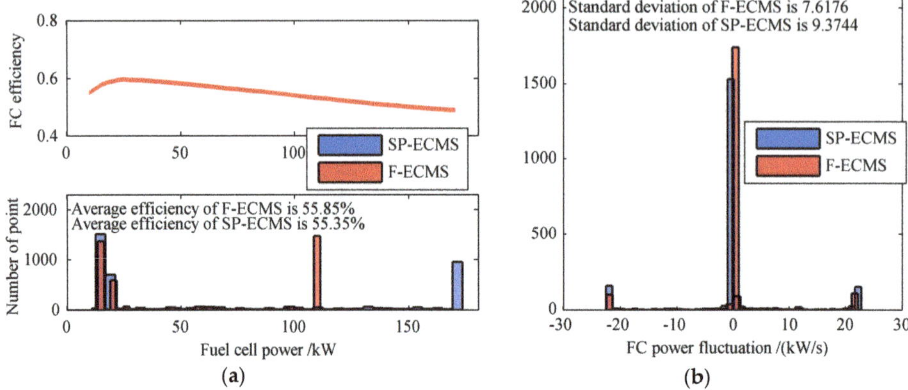

Figure 9. The results of fuel cell (FC) System: (**a**) distribution comparison; (**b**) power fluctuations.

As the main objective of this paper is to minimize the operation cost, the F-ECMS is finally selected as the control strategy for the LF-LRV. The power curves of the power sources for F-ECMS are illustrated in Figure 10. It can be seen that the F-ECMS can ensure that the hybrid system works steadily, and that the output power among the power sources rational distribution. Table 3 summarizes the improvement of F-ECMS compared with the OMC, F-OMC, and SP-ECMS.

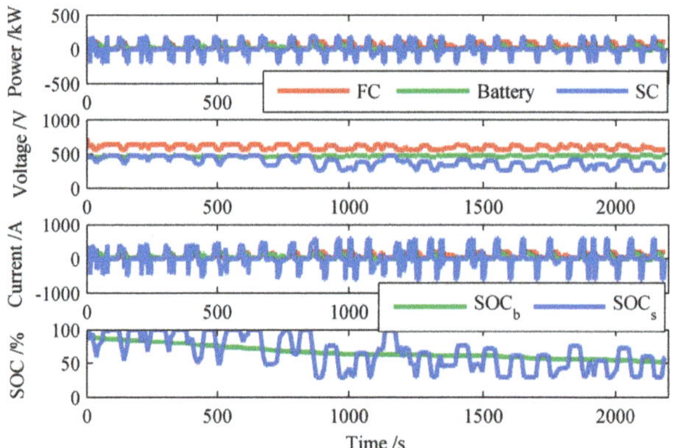

Figure 10. Performance of power sources for F-ECMS.

Table 3. The improvement of F-ECMS.

Indicators	OMC	F-OMC	SP-ECMS
Methods	Introduces ECMS and FA	Introduces ECMS	Optimized by FA
Operating cost	Save 39.62%	Save 18.28%	Save 13.81%
Fuel consumption	Save 51.02%	Save 31.98%	Save 20.75%
FC system efficiency	Improve 3.34%	Improve 0.88%	Improve 0.5%

6. Conclusions

This paper presents the optimal control strategy for a light rail vehicle, which is powered by the fuel cell, battery, and supercapacitor. The aim is to improve the global performance of the equivalent consumption minimization strategy (ECMS).

As the conventional ECMS cannot be directly applied to the fuel cell/battery/supercapacitor hybrid system, a strategy combined ECMS and operating mode control is presented. Then, the firefly algorithm is introduced to optimize the strategy, in order to enhance the global performance. The optimization objective is to minimize the operating costs of the vehicle, which contains the total fuel consumption, initial investment, and replacement costs of power sources. The balance factor and power limitations are considered as the candidate of the optimized variables.

A 100% low-floor light rail vehicle is selected as the case study. The operation costs of the vehicle based on battery-prior ECMS, supercapacitor-prior ECMS, and the proposed strategy are obtained. The operating mode control and firefly algorithm-based operating mode control are used as benchmarks. The simulation results indicate that the proposed strategy offers the lowest operating cost. Compared with operating mode control, firefly algorithm-based operating mode control, and supercapacitor-prior ECMS, the proposed strategy can save 39.62%, 18.28%, and 13.81% cost, respectively. Moreover, the strategy can reduce fuel consumption and increase the efficiency of the fuel cell system. It can be concluded that among all of the strategies mentioned above, the proposed strategy not only satisfies the power/energy demand, but also offers the best global performance for the hybrid vehicle.

Author Contributions: Conceptualization, H.Z.; methodology, H.Z.; software, H.Z. and J.Y.; validation, J.Y. and P.S.; formal analysis, H.Z.; investigation, H.Z. and J.Y.; resources, J.Z. and P.S.; data curation, H.Z.; writing—original draft preparation, H.Z.; writing—review and editing, H.Z., J.Y. and J.Z.; visualization, H.Z. and J.Y.; supervision, J.Z., P.S. and X.X.; project administration, J.Z., P.S. and X.X.; funding acquisition, J.Z. and J.Y.

Funding: This research was funded by the National Natural Science Foundation of the People's Republic of China, grant number 11572264; the Science and Technology Major Project of Sichuan Province, grant number 2019ZDZX0002; the Open Research Subject of Key Laboratory of Fluid and Power Machinery (Xihua University), Ministry of Education, grant number szjj2019-015.

Conflicts of Interest: The author declares no conflict of interest.

References

1. Yue, M.; Jemei, S.; Gouriveau, R.; Zerhouni, N. Review on Health-Conscious Energy Management Strategies for Fuel Cell Hybrid Electric Vehicles: Degradation Models and Strategies. *Int. J. Hydrogen Energy* **2019**, *44*, 6844–6861. [CrossRef]
2. Hidalgo-Reyes, J.I.; Gomez-Aguilar, J.F.; Escobar-Jimenez, R.F.; Alvarado-Martinez, V.M.; Lopez-Lopez, M.G. Classical and Fractional-order Modeling of Equivalent Electrical Circuits for Supercapacitors and Batteries, Energy Management Strategies for Hybrid Systems and Methods for the State of Charge Estimation: A State of the Art Review. *Microelectron. J.* **2019**, *85*, 109–128. [CrossRef]
3. Hong, Z.; Li, Q.; Han, Y.; Shang, W.; Zhu, Y.; Chen, W. An Energy Management Strategy based on Dynamic Power Factor for Fuel Cell/battery Hybrid Locomotive. *Int. J. Hydrogen Energy* **2018**, *43*, 3261–3272. [CrossRef]
4. Cheng, J.; Yang, G.; Zhang, K.; He, G.; Jia, J.; Yu, H.; Gai, F.; Li, L.; Hao, C.; Zhang, F. Guanidimidazole-quanternized and Cross-linked Alkaline Polymer Electrolyte Membrane for Fuel Cell Application. *J. Membr. Sci.* **2016**, *501*, 100–108. [CrossRef]
5. Kamal, T.; Hassan, S.Z.; Espinosa-Trujillo, M.J.; Li, H.; Flota, M. An Optimal Power Sharing and Power Control Strategy of Photovoltaic/Fuel Cell/Ultra-capacitor Hybrid Power System. *J. Renew. Sustain. Energy* **2016**, *8*, 035301. [CrossRef]
6. Marzougui, H.; Amari, M.; Kadri, A.; Bacha, F.; Ghouili, J. Energy Management of Fuel Cell/Battery/Ultracapacitor in Electrical Hybrid Vehicle. *Int. J. Hydrogen Energy* **2017**, *42*, 8857–8869. [CrossRef]
7. Fathabadi, H. Fuel Cell Hybrid Electric Vehicle (FCHEV): Novel Fuel Cell/SC Hybrid Power Generation System. *Energy Convers. Manag.* **2018**, *156*, 192–201. [CrossRef]
8. Das, H.; Tan, C.; Yatim, A. Fuel Cell Hybrid Electric Vehicles: A Review on Power Conditioning Units and Topologies. *Renew. Sustain. Energy Rev.* **2017**, *76*, 268–291. [CrossRef]
9. Garcia, P.; Fernandez, L.; Torreglosa, J.; Jurado, F. Comparative Study of Four Control Systems for a 400-kW Fuel Cell Battery-Powered Tramway with Two DC/DC Converters. *Int. Trans. Electr. Energy Syst.* **2013**, *23*, 1028–1048. [CrossRef]

10. Behdani, A.; Naseh, M. Power Management and Nonlinear Control of a Fuel Cell-Supercapacitor Hybrid Automotive Vehicle with Working Condition Algorithm. *Int. J. Hydrogen Energy* **2017**, *42*, 24347–24357. [CrossRef]

11. Zhou, D.; Al-Durra, A.; Gao, F.; Ravey, A.; Matraji, I.; Simoes, M. Online Energy Management Strategy of Fuel Cell Hybrid Electric Vehicles Based on Data Fusion Approach. *J. Power Sources* **2017**, *366*, 278–291. [CrossRef]

12. Bauman, J.; Kazerani, M. A Comparative Study of Fuel-Cell-Battery, Fuel-Cell-Ultracapacitor, and Fuel-Cell-Battery-Ultracapacitor Vehicles. *IEEE Trans. Veh. Technol.* **2008**, *57*, 760–769. [CrossRef]

13. Li, Q.; Yang, H.; Han, Y.; Li, M.; Chen, W. A State Machine Strategy Based on Droop Control for an Energy Management System of PEMFC-Battery-Supercapacitor Hybrid Tramway. *Int. J. Hydrogen Energy* **2016**, *41*, 16148–16159. [CrossRef]

14. Peng, F.; Zhao, Y.; Li, X.; Liu, Z.; Chen, W.; Liu, Y.; Zhou, D. Development of Master-Slave Energy Management Strategy Based on Fuzzy Logic Hysteresis State Machine and Differential Power Processing Compensation for a PEMFC-LIB-SC Hybrid Tramway. *Appl. Energy* **2017**, *206*, 346–363. [CrossRef]

15. Ali, A.M.; Soffker, D. Towards Optimal Power Management of Hybrid Electric Vehicles in Real-Time: A Review on Methods, Challenges, and State-of-the-Art Solutions. *Energies* **2018**, *11*, 476.

16. Ahmadi, S.; Bathaee, S.M.T. Multi-Objective Genetic Optimization of the Fuel Cell Hybrid Vehicle Supervisory System: Fuzzy Logic and Operating Mode Control Strategies. *Int. J. Hydrogen Energy* **2015**, *40*, 12512–12521. [CrossRef]

17. Caux, S.; Hankache, W.; Fadel, M.; Hissel, D. On-Line Fuzzy Energy Management for Hybrid Fuel Cell Systems. *Int. J. Hydrogen Energy* **2010**, *35*, 2134–2143. [CrossRef]

18. Bizon, N. Load-Following Mode Control of a Standalone Renewable/Fuel Cell Hybrid Power Source. *Energy Convers. Manag.* **2014**, *77*, 763–772. [CrossRef]

19. Yu, K.; Tan, X.; Yang, H.; Liu, W.; Cui, L.; Liang, Q. Model Predictive Control of Hybrid Electric Vehicles for Improved Fuel Economy. *Asian J. Control* **2016**, *18*, 1–14. [CrossRef]

20. Ayad, M.Y.; Becherif, M.; Henni, A. Vehicle Hybridization with Fuel Cell, Supercapacitors and Batteries by Sliding Mode Control. *Renew. Energy* **2011**, *36*, 2627–2634. [CrossRef]

21. Zhang, W.; Li, J.; Xu, L.; Ouyang, M. Optimization for a Fuel Cell/Battery/Capacity Tram with Equivalent Consumption Minimization Strategy. *Energy Convers. Manag.* **2017**, *134*, 59–69. [CrossRef]

22. Hu, Z.; Li, J.; Xu, L.; Song, Z.; Fang, C.; Ouyang, M.; Dou, G.; Kou, G. Multi-Objective Energy Management Optimization and Parameter Sizing for Proton Exchange Membrane Hybrid Fuel Cell Vehicles. *Energy Convers. Manag.* **2016**, *129*, 108–121. [CrossRef]

23. Xu, L.; Yang, F.; Li, J.; Ouyang, M.; Hua, J. Real Time Optimal Energy Management Strategy Targeting at Minimizing Daily Operation Cost for a Plug-in Fuel Cell City Bus. *Int. J. Hydrogen Energy* **2012**, *37*, 15380–15392. [CrossRef]

24. Li, M.; Li, M.; Han, G.; Liu, N.; Zhang, Q.; Wang, Y. Optimization Analysis of the Energy Management Strategy of the New Energy Hybrid 100% Low-Floor Tramcar Using a Genetic Algorithm. *Appl. Sci.* **2018**, *8*, 1144. [CrossRef]

25. Olivier, J.C.; Wasselynck, G.; Chevalier, S.; Auvity, B.; Josset, C.; Trichet, D.; Squadrito, G.; Bernard, N. Multiphysics Modeling and Optimization of the Driving Strategy of a Light Duty Fuel Cell Vehicle. *Int. J. Hydrogen Energy* **2017**, *42*, 26943–26955. [CrossRef]

26. Zhang, H.; Yang, J.; Zhang, J.; Song, P.; Li, M. Optimal Energy Management of a Fuel Cell-Battery-Supercapacitor-Powered Hybrid Tramway Using a Multi-Objective Approach. *Proc. Inst. Mech. Eng. Part F J. Rail Rapid Transit* **2019**. [CrossRef]

27. Zhang, F.; Xu, K.; Li, L.; Langari, R. Comparative Study of Equivalent Factor Adjustment Algorithm for Equivalent Consumption Minimization Strategy for HEVs. In Proceedings of the IEEE Vehicle Power and Propulsion Conference, Chicago, IL, USA, 27–30 August 2018.

28. Li, H.; Ravey, A.; N'Diaye, A.; Djerdir, A. A Novel Equivalent Consumption Minimization Strategy for Hybrid Electric Vehicle Powered by Fuel Cell, Battery and Supercapacitor. *J. Power Sources* **2018**, *395*, 262–270. [CrossRef]

29. Lin, X.; Feng, Q.; Mo, L.; Li, H. Optimal Adaptation Equivalent Factor of Energy Management Strategy for Plug-in CVT HEV. *Proc. Inst. Mech. Eng. Part D J. Automob. Eng.* **2019**, *233*, 877–889. [CrossRef]

30. Torreglosa, J.P.; Jurado, F.; Garcia, P.; Fernandez, L.M. Hybrid Fuel Cell and Battery Tramway Control Based on an Equivalent Consumption Minimization Strategy. *Control Eng. Pract.* **2011**, *19*, 1182–1194. [CrossRef]
31. Zhang, G.; Li, Q.; Han, Y.; Meng, X.; Chen, W. Study on Equivalent Consumption Minimization Strategy Based on Operation Mode and DDOH for Fuel Cell Hybrid Tramway. *Proc. CSEE* **2018**, *23*, 6905–6914.
32. Garcia, P.; Torreglosa, J.P.; Fernandez, L.M.; Jurado, F. Viability Study of a FC-Battery-SC Tramway Controlled by Equivalent Consumption Minimization Strategy. *Int. J. Hydrogen Energy* **2012**, *37*, 9368–9382. [CrossRef]
33. Yan, Y.; Li, Q.; Chen, W.; Su, B.; Liu, J.; Ma, L. Optimal Energy Management and Control in Multimode Equivalent Energy Consumption of Fuel Cell/Supercapacitor of Hybrid Electric Tram. *IEEE Trans. Ind. Electron.* **2019**, *66*, 6065–6076. [CrossRef]
34. Han, Y.; Li, Q.; Wang, T.; Chen, W.; Ma, L. Multisource Coordination Energy Management Strategy Based on SOC Consensus for a PEMFC–Battery–Supercapacitor Hybrid Tramway. *IEEE Trans. Veh. Technol.* **2018**, *67*, 296–305. [CrossRef]
35. Yang, J.; Xu, X.; Peng, Y.; Zhang, J.; Song, P. Modeling and Optimal Energy Management Strategy for a Catenary-Battery-Ultracapacitor Based Hybrid Tramway. *Energy* **2019**, *183*, 1123–1135. [CrossRef]
36. Yang, J.; Song, P.; Zhang, J.; Wang, G.; Zhang, H. Research on Simulation System of Hybrid Modern Tramway. *J. Mech. Eng.* **2017**, *53*, 161–168. [CrossRef]
37. Niu, Q.; Zhang, H.; Li, K. An Improved TLBO with Elite Strategy for Parameters Identification of PEM Fuel Cell and Solar Cell Models. *Int. J. Hydrogen Energy* **2014**, *39*, 3837–3854. [CrossRef]
38. Li, Q.; Wang, T.; Dai, C.; Chen, W.; Ma, L. Power Management Strategy Based on Adaptive Droop Control for a Fuel Cell-Battery-Supercapacitor Hybrid Tramway. *IEEE Trans. Veh. Technol.* **2018**, *67*, 5658–5670. [CrossRef]
39. Yang, J.; Zhang, J.; Song, P.; Chen, Y.; Wang, G. Research on Modeling and Operation Control of Fuel Cell Hybrid Train. *J. China Railw. Soc.* **2017**, *39*, 40–47.
40. Tremblay, O.; Dessaint, L. Experimental Validation of a Battery Dynamic Model for EV Applications. *World Electr. Veh. J.* **2009**, *3*, 1–10. [CrossRef]
41. Zhu, Y.; Murali, S.; Stoller, M.; Ganesh, K.; Cai, W.; Ferreira, P.; Pirkle, A.; Wallace, R.; Cychosz, k.; Thommes, M.; et al. Carbon-Based Supercapacitors Produced by Activation of Graphene. *Science* **2011**, *332*, 1537–1541. [CrossRef]
42. Ehsani, M.; Gao, Y.; Emadi, A. *Modern Electric, Hybrid Electric, and Fuel Cell Vehicles: Fundamentals, Theory, and Design*, 2nd ed.; CRC Press: Boca Raton, FL, USA, 2010; pp. 331–333.
43. Herrera, V.; Milo, A.; Gaztanaga, H.; Etxeberria-Otadui, I.; Villarreal, I.; Camblong, H. Adaptive Energy Management Strategy and Optimal Sizing Applied on a Battery-Supercapacitor Based Tramway. *Appl. Energy* **2016**, *169*, 831–845. [CrossRef]
44. Herrera, V.; Gaztanaga, H.; Milo, A.; Saez-de-Ibarra, A.; Etxeberria-Otadui, I.; Nieva, T. Optimal Energy Management and Sizing of a Battery-Supercapacitor-Based Light Rail Vehicle with a Multiobjective Approach. *IEEE. Trans. Ind. Appl.* **2016**, *52*, 3367–3377. [CrossRef]
45. Paganelli, G.; Delprat, S.; Guerra, T.M.; Rimaux, J.; Santin, J.J. Equivalent Consumption Minimization Strategy for Parallel Hybrid Powertrains. In Proceedings of the IEEE 55th Vehicular Technology Conference, Birmingham, AL, USA, 6–9 May 2002.
46. Xu, L.; Li, J.; Hua, J.; Li, X.; Ouyang, M. Adaptive Supervisory Control Strategy of a Fuel Cell/Battery-Powered City Bus. *J. Power Sources* **2009**, *194*, 360–368. [CrossRef]
47. Garcia, P.; Fernandez, L.M.; Torreglosa, J.P.; Jurado, F. Operation Mode Control of a Hybrid Power System Based on Fuel Cell/Battery/Ultracapacitor for an Electric Tramway. *Comput. Electr. Eng.* **2013**, *39*, 1993–2004. [CrossRef]
48. Yang, X. Firefly algorithms for multimodal optimization. In Proceedings of the International Symposium on Stochastic Algorithms, Sapporo, Japan, 26–28 October 2009.
49. Chaurasia, G.S.; Singh, A.K.; Agrawal, S.; Sharma, N.K. A Meta-Heuristic Firefly Algorithm Based Smart Control Strategy and Analysis of a Grid Connected Hybrid Photovoltaic/Wind Distributed Generation System. *Sol. Energy* **2017**, *150*, 265–274. [CrossRef]
50. Wang, D.; Luo, H.; Grunder, O.; Lin, Y.; Guo, H. Multi-Step Ahead Electricity Price Forecasting Using a Hybrid Model Based on Two-Layer Decomposition Technique and BP Neural Network Optimized by Firefly Algorithm. *Appl. Energy* **2017**, *190*, 390–407. [CrossRef]

51. Bahadormanesh, N.; Rahat, S.; Yarali, M. Constrained Multi-Objective Optimization of Radial Expanders in Organic Rankine Cycles by Firefly Algorithm. *Energy Convers. Manag.* **2017**, *148*, 1179–1193. [CrossRef]
52. Yang, J.; Xu, X.; Zhang, J.; Song, P. Multi-objective Optimization of Energy Management Strategy for Fuel Cell Tram. *J. Mech. Eng.* **2018**, *22*, 153–159. [CrossRef]

Article

Energy-Storage-Based Smart Electrical Infrastructure and Regenerative Braking Energy Management in AC-Fed Railways with Neutral Zones

Zhixuan Gao, Qiwei Lu *, Cong Wang, Junqing Fu and Bangbang He

School of Mechanical Electronic & Information Engineering, China University of Mining & Technology (Beijing), Beijing 100083, China; gaozhixuan8577@163.com (Z.G.); wangc@cumtb.edu.cn (C.W.); 18635608600@163.com (J.F.); he_bangbang@163.com (B.H.)
* Correspondence: lqw@cumtb.edu.cn; Tel.: +86-132-6038-8766

Received: 22 September 2019; Accepted: 22 October 2019; Published: 24 October 2019

Abstract: This paper presents a modified power supply system based on the current alternating current (AC)-fed railways with neutral zones that can further improve the eco-friendliness and smart level of railways. The modified system complements the existing infrastructure with additional energy-storage-based smart electrical infrastructure. This infrastructure comprises power electronic devices with energy storage system connected in parallel to both sides of each neutral zone in the traction substations, power electronic devices connected in parallel to both sides of each neutral zone in section posts, and an energy management system. The description and functions of such a modified system are outlined in this paper. The system allows for the centralized- and distributed-control of different functions via an energy management system. In addition, a control algorithm is proposed, based on the modified system for regenerative braking energy utilization. This would ensure that all the regenerative braking energy in the whole railway electrical system is used more efficiently. Finally, a modified power supply system with eight power supply sections is considered to be a case study; furthermore, the advantages of the proposed system and the effectiveness of the proposed control algorithm are verified.

Keywords: eco-friendliness; smart railway; smart electrical infrastructure; control algorithm; regenerative braking energy

---

1. Introduction

Rail transport is one of the most environmental friendly modes for transportation. Under the background of an environmentally concerned economy, it is a crucial way to satisfy increasing transport demands [1]. However, despite its high efficiency, the rail industry is still a significant energy consumer worldwide [1]. With a growing population and energy usage, the increased demand placed on rail services will further increase and the energy consumption will increase accordingly [2]. Along with the energy shortage and global warming, policy makers pay special attention to the high level of energy demand users. Due to the vast potential of energy recovery lying behind them, the targets of reduction of carbon emission and energy consumption can be achieved by regulating their energy consumption [3]. The energy savings in rail transport will play an important role in achieving this target. In rail transport, the consumption of traction energy accounts for about 60% to 80% of the total energy consumption [1]. Therefore, the regulation for traction energy will be a key area for energy savings.

To satisfy the specified energy efficiency targets, several measures have been proposed to save traction energy. Currently, the main measures for savings based on the existing technologies include renewable energy system, efficient driving, and regenerative braking.

The installation of equipment that provide clean energy from renewable sources (e.g., solar panels and wind generators) to feed the traction loads in rail may have a significant impact in the energy costs and CO_2 emissions of a railway system, increasing the environment friendliness of the railway [4–6]. The introduction of renewable energy is to partially reduce the depletion of non-renewable energy resources and thus reduce CO_2 emissions. These outcomes have been favored by scholars and many countries, such as North American, European, China, and Japan [6,7]. But in practice, the traction energy consumption of railways has not been reduced.

By contrast, the efficient driving and regenerative braking schemes differ the renewable energy scheme. They can essentially reduce the traction energy consumption of railways.

Efficient driving is an effective way to reduce the traction energy consumption of rolling stock. Typically, the running state of vehicles comprises three phases: braking, speed-holding (cruising) and acceleration. At present, many literatures have studied the efficient driving and realize the traction energy savings through the optimum combination of these phases [8–13]. To implement such trajectories and maximize savings, some auxiliary systems, such as driver advisory systems (DAS) [10,11] and automatic train operation (ATO) [12,13], must be provided to drivers. The trials of such systems have shown savings of 5-18%. However, with efficient driving, the savings of energy consumption may sacrifice service quality, such as the train speed, stopping time, etc. [1].

Regenerative braking energy (RBE) recovery is a simple and efficient way to reduce energy consumption, which cannot influence the operation of trains and service quality. Regenerative braking is a technology that can recover the mechanical energy during a locomotive braking as electricity. In some cases, over a third of the total traction energy, even a half, can be recovered using this method. If this part of energy is used effectively, the traction energy consumption of a network will be significantly reduced [1]. There are some mature RBE utilization schemes in metro, but the metro and alternating current (AC) railways are different in some respects, such as the type of power supply voltage, the amount of traction power and so on. Therefore, the RBE recovery equipment in the metro cannot be used directly in AC railway systems, but the metro schemes can be used as reference. According to experience of the metro systems and the art of research, the main regenerative energy saving measures are: a reversible inverter, optimizing operation timetable, energy storage system, and power transfer device.

Reversible inverter can directly inject the RBE into the public grid or distribution grid. Conventionally, the RBE is reversed from the traction power supply system to the public grid via traction transformers at substations. However, it is not reflected to the electric charge purchased from a utility company for railway operation [14]. Furthermore, it is not certain that the back-flown regenerative power is effectively utilized. Due to the single-phase characteristics of the back-flown power, it may be consumed in the traction transformers as a circulation current or as a heat [15]. In addition, the back-flown regenerative power, due to its stochasticity, may have impacts on the public grid.

Another option is to inject the RBE into the distribution grid via a regenerative inverter to power the distribution load in railway power supply systems, such as lighting, air conditioning, elevators, escalators, and so on. However, the distribution power of a conventional railway station on the ground is considerably smaller than the regenerative power [16]. Therefore, the RBE would be surplus, flowing back to the public grid or being stored in an energy storage system.

Optimizing operation timetable is a low-cost solution for reducing energy consumption of railways by increasing the coexistence time of the traction and braking trains in the same electric section without auxiliaries [1,17,18]. However, owing to unanticipated occurrences often emerging during the actual operation, such as the temporary addition or cancellation of trains, equipment failure, bad weather, and so on, the trains cannot consistently run according to the optimized operation timetable. This leads to a significant reduction in energy saving efficiency. Furthermore, the flexibility of this solution is poor [19].

The other option is to store the RBE in an energy storage system (ESS), by which the RBE can be utilized or sold at some point in the future. ESS can be generally divided into on-board and wayside ESS. The first standard related to railway ESS, IEC62924, was published in 2017 and it will provide guidance for the application of ESS in railway [20]. According to the measured data in the railway systems in Japan, the total energy saving effect can amount to several hundred MWh per year after applying ESS. If the ESS is installed in the appropriate location, the total traction energy consumption can be reduced by about 4-8% [21]. In realizing railway energy saving and consumption reduction, it will bring huge electricity savings to railway operation. However, this data was obtained from the Japanese direct current (DC) railway traction system; according to present reports, the installation of ESS is usually only in DC traction systems or metros. Actual cases of installing such energy storage devices in AC systems are seldom reported.

In the AC railway system, an application of RBE utilization is the use of a railway static power conditioner (RPC). By using it, the regenerative braking power can be interchanged from one supply section to an adjacent one. At present, a RPC was installed at Ushiku section post in Japan in 2014 [14]. At first, the development of RPC was to solve the three phase imbalance problem at traction substations. Then, with the emergence of RBE utilization problem, RPCs were used to improve the utilization of RBE. The capacity of RPCs for utilizing RBE can be much smaller than that of three phase balancing, but they have much potential to increase the environment friendliness of the AC railway systems by making use of regenerative braking power [22]. Other than that, many papers also put forward proposals for using power transfer devices (PTD; RPC [14,22], railway power quality compensator (RPQC) [23], or energy optimization controller (EOC) [24]) to utilize the RBE.

However, the above schemes are independent distributed control and cannot cooperate with each other, which causes each device to not effectively play its maximum potential and may even cause energy waste. The railway smart grid paradigm provides a chance to integrate each component, including renewable energy systems, energy storage systems, RBE utilization systems, and so on, using bidirectional communication equipment and advanced computational features [25–29]. Different optimal strategies are proposed to decrease the energy consumption and electricity bills of electrical railway systems by controlling cooperative operations between each component [5,6,30–32]. These methods achieve the optimal operation of the electric railway energy systems using existing electrical railway infrastructure. However, there is a neutral zone (NZ) in the existing traction power supply system, which limits energy optimization capability. Thus, [33] introduced the smart electrical infrastructure into railway systems to complement the existing electrical railway infrastructure, creating more opportunities and controllability for smart railway systems, such as power flow in the whole railway system (similar to power routing). The smart grid concept presents more efficient, safe, reliable, and modernized railway systems and offers a new perspective to the eco-friendliness of railway systems and low carbon societies in the future.

This paper presents the energy-storage-based smart electrical infrastructure to modify current AC-fed railways with neutral zones. Compared with the smart electrical infrastructure in [33], the proposed modified system increases the dimensions of the controlled variables of power flow under the action of ESS. It brings more benefits, such as further enhancing the ability to control power flow and reducing more cost; achieving full utilization of all RBE, which is crucial in further improving the eco-friendliness of railway transportation; and better regulating the clean power from renewable energy sources (photovoltaic (PV) panels and wind turbines) to achieve more efficient utilization of clean energy in railways. In addition, the proposed modified system also has other functions, such as cutting peak and filling valley, active control for line voltage and emergency power supply in case of power failure. Then, in order to utilize the RBE, this paper also proposed a control algorithm to achieve the full utilization of all RBE by the use of the proposed energy-storage-based smart electrical infrastructure. It should be emphasized that when realizing the utilization of RBE, the capacity and cost of the energy storage system in the proposed modified system will be much less than that of the system adopting ESS alone; this is verified in the validation section of the paper. The work done in this paper will provide options for railway smart control and energy saving and consumption reduction.

This paper is organized as follows: The regenerative braking energy utilization schemes are described in Section 2. The description and functions of the modified system are outlined in Section 3. The proposed control algorithm based on the modified system for regenerative braking energy utilization is presented in Section 4. The verification results and analysis are represented in Section 5. Finally, conclusions are drawn in Section 6.

2. Regenerative Braking Energy Utilization Schemes

Figure 1 summarizes some possible solutions for RBE utilization that have already used or proposed. As shown in Figure 1a, regenerative resistance has been one of the most effective methods in the past, but the RBE is not utilized in this solution. A simple scheme is reversing the regenerative power from a railway power supply system to the utility grid directly, as shown in Figure 1b. However, it is not known that such back-flowed single phase regenerative power is utilized properly in the utility grid and the back-flow of power is not reflected in electrical cost. Back-flow to the distribution grid shown in Figure 1c is one of the most effective methods to utilize RBE in metros because the load of metro stations is large enough to utilize the regenerative power simultaneously. However, in AC railways such as high-speed or burden, typically the regenerative power is more than the power demand of distribution grids such as railway stations and trackside signaling systems. Therefore, whether this scheme can be used in AC railways still needs further discussion. Compared with the above solutions, the installation of ESS could be effective because it is independent of operation schedules of trains and other electrical loads, as shown in Figure 1d. The ESS can be installed along the railway trackside to store RBE. Then, the RBE stored in ESS during train braking will be utilized for the powering of systems later. However, the cost of ESS needs to be concerned in AC railways. Figure 1e shows an RBE utilization scheme via the use of a PTD (such as RPC [14] or EOC [24]) connected in parallel to both sides of each NZ in section posts (SPs). PTD can interchange regenerative power from one feeder catenary to another and it contributes to realize the simultaneous utilization of RBE within the railway systems. Figure 1f illustrates the RBE utilization scheme via the use of a PTD (such as RPQC [23] or RPC [22]) connected in parallel to both sides of each NZ in traction substations (TSs). Then, part of the traction power in one feeding section can be supplied by the RBE recovered from the other adjacent feeding section, similar to Figure 1e. Figure 1g illustrates the scheme that complements the existing railway power supply system with additional PTDs [33]. This makes it possible for control of power flow through the whole railway system. In AC railways, PTDs are able to transfer the power specified by a control system from one feeding section to the other nonadjacent electrical sections. Since the supplied power voltage is high enough, 20 kV or 25 kV, the regenerative power can be delivered over long distances, which is impossible with DC railway or metro [15].

However, as shown in Figure 1e,f, if there are no traction locomotives in the adjacent feeding section, the regenerative power flows back into the utility grid. The schemes in Figure 1g can solve this problem to a certain degree by transferring the surplus regenerative power to the traction locomotives in the other nonadjacent electrical sections. However, if the amount of regenerative power is more than the traction power in the whole railway system, the surplus regenerative power flows back into the utility grid. Therefore, to solve these problems, the installation of PTD with the ESS (EPTD) is a feasible option [34–38]. Via EPTD, it is possible to interchange RBE with the adjacent feeding sections when there are traction locomotives present and store the RBE when there is a surplus. Figure 2 shows the proposed energy-storage-based smart electrical infrastructure for AC-fed railways with NZs. The proposed modified system links the whole railway power supply system together by PTDs and EPTDs, allowing power flow, energy storage, and release within the internal railway system. Moreover, the modified system can not only use RBE, but also can also use clean energies and control their cooperation. Other than these, the proposed modified system also has other functions as follows:

- Power quality improvement;
- Peak cutting and valley filling;
- Active control for line voltage;
- Emergency power supply.

The following discussion covers the analysis of each function of the proposed system.

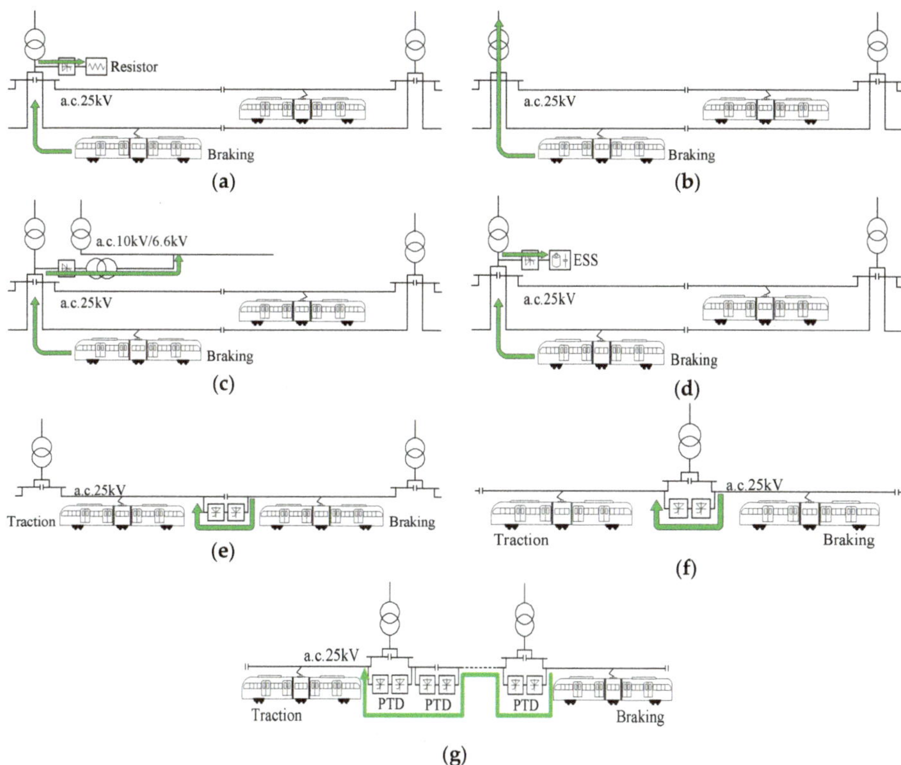

Figure 1. Regenerative braking energy (RBE) utilization schemes (**a**) regenerative resistance; (**b**) feeding into grid; (**c**) feeding into distribution grid; (**d**) storage energy with an energy storage system (ESS); (**e**) tie feeding in a section post (SP); (**f**) tie feeding in a traction substation (TS); and (**g**) smart electrical infrastructure.

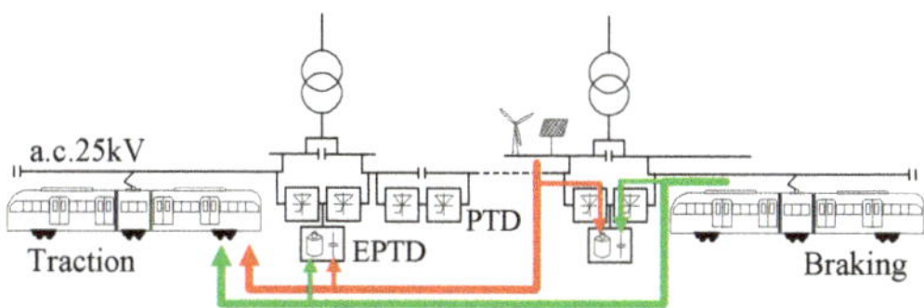

Figure 2. Energy-storage-based smart electrical infrastructure.

3. Description of the Modified System

The railway traction system with energy-storage-based smart electrical infrastructure is shown in Figure 3. The red dashed line is the existing infrastructure in AC-fed railways with NZs. For the sake of analysis, all locomotives on the same power supply section are equivalent to one electrical locomotive load. The existing system consists of n TSs, where TS_m ($m = 1, 2, \ldots, n$) is used to supply trains L_k ($k = 1, 2, \ldots, 2n$) in the power supply sections S_k. The NZ is used to ensure electrical insulation between two adjacent power supply sections. The green dashed line is the proposed energy-storage-based smart electrical infrastructure. The modified system consists of the additional PTDs and EPTDs connected in parallel to both sides of each NZ and an energy management system (EMS). The acronym TS-EPTD refers to the PTDs with the ESS located in the NZs of the TSs. The acronym SP-PTD refers to the PTDs located in the other NZs of the SPs. The installation of TS-EPTDs will connect all power supply sections to ESSs, so there is no need to install EESs in SP-PTDs. The active and reactive power transferred by PTDs and EPTDs is specified by the EMS via the bidirectional communication equipment. The modified system can perform a centralized and a distributed-control. The centralized architecture can perform a global optimization to achieve the cooperative operations between each component, significantly enhancing the energy-management performance of the systems. The distributed control can further enhance redundancy and reliability of the railway systems [31,33]. Each EPTD and PTD can realize data acquisition (DA) of voltage and current on each power supply section and transmit the data to the EMS via the bidirectional communication equipment. The EMS performs a global optimization by the use of the data of voltage and current and has power control (PC) over the EPTDs and PTDs via the bidirectional communication equipment. It should be pointed out that the proposed system is also applicable for autotransformer (AT)-based railway systems.

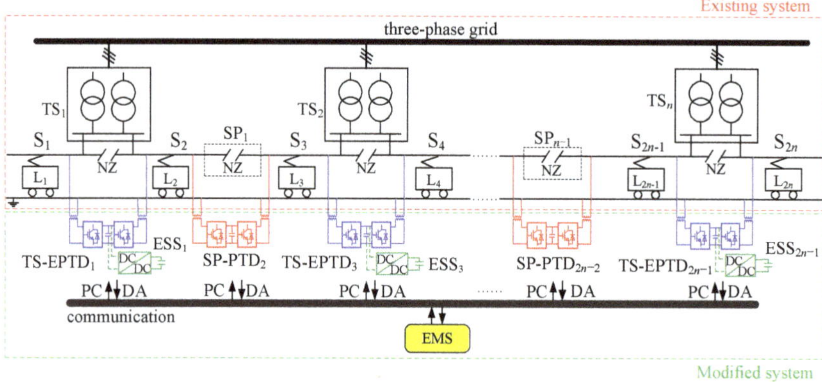

Figure 3. The railway traction system with energy-storage-based smart electrical infrastructure.

3.1. Regenerative Braking Energy Utilization

The simplified system and its electrical description are shown in Figure 4. u_k and i_k are the voltage and current on each power supply section provided by traction transformers, $i_{L,k}$ is the load current produced by locomotives (traction or regeneration), $i_{E,k,1}$ and $i_{E,k,2}$ are the transferring current controlled by EPTDs, $i_{ES,k}$ is the discharge or charge current controlled by ESSs in EPTDs, and $i_{P,k,1}$ and $i_{P,k,2}$ are the transferring current controlled by PTDs. The total cost of the whole railway system P_{cost} can be expressed as:

$$P_{cost} = \sum_{k=1}^{2n} i_k u_k. \tag{1}$$

Assuming that X1 locomotives are in the traction state; R1 locomotives are in the regenerative state; and X1 + R1 = 2n. Then, the locomotive traction power and the locomotive regenerative power can be expressed as:

$$P_{traction} = \sum_{k=1}^{X1} i_{L,k} u_k. \tag{2}$$

$$P_{\text{regeneration}} = \sum_{k=1}^{R1} i_{L,k} u_k. \tag{3}$$

If there is no modified system in the railway, the feeder power of the traction transformer is the locomotive power. Due to the existence of NZ, the regenerative power cannot be transferred to the other traction electrical sections, and can only be fed back to the utility grid. However, the utility companies do not pay for regenerative power, so the railway operators will pay the charge consumed by all traction rolling stocks. Neglect all losses in the system, then, the total cost of the whole railway system is:

$$P_{\cos t} = P_{\text{traction}}. \tag{4}$$

After adding the modified system, the feeder power is affected by the locomotive power and the transmission power in EPTDs and PTDs, as shown in Figure 4. Their relationship can be obtained as:

$$\begin{cases} i_{E,k,1} - i_{E,k,2} - i_{ES,k} = 0 \\ i_{P,k,1} - i_{P,k,2} = 0 \end{cases}. \tag{5}$$

(i) If k is an odd number:

$$i_k = \begin{cases} i_{L,1} + i_{E,1,1}, & k = 1 \\ i_{L,k} - i_{P,k-1,2} + i_{E,k,1}, & k > 1 \end{cases}. \tag{6}$$

(ii) If k is an even number:

$$i_k = \begin{cases} i_{L,k} + i_{P,k+1,1} - i_{E,k,2}, & k < n \\ i_{L,2n} - i_{E,2n-1,2}, & k = 2n \end{cases}. \tag{7}$$

Submitted (5)–(7) into (1):

$$P_{\cos t} = P_{\text{traction}} - P_{\text{regeneration}} + P_{ES}, \tag{8}$$

where

$$P_{ES} = \sum_{k=1}^{n} i_{ES,k} v_{dc}, \tag{9}$$

where v_{dc} is the dc bus voltage in EPTD_k.

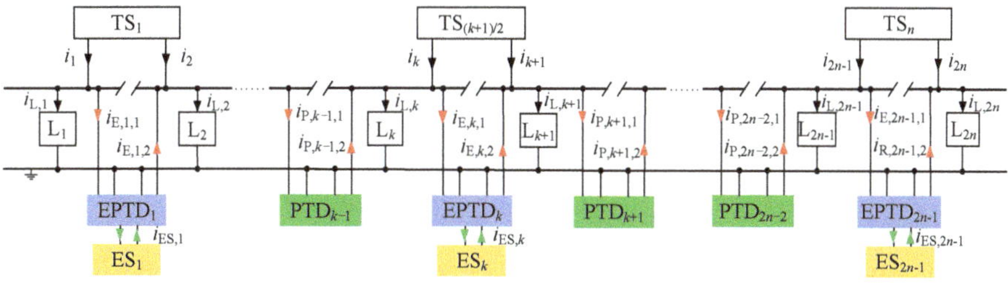

Figure 4. The simplified system and its electrical description.

A comparison of (4) and (8) illustrates that after the action of the modified system, the regenerative power can be transferred to the other traction electrical sections, and both the traction energy consumption and the electrical bills for railway operators will be reduced. Furthermore, it can be known from (8) that if the regenerative power is surplus after transferring, the surplus regenerative power can be stored in ESS instead of feeding back to the grid, and will be used for traction later. Apparently, all regenerative power can be used within the internal railway system, further improving the utilization rate of the regenerative power. The modified system enhances the means of regulating RBE utilization in traction power supply system.

The energy distribution of the whole railway system is shown in Figure 5. $P_{\_tra}$ is the judgment threshold of the traction condition; $P_{\_reg}$ is the judgment threshold of the regeneration condition; P_{grid} is the power supplied by the utility grid; P_{PTD} is the transferring power by EPTD and PTD; P_{ES_disch} is the discharge power of EES; P_{ES_ch} is the charge power of EES. $P_{\_tra}$ and $P_{\_reg}$ can divide power allocation into three states: traction, non-load, and regeneration. In the traction condition, the traction power is supplied together by the grid, the regenerative power and the stored regenerative power in EES. In the non-load condition, there are no trains running or in the cruising state. However in practice, there will be losses in transformers, lines, etc. At this time, the line loss, transformer loss and other losses are powered by the grid. Furthermore, the ESS can be charged for other functions in this condition. In the regeneration condition, the regenerative power supplies the traction power demanded by the locomotives, and the surplus part is stored in ESS.

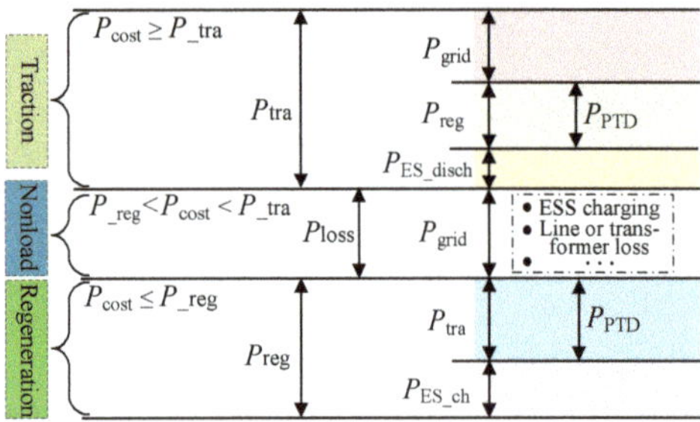

Figure 5. Energy distribution of the whole railway system.

3.2. Power Quality Improvement

The output powers of two-phase windings in single-phase railway networks have significant differences due to the imbalance of traction loads, causing negative sequence, reactive power, and other power quality problems in the upstream three-phase utility grid. The modified system also has the ability to improve power quality. An EPTD comprises two single-phase back-to-back AC/DC/AC converters and one DC/DC converter with a bus dc link in between. It can control active power exchange and compensate reactive power to present the upstream three phase power as balanced and free of reactive power. By compensating the reactive loads ($Q_{EPTDk,1}$ and $Q_{EPTDk,2}$) and balancing the active load power (P_{EPTDk}) between the two railway power supply sections (take V/V transformer as an example, see Figure 6), the problem of negative sequence and reactive power pollution to the upstream three-phase utility grid can be eliminated. The EPTD can also be used as a three phase balancer for Scott transformer and other transformer connections. At present, many control methods of EPTD (RPC or ES-RPC) have been proposed [35–41]. This paper did not elaborate too much on it. However, the capacity of the EPTD for three phase balancer can be much larger than that of the EPTD used for the regenerative power user. The capacity reduction methods of EPTD (mainly RPC) have also been investigated [42–44].

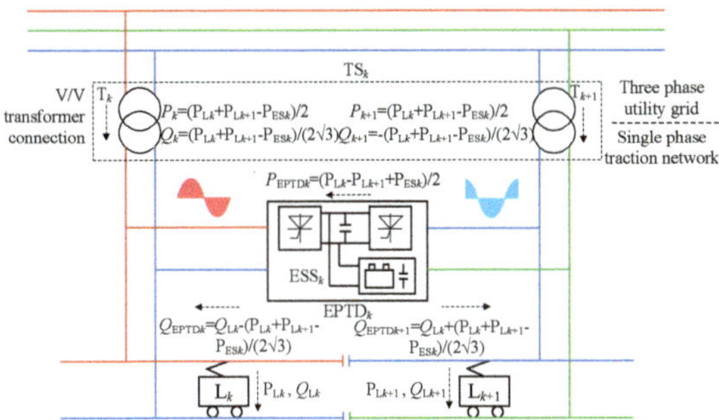

Figure 6. Power quality improvement by PTD with the ESS (EPTD).

3.3. Peak Cutting and Valley Filling Control

Due to the influence of train running intervals and train carrying capacities, there is peak power in the traction substation. The traction transformer will experience huge power at peak positions, when power surges to traction transformer. In addition, the operation cost and the investment cost are large if the traction transformers are designed according the peak power in convention [22,28,33,35,36]. The effect of peak power surges can be reduced by the modified system. EPTD and PTD can control power flow on both sides connected to it and EPTD can also store up energy as well as release energy, realizing the peak shift control.

Figure 7 shows an example of a peak load waveform caused by the coincidence of some train loads. In a certain power supply section, peak shaving can be realized by releasing the energy stored in EPTD and transferring the power from other adjacent power supply intervals [35,36]. In addition, the cost of the whole railway will be reduced via peak-shift control adopting the smart electrical infrastructure in [33]. The modified system in this paper will further enhance the flexibility of peak-shift control, and the cost of the whole railway will be reduced even more. It should be pointed out that the storage energy in EES will come from regenerative braking energy (see Figure 2) on the one hand and charging power from grid (see Figure 5) on the other hand. By realizing the function of peak shaving, the economic cost of the whole railway will be reduced.

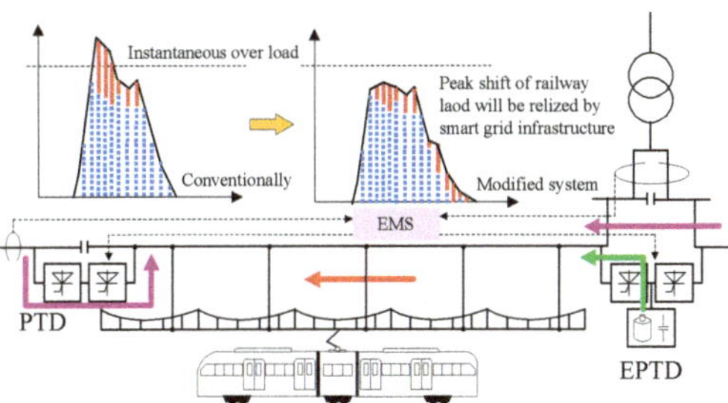

Figure 7. Peak shaving based on the modified system.

Similarly, there is valley power in the traction substation and valley filling control also needs to be considered. Above all, the filling valley can increase the utilization rate of traction transformers [35,36]. Moreover, valley filling can help balance the supply and demand in an electrical system and reduce cost of electrical railway system. According to market price fluctuations in the electricity market, buying energy at low prices during valley power duration by storing it in ESSs, and when the price is higher, using it to power. In this way, the cost can be reduced even without reducing the total energy consumption [31].

3.4. Management and Control for Clean Energy

Renewable sources such as PV panels and wind generators may be installed to feed railway loads, and this will become a trend in the future [45]. The aim of the measure is to partially power railway loads with clean energy and thus reduce CO_2 emissions. Covering part of the energy needs of a railway system through local renewable sources may significantly reduce the amount of energy imported and reduce the ratio of emitted per kWh consumption. However, there are two main problems after introducing renewable sources:

(1) Locational mismatch:

There are distributed usable spaces along railroad tracks for renewable sources in some supply sections of railway systems. In some cases, the traction power in the corresponding electrical section is less than the clean energy, but other electrical sections without renewable sources require power. That means that in some cases the consumption location is away from the generation area, causing locational mismatch (see the location of the regeneration sources in Figure 2).

(2) Temporal mismatch:

In many stations, huge amounts of electric power are consumed during rush hours. However, the hours of PV generation are fixed, i.e., PV generation is active in the daytime and inactive at night. When generating more, the power consumption may not be high. Wind generators experience a similar effect. Therefore, there are temporal mismatches between railway consumption and renewable generation [22].

The modified system can eliminate the locational and temporal mismatches by controlling EPTDs and PTDs so that they can transfer the clean energy to other sections and store as well as release the clean energy to increase the buffer time. By this approach, clean energy can be reasonably controlled through EMS improving the utilization of clean energy, as shown in Figure 2.

3.5. Active Control for Line Voltage

Traction and regenerative braking of locomotives will cause voltage fluctuations on lines, and the introduction of regenerative sources will also cause voltage fluctuations on lines [20,28]. The effects of the equivalent parameters of the line, such as the reactive power generated by parasitic capacitor, can also cause line voltage fluctuations [46,47]. The range of voltage fluctuation can be suppressed by introducing the modified system. Figure 8 shows the concept of line voltage fluctuation at line ends and the concept of voltage control by use of the modified system. When a locomotive brakes, the regenerative braking power may cause over-voltage in the contact line. The EMS can carry out active voltage control by transferring power via PTD and EPTD and storing the regenerative power in ESS. Similarly, when a locomotive accelerates, the regenerative braking power may cause under-voltage in the contact line. The voltage in the contact line can be kept at a normal via active voltage control. In addition, the modified system can also compensate reactive power to control voltage via PTD and EPTD.

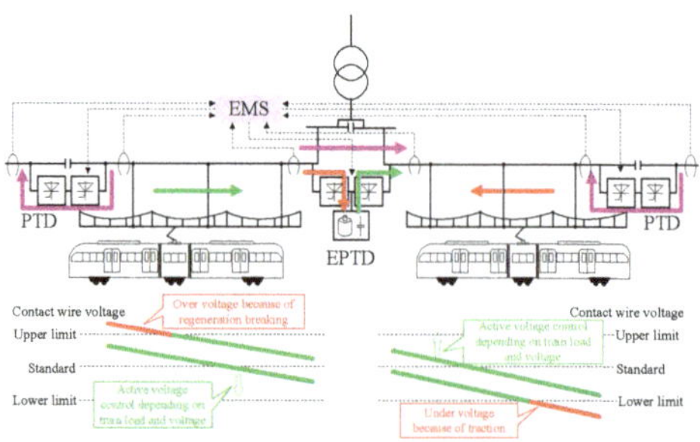

Figure 8. Active voltage control based on the modified system.

3.6. Emergency Power Supply

In electrified railways, power failures will have important impacts on the operation and the service quality of railways. During blackout conditions, trains will stop running; there will be no ventilation and mugginess in trains, which greatly reduces the comfort of passengers. The modified system will alleviate this problem. In blackout conditions, the EMS can carry out emergency control by the use of ESS and power of other electrical sections, as shown in Figure 9. The emergency control can maintain a train's operation during blackout conditions at least until it reaches the nearest station. This will avoid long stops between stations on the canal [20], and keep air-conditioning in operation during the remainder of the train journey to ensure the quality of service provided to passengers.

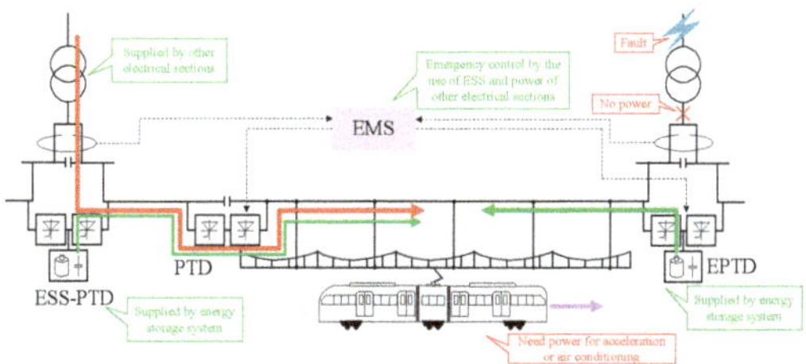

Figure 9. Emergency control based on the modified system.

4. Control Algorithm for Utilization of RBE

This paper focused on the realization of eco-friendliness, energy saving, and low carbon of a railway. In order to achieve this goal, a control algorithm for the utilization of RBE based on the modified system was proposed in this paper.

4.1. Algorithm Principles

In the control process for the utilization of RBE in the whole railway line using the modified system, the loss of the RBE in the lines and the power converters is unavoidable. The line loss of the RBE in the flow process is small, accounting for approximately 0.2% [48]. Therefore, the line loss was ignored and the efficiency of the converters alone was considered in this paper. In addition, the capacity of the ESSs was considered to be unlimited.

To utilize the RBE using the modified system, a control algorithm was designed in this paper. The control algorithm principles is as follows:

(1)   Adjacent transmission by EPTD and PTD (see Figure 10a,b).
(2)   Transmission between two power supply sections (see Figure 10c).
(3)   Storage the surplus RBE in ESS (see Figure 10d).
(4)   Release the energy in ESS (see Figure 10e).

The efficiency of the converter is the most important factor affecting the transmission loss of regenerative braking power. The typical topology of an EPTD is shown in Figure 11, and a PTD was similar, including only AC−DC, and DC−AC. In Figure 11, there were two efficiency paths. Path 1 was transferring the regenerative power by the use of AC−DC, DC−AC converter, and its transmission efficiency was $\eta_{(AC-DC)} \cdot \eta_{(DC-AC)}$. Path 2 was storing or releasing regenerative power by the use of AC−DC, DC−DC converter, and its transmission efficiency was $\eta_{(AC-DC)} \cdot \eta_{(DC-DC)}$. At present, the efficiency of power electronic converters was very high, above 90%, so the efficiencies of $\eta_{(AC-DC)}$, $\eta_{(DC-AC)}$, $\eta_{(DC-DC)}$ were considered to be similar in this paper.

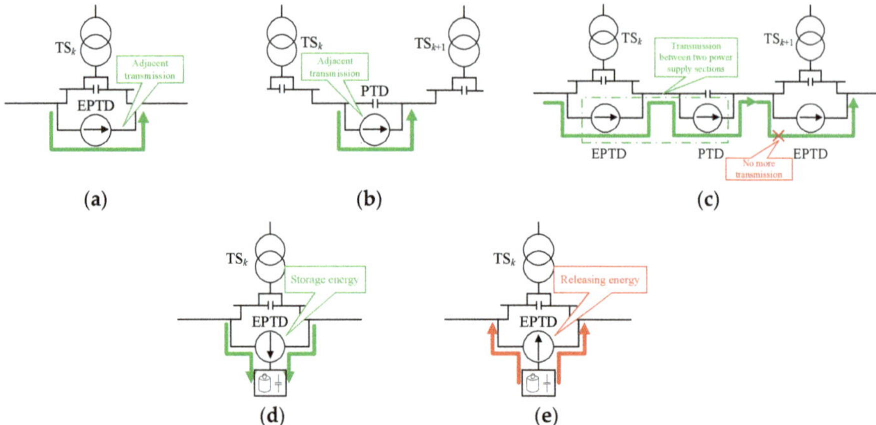

Figure 10. Control algorithm principle: (**a**) adjacent transmission; (**b**) adjacent transmission; (**c**) transmission between two power supply sections; (**d**) storage energy; and (**e**) release energy.

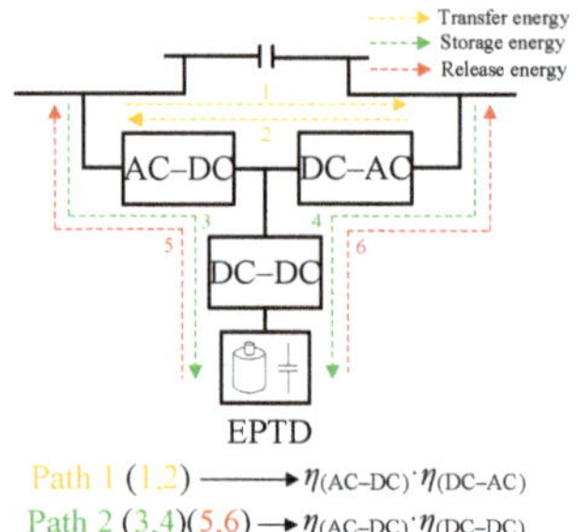

$$\text{Path 1 (1,2)} \longrightarrow \eta_{(AC-DC)} \cdot \eta_{(DC-AC)}$$
$$\text{Path 2 (3,4)(5,6)} \longrightarrow \eta_{(AC-DC)} \cdot \eta_{(DC-DC)}$$

Figure 11. Typical topology and operation paths of EPTD.

The regenerative power is charged to the ESS during train braking and will be utilized for the powering of systems later. Therefore, the transmission efficiency of RBE via ESS is:

$$\eta_{ES} = \eta_{(AC-DC)} \cdot \eta_{(DC-DC)} \cdot \eta_{(DC-DC)} \cdot \eta_{(AC-DC)}. \tag{10}$$

When the surplus regenerative power is transferred to other power sections, the transmission efficiency of RBE via PTD and EPTD is:

$$\eta_{T} = \left(\eta_{(AC-DC)} \cdot \eta_{(DC-AC)} \right)^{m}, \tag{11}$$

where m is the number of power supply sections during the process of regenerative power transmission.

When taking $m = 1$, it can be seen by contrasting (10) and (11) that the transmission efficiency, when transferring regenerative power to the adjacent section via PTD or EPTD, is obviously higher than when it is storing and releasing regenerative power via ESS. Therefore, the most efficient way is to transfer regenerative braking power to the adjacent power supply sections. After principle (1), if there is regenerative power surplus in the whole railway system, this part power will be transferred to other non-adjacent power sections or stored in the ESS. When taking $m = 2$, the transmission efficiency of transferring regenerative power between two sections via PTD and EPTD is similar to storing and releasing regenerative power via ESS. In order to reduce the capacity and cost of ESS, it allows the transfer of the regenerative power between two power sections via PTD and EPTD. After principle (2), if there is still regenerative power surplus, the surplus power continues to be transferred to other non-adjacent power sections or stored in the ESS. When taking $m = 3$, the transmission efficiency transferring regenerative power between the three sections via PTDs and EPTDs is lower than storing and releasing regenerative power via ESS. Therefore it is not economical to transfer the regenerative power between three sections, as shown in Figure 10c. Instead, all surplus regenerative power is stored in ESS. Obviously, the RBE utilization efficiency via the modified system in this paper was higher than that via the system in [33]. After principles (1), (2), and (3), if there is traction power needed in some power sections, the previously stored regenerative power will be released to supply the traction power.

4.2. Algorithm Steps

In accordance with the control algorithm principle, the algorithm steps and computational process were designed.

According to the Figure 4, the feeder power satisfied:

$$P_k = i_k u_k. \tag{12}$$

Assuming that transferring regenerative power by EPTD and PTD is $P_{\text{PTD},k}$; the exchange power in ESS is $P_{\text{ES},k}$. A simplified diagram of power flow in the whole railway system for RBE utilization can be obtained, as shown in Figure 12. The green arrow and the red arrow are the positive direction of transmission power and energy storage power respectively, and the power is negative when they are pointing in the opposite direction. Therefore, transferring power to the right is positive and to left is negative; storing power in an ESS is positive and releasing power from an ESS is negative. Defining the transmission efficiency of each converter is $\eta_k = \eta_{(\text{AC–DC})}\cdot\eta_{(\text{DC–AC})}$; the storage and release efficiency of an ESS from its left port is $\eta_{Lk} = \eta_{(\text{AC–DC})}\cdot\eta_{(\text{DC–DC})}$ (arrow 3 and 5 in Figure 11); and the storage or release efficiency of an ESS from its right port is $\eta_{Rk} = \eta_{(\text{DC–AC})}\cdot\eta_{(\text{DC–DC})}$ (arrow 4 and 6 in Figure 11). Before the modified system operates, each feeder power is equal to the train power in the corresponding power supply sections.

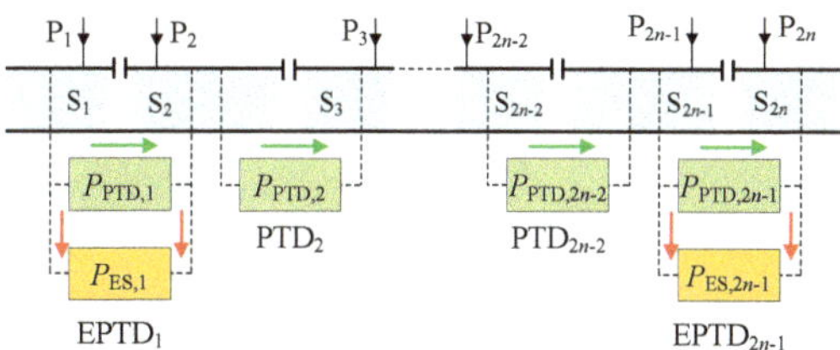

Figure 12. The simplified electrical description for RBE utilization.

(1) Step 1: According to the power data of $P_1, P_2, \dots, P_{2n}$, screening out the power supply sections where the RBE is located and storing the number of the power supply sections in variable j. Firstly, the regenerative power is transmitted to the adjacent power supply sections, as shown in Figure 13.

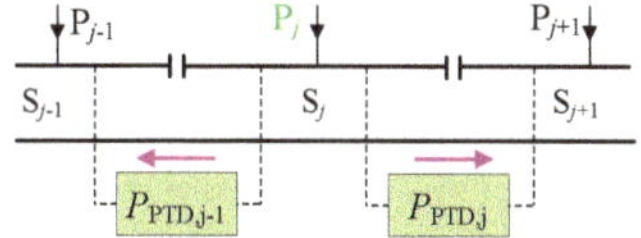

Figure 13. Adjacent transmission.

Case 1: $P_{j-1} \leq 0$, $P_{j+1} \leq 0$.

No transmission.

Case 2: $P_{j-1} > 0$, $P_{j+1} \leq 0$.

The regenerative power in the j power supply section is transferred to the $j - 1$ power supply section. The RBE should be maximized transmission. Therefore, the transmission power $P_{\text{PTD\_1},j-1}$ in step 1 can be calculated by:

$$P_{\text{PTD\_1},j-1} = -\min\{|P_{j-1}/\eta_{j-1}|, |P_j|\}. \tag{13}$$

It should be noted that when taking $j = 2n$, the regenerative power in the $2n$th power supply section is transferred to the $(2n - 1)$th power supply section only.

Case 3: $P_{j-1} \leq 0$, $P_{j+1} > 0$.

The transmission power $P_{\text{PTD\_1},j}$ in step 1 can be calculated by:

$$P_{\text{PTD\_1},j} = \min\{|P_{j+1}/\eta_j|, |P_j|\}. \tag{14}$$

It should be noted that when taking $j = 1$, the regenerative power in the 1st power supply section is transferred to the 2nd power supply section only.

Case 4: $P_{j-1} > 0$, $P_{j+1} > 0$.

(i) if $|P_j| \geq |P_{j-1}/\eta_{j-1}| + |P_{j+1}/\eta_j|$.

The regenerative power in the j power supply section is transferred to the $j - 1$ power supply section and the $j + 1$ power supply section. The regenerative power in the j power supply section can supply all traction power required by the $j - 1$ and $j+1$ power supply sections. The transmission power in step 1 can be calculated by:

$$\left\{ \begin{array}{l} P_{\text{PTD\_1},j-1} = -|P_{j-1}/\eta_{j-1}| \\ P_{\text{PTD\_1},j} = |P_{j+1}/\eta_j| \end{array} \right. . \tag{15}$$

(ii) if $|P_j| < |P_{j-1}/\eta_{j-1}| + |P_{j+1}/\eta_j|$.

Transmission in accordance with converter efficiency to achieve efficient use of RBE.

When $\eta_{j-1} > \eta_j$:

$$\left\{ \begin{array}{l} P_{\text{PTD\_1},j-1} = -\min\{|P_{j-1}/\eta_{j-1}|, |P_j|\} \\ P_{\text{PTD\_1},j} = \left(|P_j| - |P_{\text{PTD\_1},j-1}|\right) \end{array} \right. . \tag{16}$$

When $\eta_{j-1} \leq \eta_j$:

$$\left\{ \begin{array}{l} P_{\text{PTD\_1},j-1} = -\left(|P_j| - |P_{\text{PTD\_1},j}|\right) \\ P_{\text{PTD\_1},j} = \min\{|P_{j+1}/\eta_j|, |P_j|\} \end{array} \right. . \tag{17}$$

(2) Step 2: After step 1, recalculating the feeder power $P_1, P_2, \ldots, P_{2n}$. Then, screening out the power supply sections where the RBE is located and storing the number of the power supply sections in variable j again. After adjacent transmission, the power supply section adjacent to the section with regenerative power has no traction power. Therefore, the regenerative power needs to be transmitted between two power supply sections, as shown Figure 14.

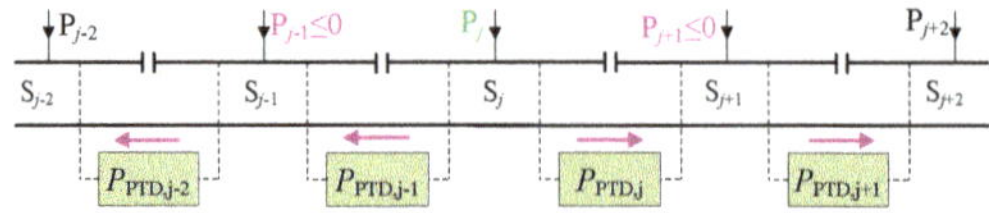

Figure 14. Transmission between two power supply sections.

Case 1: $P_{j-2} \leq 0$, $P_{j+2} \leq 0$.

No transmission.

Case 2: $P_{j-2} > 0$, $P_{j+2} \leq 0$.

The regenerative power in the j power supply section is transferred to the $j - 2$ power supply section. However, the transmission process requires two converters to complete. The transmission power $P_{PTD\_2,j-1}$ and $P_{PTD\_2,j-2}$ in step 2 both need both to be calculated by:

$$\begin{cases} P_{PTD\_2,j-1} = -\min\{|P_{j-2}/(\eta_{j-2} \cdot \eta_{j-1})|, |P_j|\} \\ P_{PTD\_2,j-2} = P_{PTD,j-1} \cdot \eta_{j-1} \end{cases} . \tag{18}$$

It should be noted that when taking $j = 2n$ or $j = 2n - 1$, the regenerative power in the $2n$th or $(2n - 1)$th power supply section is transferred to the $(2n - 2)$th or $(2n - 3)$th power supply section only.

Case 3: $P_{j-2} \leq 0$, $P_{j+2} > 0$.

The transmission power $P_{PTD\_2,j}$ and $P_{PTD\_2,j+1}$ in step 2 both need to be calculated by:

$$\begin{cases} P_{PTD\_2,j} = \min\{|P_{j+2}/(\eta_j \cdot \eta_{j+1})|, |P_j|\} \\ P_{PTD\_2,j+1} = P_{PTD,j} \cdot \eta_j \end{cases} . \tag{19}$$

It should be noted that when taking $j = 1$ or $j = 2$, the regenerative power in the 1st or 2nd power supply section is transferred to the 3rd or 4th power supply section only.

Case 4: $P_{j-2} > 0$, $P_{j+2} > 0$.

(i) if $|P_j| \geq |P_{j-2}/(\eta_{j-1} \cdot \eta_{j-2})| + |P_{j+2}/(\eta_j \cdot \eta_{j+1})|$.

The regenerative power in the j power supply section is transferred to the $j - 2$ power supply section and the $j + 2$ power supply section. The regenerative power in the j power supply section can supply all traction power required by the $j - 2$ and $j + 2$ power supply sections. The transmission power in step 2 can be calculated by:

$$\begin{cases} P_{PTD\_2,j-1} = -|P_{j-2}/(\eta_{j-1} \cdot \eta_{j-2})| \\ P_{PTD\_2,j-2} = P_{PTD\_2,j-1} \cdot \eta_{j-1} \\ P_{PTD\_2,j} = |P_{j+2}/(\eta_j \cdot \eta_{j+1})| \\ P_{PTD\_2,j+1} = P_{PTD\_2,j} \cdot \eta_j \end{cases} . \tag{20}$$

(ii) if $|P_j| < |P_{j-2}/(\eta_{j-1} \cdot \eta_{j-2})| + |P_{j+2}/(\eta_j \cdot \eta_{j+1})|$.
Transmission according to converter efficiency.
When $\eta_{j-1} \cdot \eta_{j-2} > \eta_j \cdot \eta_{j+1}$:

$$\begin{cases} P_{PTD\_2,j-1} = -\min\{|P_{j-1}/(\eta_{j-1} \cdot \eta_{j-2})|, |P_j|\} \\ P_{PTD\_2,j-2} = P_{PTD\_2,j-1} \cdot \eta_{j-1} \\ P_{PTD\_2,j} = (|P_j| - |P_{PTD\_2,j-1}|) \\ P_{PTD\_2,j+1} = P_{PTD\_2,j} \cdot \eta_j \end{cases} . \tag{21}$$

When $\eta_{j-1} \cdot \eta_{j-2} \leq \eta_j \cdot \eta_{j+1}$:

$$\begin{cases} P_{PTD\_2,j-1} = -(|P_j| - |P_{PTD\_2,j}|) \\ P_{PTD\_2,j-2} = P_{PTD\_2,j-1} \cdot \eta_{j-1} \\ P_{PTD\_2,j} = \min\{|P_{j+1}/(\eta_j \cdot \eta_{j+1})|, |P_j|\} \\ P_{PTD\_2,j+1} = P_{PTD\_2,j} \cdot \eta_j \end{cases} . \tag{22}$$

(3) Step 3: After step 2, Recalculating the feeder power $P_1, P_2, \ldots, P_{2n}$. Then, screening out the power supply sections where the RBE is located and storing the number of the power supply sections in variable j again. After step 2, if the regenerative power is still surplus, the surplus regenerative power will be stored in ESS.

(i) if j is an odd number.

The regenerative energy power will be stored in ESS by arrow 3 in Figure 11. The storage energy power $P_{ES\_3,j}$ in step 3 can be calculated by:

$$P_{ES\_3,j} = \left| P_j \cdot \eta_{Lj} \right|. \tag{23}$$

(ii) if j is an even number.

The storage energy power will be stored in ESS by arrow 4 in Figure 11. The storage energy power in step 3 can be calculated by:

$$P_{ES\_3,j-1} = \left| P_j \cdot \eta_{Rj-1} \right|. \tag{24}$$

After storage, all regenerative power is utilized in railway system.

(4) Step 4: Screening out the ESS with storage energy and storing the number of the ESSs in variable j. j is an odd number (see Figure 3).

Case 1: $P_j = 0$, $P_{j+1} = 0$.

No release energy.

Case 2: $P_j > 0$, $P_{j+1} = 0$.

The regenerative power in the j ESS is transferred to the j power supply section by the path 5 in Figure 11. The release energy power $P_{ES\_4,j}$ in step 4 can be calculated by:

$$P_{ES\_4,j} = -\min\left\{ \left| P_j / \eta_{Lj} \right|, \left| P_{ESj} \right| \right\}. \tag{25}$$

Case 3: $P_j = 0$, $P_{j+1} > 0$.

The regenerative power in the j ESS is transferred to the j power supply section by the path 6 in Figure 11. The release energy power in step 4 can be calculated by:

$$P_{ES\_4,j} = -\min\left\{ \left| P_{j+1} / \eta_{Rj} \right|, \left| P_{ESj} \right| \right\}. \tag{26}$$

Case 4: $P_j > 0$, $P_{j+1} > 0$.

(i) if $|P_{ESj}| \geq |P_j/\eta_{Lj}| + |P_{j+1}/\eta_{Rj}|$.

The regenerative power in the j ESS is transferred to the j power supply section and the $j + 1$ power supply section. The release energy power in step 4 can be calculated by:

$$P_{ES\_4,j} = -\left(\left| P_j / \eta_{Lj} \right| + \left| P_{j+1} / \eta_{Rj} \right| \right). \tag{27}$$

(ii) if $|P_{ESj}| < |P_j/\eta_{Lj}| + |P_{j+1}/\eta_{Rj}|$.

When $\eta_{Lj} > \eta_{Rj}$, the storage energy in ESS_j first releases along arrow 5 in Figure 11, achieving efficient utilization. When $\eta_{Lj} \leq \eta_{Rj}$, the storage energy in ESS_j first releases along arrow 6 in Figure 11, achieving efficient utilization. However, it should be noted that all storage energy in ESS_j is fully utilized in above two cases, and the release energy power can be calculated by:

$$P_{ES\_4,j} = -\left| P_{ESj} \right|. \tag{28}$$

Through step 4, the regenerative power stored in the energy storage system is utilized.

Finally, the transmission power references of EPTDs and PTDs and the storage as well as release power references can be calculated by the use of P_{PTD_1}, P_{PTD_2}, P_{ES_3}, and P_{ES_4}. These references are calculated by the EMS and the relevant data is delivered to the PTDs and EPTDs via bidirectional communication equipment.

In summary, all regenerative power was used within the internal railway system more efficiently by the modified smart system proposed in this paper.

5. Results Verification and Analysis

To evaluate the effectiveness of the modified system and the proposed control algorithm, a modified power supply system consisting of four TSs and eight power supply sections was considered. The diagram of the traction power supply system is shown in Figure 15. The variable data of EPTDs and PTDs and 100 groups of locomotive power data in eight supply power sections were randomly generated in Matlab. The efficiency of AC–DC, DC–AC, and DC–DC converters in EPTDs and PTDs is shown in Table 1. The efficiency η_j, η_{Lj}, and η_{Rj} could be calculated by the use of the data in Table 1. The control algorithm was realized in Matlab.

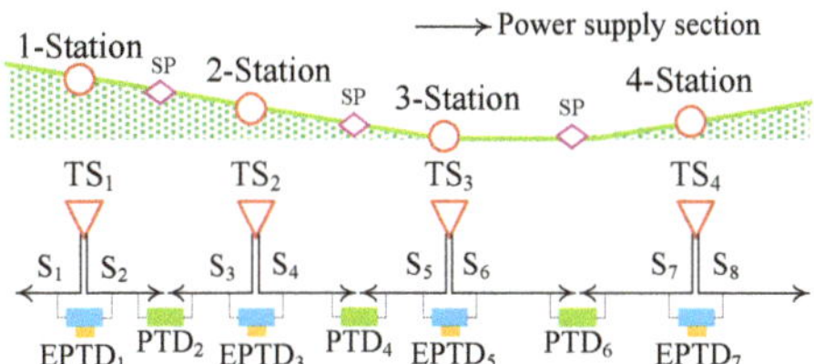

Figure 15. The diagram of the modified power supply system.

Table 1. Converter efficiency.

Type	$\eta_{(AC–DC)}$	$\eta_{(DC–AC)}$	$\eta_{(DC–DC)}$
EPTD1	0.9690	0.9756	0.9753
PTD2	0.9609	0.9665	–
EPTD3	0.9861	0.9501	0.9736
PTD4	0.9886	0.9790	-
EPTD5	0.9847	0.9586	0.9721
PTD6	0.9810	0.9597	-
EPTD7	0.9881	0.9575	0.9786

5.1. Feasibility Evaluation of the Modified System

Based on the aforementioned data, 100 groups of train power consumption of no control results in the original system and control results in the modified system were obtained, as shown in Figure 16. With the proposed system and the control algorithm, the regenerative power was effectively utilized within the internal railway system and the power consumption of the whole railway system was significantly reduced. Moreover, when there are surplus regenerative power in the railway system, the surplus power is stored in ESSs and will be utilized for the powering in the other groups later. The storage energy results in all the ESSs are shown in Figure 17. It can be seen from Figure 17 that the ESSs were not only charged but also discharged, realizing full utilization of all the regenerative power within the internal railway system.

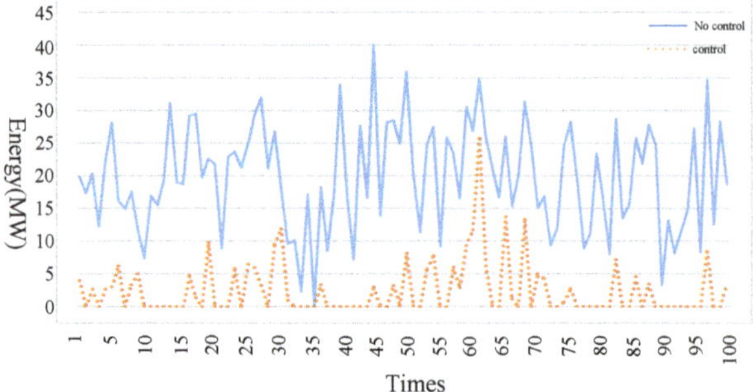

Figure 16. Results before and after control.

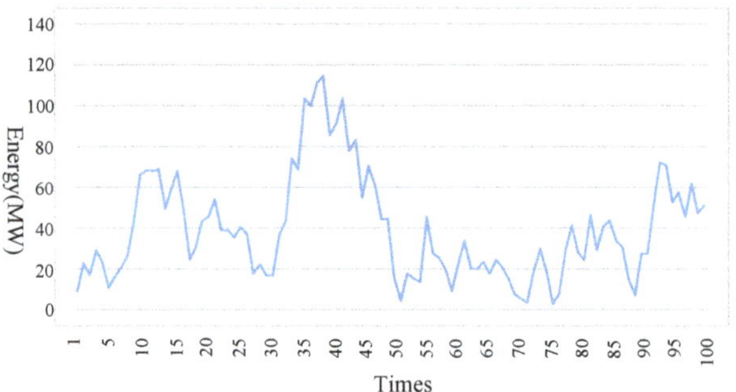

Figure 17. Energy storage results.

In addition, six groups of the locomotive power data in eight supply power sections of no control results in the original system and control results in the modified system were compared. The comparison results are presented in Figure 18. The power in the corresponding six groups of ESSs is presented in Figure 19. It can be seen from the results that the power consumption in most of supply power sections was reduced to zero after utilizing the regenerative power in trains and the energy in ESSs via modified system, realizing the eco-friendliness and low carbon of railway transportation. Specially, In Figure 18f, the power consumption in all the supply power sections was reduced to zero, realizing the zero-emission of railway transportation. Furthermore, the ESS in $EPTD_7$ had no storage power because the regenerative power in 7th and 8th sections was not surplus after regulating by the control algorithm.

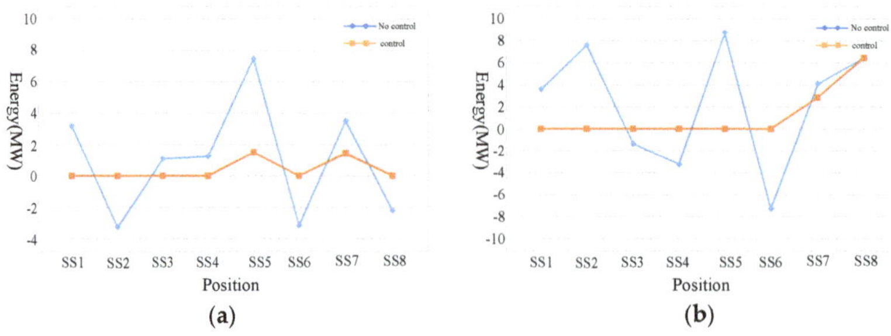

Figure 18. *Cont.*

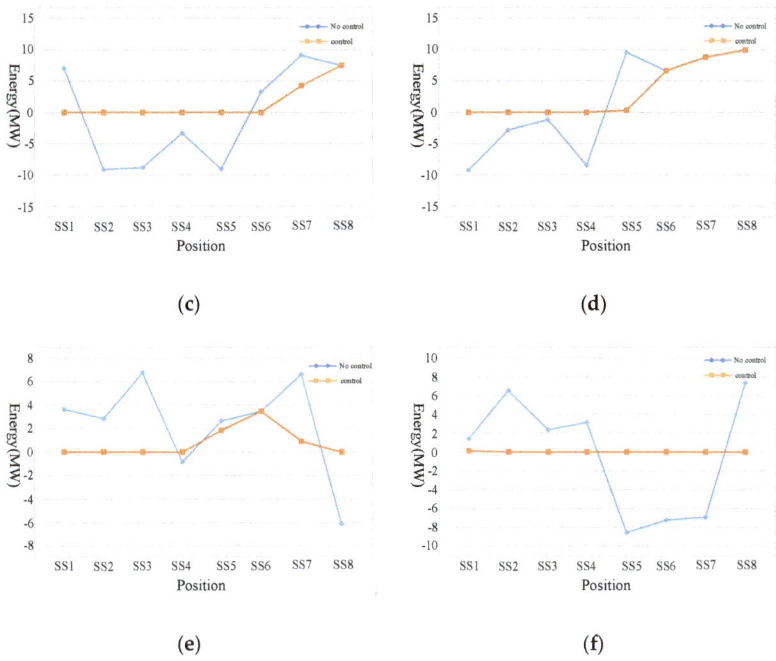

(c)　　　　　　　　　　　　　　(d)

(e)　　　　　　　　　　　　　　(f)

Figure 18. Partial comparison results before and after control: (**a**) 60th group; (**b**) 61th group; (**c**) 62th group; (**d**) 63th group; (**e**) 64th group; and (**f**) 65th group.

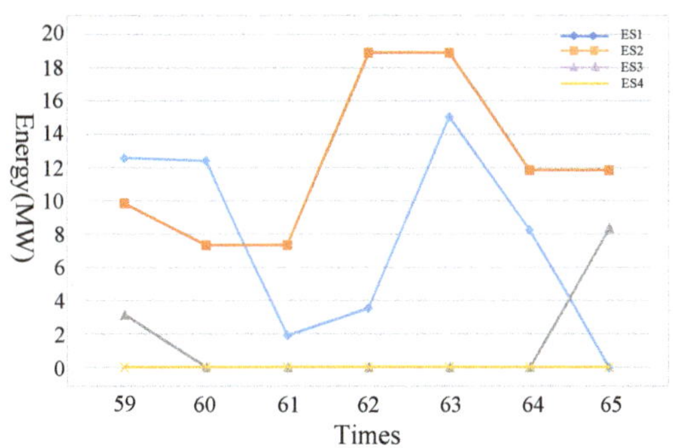

Figure 19. The results in six groups of ESSs.

5.2. Advantage Evaluation of the Modified System

Furthermore, a comparison between the energy storage power of the proposed modified system and that of the system with ESSs alone (Figure 1d) was made in Matlab to verify the advantage of the proposed modified system, as shown in Figure 20. It can be seen that the capacity of the energy storage system adopting the proposed system was greatly reduced. When the system with ESSs alone was also adopted, the regenerative power could be utilized through four converter links AC−DC/DC−DC/DC−DC/DC−AC. Compared with the system with ESSs alone, the modified system only needs to pass through two converter links AC−DC/DC−AC to utilize regenerative power when adjacent transmission, which enhances the utilization rate of the regenerative braking energy. In addition, compared with the scheme in [33], the proposed system can utilize all the regenerative power and avoid regenerative power waste. For example, In Figure 18c, the total regenerative power was 30.347 MW, and the total traction power was 26.973 MW. The regenerative power was more than the traction power. If the scheme in [33] is adopted to utilize regenerative power, it will cause regenerative power waste. Instead, the proposed system will solve this problem. Moreover, the proposed system could also enhance the utilization rate of regenerative power compared with the scheme in [33], as analyzed in Section 4.2, avoiding regenerative power transmission between three sections.

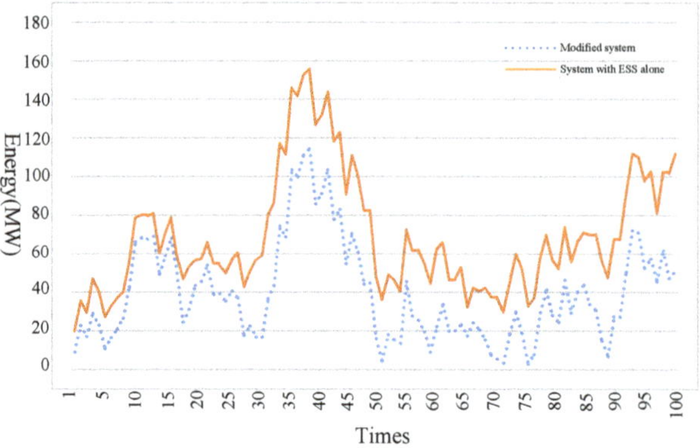

Figure 20. A comparison results of energy storage power between the proposed modified system and the energy storage system.

However, the cost of energy-storage-based smart electrical infrastructure needs to be considered. The cost of batteries is four times that of converters [49], but there may be discrepancies in practice. If only from the perspective of regenerative braking energy utilization, the economy of the proposed system compared with the system with ESSs alone needs to be discussed, because some literatures consider that the systems with ESSs alone are more economical [20,22]. However, considering that the system has other functions, it will bring more benefits at the same cost. From this point of view, the proposed system will be more competitive economically. In addition, with the development of the technology and the increased demand of batteries, the cost of the batteries will be gradually reduced, thereby greatly improving the application potential of the proposed system. Moreover, in some railways with large slopes, more regenerative braking energy will be generated [17]. In these railways, the economic benefits of the proposed system will be more obvious. The scheme of adding ESS in AC-fed railway system has also received attention and been studied [34–36,50], which proves the feasibility of the application of the proposed system based on the energy storage scheme in the railway system. Last but not least, the capacity of the ESSs in the proposed system can be optimized, thereby reducing the cost of the ESSs. Then, the regenerative braking energy can be effectively utilized while ensuring the minimum capacity and cost of the ESSs, thereby further improving the economics of the proposed modified system. It also should be noted that this paper did not consider the issue of the capacity constraints of the ESSs. However in the future research work, the author will further optimize the capacity of the ESSs and design the control strategy under the capacity constraints of the ESSs, thereby enhancing the economic advantage of the proposed system.

In addition, it should be noted that the data was randomly obtained by Matlab, and whether the result of zero-emission of railway transportation in Figure 18f will be achieved in practice still needs to be discussed. This mainly depends on the regenerative braking energy produced by locomotives in actual. However, if clean energy is added for powering traction in the railway system in the future, combined with the energy-storage-based smart electrical infrastructure proposed in this paper, the zero-emission of railway transportation could be realized in the future.

The feasibility of the additional functions of power quality improvement, peak cutting and valley filling, management and control for clean energy, active control for line voltage, and emergency power supply are widely discussed in literatures [5,6,14,20–24,28,30–32,35–44]. For the realization process and verification results of those functions, readers can see the above-mentioned literatures. This paper would no longer validate those functions. The energy-storage-based smart electrical system proposed in this paper could integrate these functions instead of allowing them to operation independently, providing options for smart and eco-friendliness of railway in the future.

This paper focused on regenerative braking energy utilization via the proposed system, realizing eco-friendly railway transportation system. In the future, the author will further investigate the other functions of the proposed system, so as to maximize the benefits of the proposed system.

6. Conclusions

This paper presented a modified system to improve the current AC-fed railway power supply systems, which could enhance the eco-friendliness and smart level of railway. The modified system further improved the degree of controllability of the infrastructure, which allowed for many functions as follows:

- Utilization of regenerative braking energy;
- Power quality improvement;
- Peak cutting and valley filling;
- Management and control for clean energy;
- Active control for line voltage;
- Emergency power supply.

In addition, to utilize the regenerative braking energy, a control algorithm based on the modified system was proposed to realize efficient utilization of all the regenerative power within the internal railway system, making traction consumption of railway systems reduction greatly and more eco-friendliness.

Finally, the advantages of the proposed system and the effectiveness of the proposed control algorithm were verified in a case of a modified power supply system consisting of four traction substations and eight power supply sections. The results showed that the regenerative braking energy in railway could be efficiently utilized by using the improved system and the control algorithm. The work conducted in this paper would provide alternative methods for railway smart control and energy saving and consumption reduction.

Author Contributions: Z.G. and Q.L. designed the research; J.F. and C.W. gave help of validation and formula derivation; Z.G. and B.H. finished the verification; Z.G. wrote the paper.

Funding: This research was funded in part by [the Shenhua Group Co., Ltd., Science and Technology Innovation Projects] grant number [CSIE16024877], and in part by [the National Natural Science Foundation of China] grant number [51577187].

Conflicts of Interest: The authors declare no conflict of interest.

References

1. Douglas, H.; Roberts, C.; Hillmansen, S.; Schmid, F. An assessment of available measures to reduce traction energy use in railway networks. *Energy Conv. Manag.* **2015**, *106*, 1149–1165. [CrossRef]
2. The International Energy Agency (IEA) and the International Union of Railways (UIC). Railway Handbook: Energy Consumption and CO_2 Emissions. Available online: https://www.iea.org/topics/transport/railwayhandbook (accessed on 3 October 2019).
3. Bukarica, V.; Tomsic, Z. Design and evaluation of policy instruments for energy efficiency market. *IEEE Trans. Sustain. Energy* **2017**, *8*, 354–362. [CrossRef]
4. Pankovits, P.; Pouget, J.; Robyns, B.; Delhaye, F.; Brisset, S. Towards railway-smartgrid: Energy management optimization for hybrid railway power substations. In Proceedings of the 5th IEEE PES Innovative Smart Grid Technologies Conference Europe (ISGT Europe), Istanbul, Turkey, 12–15 October 2014.

5. Hernandez, J.C.; Sutil, F.S. Electric vehicle charging stations feeded by renewable: PV and train regenerative braking. *IEEE Latin Am. Trans.* **2016**, *14*, 3262–3269. [CrossRef]

6. Aguado, J.A.; Racero, A.J.S.; De La Torre, S. Optimal operation of electric railways with renewable energy and electric storage systems. *IEEE Trans. Smart Grid.* **2018**, *9*, 993–1001. [CrossRef]

7. Hayashiya, H.; Itagaki, H.; Morita, Y.; Mitoma, Y.; Furukawa, T.; Kuraoka, T.; Fukasawa, Y.; Oikawa, T. Potentials, peculiarities and prospects of solar power generation on the railway premises. In Proceedings of the 2012 International Conference on Renewable Energy Research and Applications (ICRERA), Nagasaki, Japan, 11–14 November 2012.

8. Bocharnikov, Y.V.; Tobias, A.M.; Roberts, C.; Hillmansen, S.; Goodman, C.J. Optimal driving strategy for traction energy saving on DC suburban railways. *IET Electr. Power Appl.* **2007**, *5*, 675–682. [CrossRef]

9. Li, L.; Dong, W.; Ji, Y.D.; Zhang, Z.K.; Tong, L. Minimal-energy driving strategy for high-speed electric train with hybrid system model. *IEEE Trans. Intell. Transp. Syst.* **2013**, *14*, 1642–1653. [CrossRef]

10. Johnson, P.; Brown, S. A simple in-cab schedule advisory system to save energy and improve on-time performance. In Proceedings of the IET Conference on Railway Traction Systems (RTS 2010), Birmingham, UK, 13–15 April 2010.

11. Li, Y.; Tomas, L.; Per, L. Achieving energy-efficiency and on-time performance with Driver Advisory Systems. In Proceedings of the 2013 IEEE International Conference on Intelligent Rail Transportation Proceedings, Beijing, China, 30 August–1 September 2013.

12. Wang, Y.; Hou, Z.S.; Li, X.Y. A novel automatic train operation algorithm based on iterative learning control theory. In Proceedings of the 2008 IEEE International Conference on Service Operations and Logistics, and Informatics, Beijing, China, 12–15 October 2008.

13. Li, Z.X.; Yin, C.K.; Jin, S.T.; Hou, Z.S. Iterative learning control based automatic train operation with iteration-varying parameter. In Proceedings of the 2013 10th IEEE International Conference on Control and Automation (ICCA), Hangzhou, China, 12–14 June 2013.

14. Hayashiya, H.; Yokokawa, S.; Iino, Y.; Kikuchi, S.; Suzuki, T.; Uematsu, S.; Sato, N.; Usui, T. Regenerative energy utilization in a.c. traction power supply system. In Proceedings of the 2016 IEEE International Power Electronics and Motion Control Conference (PEMC), Varna, Bulgaria, 25–28 September 2016.

15. Hayashiya, H.; Kikuchi, S.; Matsuura, K.; Hino, M.; Tojo, M.; Kato, T.; Ando, M.; Oikawa, T.; Kamata, M.; Munakata, H. Possibility of energy saving by introducing energy conversion and energy storage technologies in traction power supply system. In Proceedings of the 2013 15th European Conference on Power Electronics and Applications (EPE), Lille, France, 2–6 September 2013.

16. Hayashiya, H.; Nakao, Y.; Aoki, Y.; Kobayashi, S.; Ogihara, M. Comparison between energy storage system and regenerative inverter in D.C. traction power supply system for regenerative energy utilization. In Proceedings of the 2017 19th European Conference on Power Electronics and Applications (EPE'17 ECCE Europe), Warsaw, Poland, 11–14 September 2017.

17. Lu, Q.W.; He, B.B.; Wu, M.Z.; Zhang, Z.C.; Luo, J.T.; Zhang, Y.K.; He, R.K.; Wang, K.Y. Establishment and analysis of energy consumption model of heavy-haul train on large long slope. *Energies* **2018**, *11*, 965. [CrossRef]

18. Yang, X.; Li, X.; Gao, Z.Y.; Wang, H.W.; Tang, T. A cooperative scheduling model for timetable optimization in subway systems. *IEEE Trans. Intell. Transp. Syst.* **2013**, *14*, 438–447. [CrossRef]

19. González-Gil, A.; Palacin, R.; Batty, P. Sustainable urban rail systems: Strategies and technologies for optimal management of regenerative braking energy. *Energy Convers. Manag.* **2013**, *75*, 374–388. [CrossRef]

20. Hayashiya, H.; Iino, Y.; Takahashi, H.; Kawahara, K.; Yamanoi, T.; Sekiguchi, T.; Sakaguchi, H.; Sumiya, A.; Kon, S. Review of regenerative energy utilization in traction power supply system in Japan: Applications of energy storage systems in d.c. traction power supply system. In Proceedings of the IECON 2017–43rd Annual Conference of the IEEE Industrial Electronics Society, Beijing, China, 29 October–1 November 2017.

21. Hayashiya, H.; Suzuki, T.; Kawahara, K.; Yamanoi, T. Comparative study of investment and efficiency to reduce energy consumption in traction power supply: A present situation of regenerative energy utilization by energy storage system. In Proceedings of the 2014 16th International Power Electronics and Motion Control Conference and Exposition (PEMC), Antalya, Turkey, 21–24 September 2014.

22. Hayashiya, H.; Watanabe, Y.; Fukasawa, Y.; Miyagawa, T.; Egami, A.; Iwagami, T.; Kikuchi, S.; Yoshizumi, H. Cost impacts of high efficiency power supply technologies in railway power supply–Traction and Station. In Proceedings of the 2012 15th International Power Electronics and Motion Control Conference (EPE/PEMC), Novi Sad, Serbia, 4–6 September 2012.

23. Kaleybar, H.J.; Kojabadi, H.M.; Brenna, M.; Foiadelli, F.; Zaninelli, D. An intelligent strategy for regenerative braking energy harvesting in AC electrical railway substation. In Proceedings of the 2017 5th IEEE International Conference on Models and Technologies for Intelligent Transportation Systems (MT-ITS), Naples, Italy, 26–28 June 2017.

24. Wang, Y.; Chen, M.; Lei, G.; Luo, J. Flexible traction power system adopting energy optimisation controller for AC-fed railway. *Electron. Lett.* **2017**, *53*, 554–556. [CrossRef]

25. Pilo De La Fuente, E.; Mazumder, S.K.; Franco, I.G. Railway Electrical Smart Grids: An introduction to next-generation railway power systems and their operation. *IEEE Electrif. Mag.* **2014**, *2*, 49–55. [CrossRef]

26. Palfreyman, T.; Hewings, D. The smart grid applied to railway traction systems: A future vision and integrated protection & control. In Proceedings of the 6th IET Professional Development Course on Railway Electrification Infrastructure and Systems (REIS 2013), London, UK, 3–6 June 2013.

27. Bartłomiejczyk, M. Smart grid technologies in electric traction: Mini inverter station. In Proceedings of the 2017 Zooming Innovation in Consumer Electronics International Conference (ZINC), Novi Sad, Serbia, 31 May–1 June 2017.

28. Hayashiya, H.; Yoshizumi, H.; Suzuki, T.; Furukawa, T.; Kondoh, T.; Kitano, M.; Aoki, T.; Ishii, T.; Kurosawa, N.; Miyagawa, T. Necessity and possibility of smart grid technology application on railway power supply system. In Proceedings of the 14th European Conference on Power Electronics and Applications, Birmingham, UK, 30 August–1 September 2011.

29. Nasr, S.; Iordache, M.; Petit, M. Smart micro-grid integration in DC railway systems. In Proceedings of the IEEE PES Innovative Smart Grid Technologies, Europe, Istanbul, Turkey, 12–15 October 2014.

30. Şengör, I.; Kiliçkiran, H.C.; Akdemir, H.; Kekezoğlu, B.; Erdinç, O.; Catalão, J.P.S. Energy management of a smart railway station considering regenerative braking and stochastic behaviour of ESS and PV generation. *IEEE Trans. Sustain. Energy* **2018**, *9*, 1041–1050. [CrossRef]

31. Khayyam, S.; Ponci, F.; Goikoetxea, J.; Recagno, V.; Bagliano, V.; Monti, A. Railway Energy management system: Centralized-decentralized automation architecture. *IEEE Trans. Smart Grid.* **2016**, *7*, 1164–1175. [CrossRef]

32. Novak, H.; Vasak, M.; Lesic, M. Hierarchical energy management of multi-train railway transport system with energy storages. In Proceedings of the IEEE International Conference on Intelligent Rail Transportation (ICIRT), Birmingham, UK, 23–25 August 2016.

33. Pilo, E.; Mazumder, S.K.; González-Franco, I. Smart electrical infrastructure for AC-fed railways with neutral zones. *IEEE Trans. Intell. Transp. Syst.* **2015**, *16*, 642–652. [CrossRef]

34. Kazunori, T.; Masaaki, T.; Hitoshi, H.; Yumiko, I. Proposal and effect evaluation of RPC application with energy storage system for regenerative energy utilization of high speed railway. *J. Int. Counc. Electr. Eng.* **2017**, *7*, 227–232.

35. Ma, Q.; Guo, X.; Luo, P.; Zhang, Z.W. Coordinated control strategy design of new type railway power regulator based on super capacitor energy storage. *Trans. Chin. Electr. Soc.* **2018**, *34*, 765–776. (In Chinese)

36. Ma, Q.; Guo, X.; Luo, P.; Zhang, Z.W. A novel railway power conditioner based on super capacitor energy storage system. *Trans. Chin. Electr. Soc.* **2018**, *33*, 1208–1218. (In Chinese)

37. Wei, W.J.; Hu, H.T.; Wang, K.; Chen, J.Y.; He, Z.Y. Energy storage scheme and control strategies of high-speed railway based on railway power conditioner. *Trans. Chin. Electr. Soc.* **2019**, *34*, 1290–1299. (In Chinese)

38. Deng, W.L.; Dai, C.H.; Han, C.B.X.; Chen, W.R. Back-to-back hybrid energy storage system of electric railway and its control method considering regenerative braking energy recovery and power quality improvement. *Proc. CSEE* **2019**, *39*, 2914–2923. (In Chinese)

39. Luo, A.; Wu, C.P.; Shen, J.; Shuai, Z.K.; Ma, F.J. Railway static power conditioners for high-speed train traction power supply systems using three-phase V/V transformers. *IEEE Trans. Power Electron.* **2011**, *26*, 2844–2856. [CrossRef]

40. Chen, M.W.; Chen, Y.Y.; Wei, M.C. Modeling and control of a novel hybrid power quality compensation system for 25-kV electrified railway. *Energies* **2019**, *12*, 3303. [CrossRef]

41. Roudsari, H.M.; Jalilian, A.; Jamali, S. Flexible fractional compensating mode for railway static power conditioner in a V/v traction power supply system. *IEEE Trans. Ind. Electron.* **2018**, *65*, 7963–7974. [CrossRef]

42. Zhang, D.H.; Zhang, Z.X.; Wang, W.; Yang, Y.L. Negative sequence current optimizing control based on railway static power conditioner in V/v traction power supply system. *IEEE Trans. Power Electron.* **2016**, *31*, 200–212. [CrossRef]

43. Langerudy, A.T.; Mousavi, G.S.M. Hybrid railway power quality conditioner for high-capacity traction substation with auto-tuned DC-link controller. *IET Electr. Syst. Transp.* **2016**, *6*, 207–214. [CrossRef]

44. Zhu, Z.; Ma, F.J.; Wang, X.; Deng, L.F.; Li, G.X.; Wei, X.W.; Tang, Y.X.; Liu, S.Y. Operation analysis and a game theoretic approach to dynamic hybrid compensator for the V/v traction system. *IEEE Trans. Power Electron.* **2019**, *34*, 8574–8587. [CrossRef]

45. The International Energy Agency (IEA). The Future of Rail: Opportunities for Energy and the Environment. Available online: https://www.iea.org/futureofrail (accessed on 23 October 2019).

46. Perin, I.; Walker, G.R.; Ledwich, G. Load sharing and wayside battery storage for improving AC railway network performance, with generic model for capacity estimation, part 1. *IEEE Trans. Ind. Electron.* **2019**, *66*, 1791–1798. [CrossRef]

47. Perin, I.; Walker, G.R.; Ledwich, G. Load sharing and wayside battery storage for improving AC railway network performance, with generic model for capacity estimation, part 2. *IEEE Trans. Ind. Electron.* **2018**, *65*, 9459–9467. [CrossRef]

48. Ma, F.J.; Luo, A.; Xu, X.Y.; Xiao, H.G.; Wu, C.P.; Wang, W. A simplified power conditioner based on half-bridge converter for high-speed railway system. *IEEE Trans. Power Electron.* **2013**, *60*, 728–738. [CrossRef]

49. Gelman, V. Energy storage that may be too good to be true: Comparison between wayside storage and reversible thyristor controlled rectifiers for heavy rail. *IEEE Veh. Technol. Mag.* **2013**, *8*, 70–80. [CrossRef]

50. Gui, G.P.; Luo, L.F.; Liang, C.G.; Hu, S.J.; Li, Y.; Cao, Y.J.; Xie, B.; Xu, J.Z.; Zhang, Z.W.; Liu, Y.X.; et al. supercapacitor integrated railway static power conditioner for regenerative braking energy recycling and power quality improvement of high-speed railway system. *IEEE Trans. Transp. Electrif.* **2019**, *5*, 702–714.

Article

Operational Simulation Environment for SCADA Integration of Renewable Resources

Diego Francisco Larios *, Enrique Personal, Antonio Parejo, Sebastián García, Antonio García and Carlos Leon

Department of Electronic Technology, Escuela Politécnica Superior, University of Seville, Seville 41011, Spain; epersonal@us.es (E.P.); aparejo@us.es (A.P.); sgarcia15@us.es (S.G.); antgar@us.es (A.G.); cleon@us.es (C.L.)
* Correspondence: dlarios@us.es; Tel.: +34-954-557-192

Received: 29 January 2020; Accepted: 1 March 2020; Published: 13 March 2020

Abstract: The complexity of power systems is rising mainly due to the expansion of renewable energy generation. Due to the enormous variability and uncertainty associated with these types of resources, they require sophisticated planning tools so that they can be used appropriately. In this sense, several tools for the simulation of renewable energy assets have been proposed. However, they are traditionally focused on the simulation of the generation process, leaving the operation of these systems in the background. Conversely, more expert SCADA operators for the management of renewable power plants are required, but their training is not an easy task. SCADA operation is usually complex, due to the wide set of information available. In this sense, simulation or co-simulation tools can clearly help to reduce the learning curve and improve their skills. Therefore, this paper proposes a useful simulator based on a JavaScript engine that can be easily connected to any renewable SCADAs, making it possible to perform different simulated scenarios for novel operator training, as if it were a real facility. Using this tool, the administrators can easily program those scenarios allowing them to sort out the lack of support found in setting up facilities and training of novel operator tasks. Additionally, different renewable energy generation models that can be implemented in the proposed simulator are described. Later, as a use example of this tool, a study case is also performed. It proposes three different wind farm generation facility models, based on different turbine models: one with the essential generation turbine function obtained from the manufacturer curve, another with an empirical model using monotonic splines, and the last one adding the most important operational states, making it possible to demonstrate the usefulness of the proposed simulation tool.

Keywords: co-simulation; renewable resources; SCADA; operator training

---

1. Introduction

In the last few years, the presence of renewable energy sources has increased significantly within the power systems. These new sources, contrary to the traditional energy approach, are not only generated in huge central, but also in distributed small, mini, or micro power plants. This approach is covered under the paradigm called Distributed Generation (DG) [1]. Specifically, this paradigm is one of the keystones in Smart Grids (SG) [2–5], following an optimal control of all the connected elements, including in them an active role of consumers and producers and even proposing a new actor, the prosumer. In the DG paradigm, the resources of generation and/or storage are connected directly to the distribution network, as it is defined in the EU Directive 2019/944 [6]. In this way, they are located closer to consumption points. Those elements, which are usually renewables, are called Distributed Energy Resources (DERs) [7].

Taking under consideration these trends, it is clear that the power system is becoming more complex: the number of generation points is increasing, and the production is highly dependent on the weather. This is why different technologies are used in SG to achieve the monitoring and control levels to maintain the coordination and stability of the whole system. Those advances cover all the areas from Automatic Metering Infrastructure (AMI) [8], fault location systems [9], electric vehicle integration [10], and of course, the generation control. Above all, reliable systems for planning, coordination, and analysis are an imperative in the future of the power system.

Specifically, it is in generation control where prediction and simulation are especially important. The balancing of the whole power system depends on how accurate the predictions of generation are, an aspect that is becoming more and more difficult due to the increasing number of DG facilities.

In this sense, utilities or power companies are currently integrating, at different levels, tools to support their assets' management. This trend is included in their strategies of network digitization, in which the digital twin is the paradigm that is arousing more interest [11]. Thus, thanks to the greater knowledge of the network that provides these technologies (more detailed models and greater amount of real-time data), it is possible to define new analytic applications and diagnosis based on this network characterization (e.g., [12,13]).

Specifically, simulation software is directly affected by this improvement of the network characterization. The ability to reproduce the behavior of a system in different scenarios or configurations is very useful, becoming essential tools for planning tasks. Due to this, it is easy to find simulators for the study, modeling, or design of DER elements.

On the one hand, there is software focused on the analysis or the design of the DER elements themselves. For example, QBlade is widely used to design blades and simulate their behavior, achieving good results even in cases of vortexes' existence [14] or low wind conditions [15]. PVSyst^{TM} is a software tool to size photovoltaic (PV) facilities and to estimate their production depending on the geographic situation, solar panel technology, and other parameters [16]; likewise, Bluesol is a similar example [17].

On the other hand, general power simulators can be also used to study the DER behavior. Thus, this second group is usually divided into two types: Electromagnetic Transient (EMT) analysis and power flow analysis. In this first approach, EMT tools such as PSCAD^{TM}, DIgSILENT PowerFactory^{TM}, and the MATLAB^R SimPowerSystems^{TM} toolbox have been widely used through their model library to simulation of the transient behavior of individual assets or sets of them. As an example, PowerFactory^{TM} has been used in [18] to simulate the wind turbine model following the guidelines defined by the IEC 61400-27-1 [19]; or PSCAD^{TM}, which was used in [20] to test a new protection scheme for relay in power systems with PV generation; or in [21] to test dynamic reactive power compensations for PV inverters. MATLAB$^{®}$ and Simulink$^{®}$ have been also extensively used for the simulation of microgrids and hybrid systems using ad hoc models (e.g., [22]). In the second subgroup, power flow simulators are focusing on energy transfer and lose analysis along the power system. Simulation tools such as PowerWorld^{TM}, PSS$^{®}$E, and OpenDSS are commonly applied for this type of study, even being integrated into the corporate utilities' tools. As an example, PowerWorld^{TM} was used in [23] to carry out a review of complex network analysis in electric power systems using an IEEE Node Test Feeder [24]. Besides, OpenDSS is a free software by EPRI$^{®}$, which has been used in Demand Response (DR) analyses [25], Volt-Var Optimization (VVO) [26], etc. Moreover, OpenDSS is also suitable for evaluating the analysis under different harmonic orders, making Power Quality (PQ)analyses possible (e.g., [27]). Furthermore, new tools like DRIVE [28] completes OpenDSS's capabilities, making Hosting Capacity Analysis (HCA) possible [29], which is essential for utilities in the present scenario of DER deployment and the application in new flexibility services [30].

Thus, it is important to note that, when the simulation is applied considering cyber-physical systems, different components could be interconnected to form a more complete one. For example, for a PV farm simulation, not only is the model of the solar panels necessary, but also, the models of the inverters and the control system should be studied in order to create a more realistic and complete scenario. This approach in which different simulation models are combined is called co-simulation [31].

An example of a simulator that covers this function is Mosaik, a framework for SG co-simulation [32]. This tool has been used by researchers for the study and evaluation of different energy scenarios [33].

In summary, all of these tools are focused on: (A) the analysis or design of the elements or (B) studying DER energy production and power flow for planning task. As can be seen, it is not usual that they take under consideration the operation task from the operators' point of view [34]. However, even with all the automation existing in power plants, most of the monitoring operations and maintenance are actually done by (human) workers.

In this sense, this paper proposes a novel simulation tool focusing mainly on operator training. Therefore, it is designed to replicate operation situations that are very important in training process, as for example very high wind speed or gust winds in offshore facilities.

In generation plants, systems that are responsible for monitoring and managing these kinds of facilities are commonly implemented by Supervisory Control And Data Acquisition (SCADA) systems [35]. They constitute the human-machine interface of the plant, in which the operational parameters both under normal and failure conditions are shown (e.g., warnings and alarms). In this way, some authors such as K.S.Lin [36] studied the application of complex Failure Mode and Effects Analysis (FMEA) based on Risk Priority Number (RPN), or other methods.

Once the importance of the SCADA design and its operation are established, it is easy to note that the ability and skills of the SCADA operator have no less importance for keeping them in secure and safe conditions, but as described before, most of the simulators that appear in the literature do not consider this scenario.

Some authors studied how to include new information analysis in the SCADA to support the operator in the decision process. An example of this can be observed in the work of Y. Liu et al. [37], which proposed a blade icing detection system based on machine learning and deep learning. In the same way, Schlechtingen et al. [38] proposed an Adaptive Neuro-Fuzzy Interference Systems (ANFIS) for wind turbine condition monitoring using SCADA data. Its objective was the detection of abnormal behavior and indicating component malfunctions or faults using captured signals from the SCADA. Therefore, these papers were more focused on improving the available information for operators, not training them in a secure way to operate the resources over those situations, and not offering any assistance for the learning process.

Therefore, a simulation tool that makes it possible not only to simulate the production of a renewable plant under normal and anomalous conditions, but that also considers the manual actions performed by operators and the information they require is essential. This approach is particularly interesting for operator training, where they need to learn how to face all those different operational situations.

Due to this necessity, this paper is focused on the proposal of a SCADA-integrable simulator that helps in the training of operators for renewable generation systems. Specifically, it raises a simulator focused on the human operational aspects of the facilities and not an accurate production simulation. This second scope is easier to observe in the literature, while the human factors are sometimes not even considered. In this sense, this paper proposes a simulator useful for SCADA operator training. This tool can be operated from real SCADA, acting over simulated values and events instead of real resources, without an appreciable difference from the point of view of operators. Therefore, it could be used as a training environment for novel operators, reducing the learning time and accelerating their labor incorporation, safely.

Moreover, the user interface, the programming language, and methodology are designed to be simple, allowing an easy modification of the models, even from non-expert users.

This paper is organized as follows. Section 2 exposes the general requirements for the simulator. Later, the simulator tool architecture and its main parts are described in Section 3. Section 4 shows the simulator usefulness, describing possibles models that could be used for the proposed simulator. Section 5 presents three study cases for the simulation tool. Finally, the conclusions in Section 6 are shown.

2. Simulation Requirements

Currently, as was described above, the presence and complexity of DERs are growing continuously. For this reason, more expert users are required, and they have to be trained to operate them correctly on a daily basis. Moreover, these users should not only know about normal operation procedures, but even the best ones to avoid situations that may cause a safety or security risk (for users and facilities, respectively). As K.-S. Lin stated, "safety and security are key issues in SCADA systems" [36]. On the other hand, the start up of a DER is a complex task that requires multiple verification tests of the SCADA.

Because of all the above, a clear need arises: operators need to be trained to deal with these situations. In this sense, it is essential to have a simulation tool for training SCADA users in the field of renewable resource management, following the simulation approaches proposed in Section 1. Specifically, the use of a simulation tool that allows the SCADA users to interact with different simulated scenarios (such as: historical weather conditions, failure situations, etc.) is a valuable instrument. Thanks to this, operators can be trained to react to both normal and emergency situations, without causing true risks in real facilities.

However, for a simulator like the one described above to be really useful, it should meet certain requirements. For example, it should be transparent for the user, i.e., maintaining the same interaction with both the real or the simulation plant. In this sense, as can be seen in Figure 1, users need to be able to switch between (or even combine) real and simulated DER if necessary, but always maintaining the same SCADA interface.

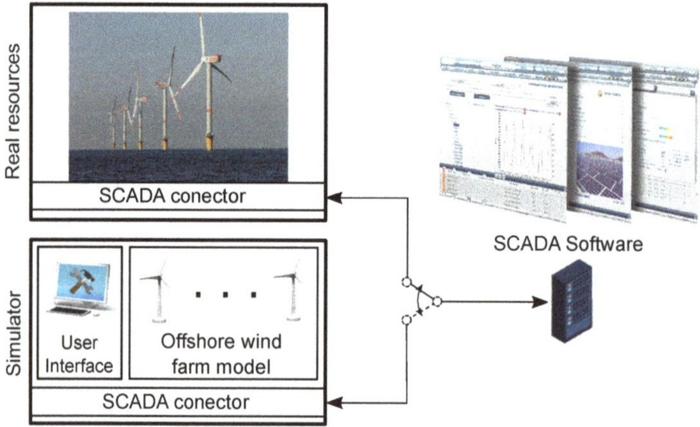

Figure 1. Connection between SCADA and the simulator.

In a regular operation, SCADA shows the real resources' information. However, when new operators should be trained for a specific issue, the simulation of this scenario could be executed, always maintaining the same SCADA interface.

Additionally, the simulation must also be executed in the background, evolving independently. Although SCADA could not obtain values from the simulation, the simulator would continue evolving its simulation variables. Therefore, when SCADA asks for the value of a variable, it will always get the most updated value, as when this happens in a real facility.

Finally, the simulator should be friendly for the users, providing a simple interface to configure the simulation parameters. Moreover, common SCADA users do not need to have advanced programming skills. Therefore, the simulator should provide the necessary tools to create, modify, and run DER models easily, especially designed for training purposes.

3. Simulation Tool Description

To accomplish the requisites exposed in Section 2, the proposed simulation tool was designed with the architecture depicted in Figure 2. As can be seen in this figure, the simulator was based on three different blocks that exchanged information among them. Specifically, it mainly consisted of a data repository (the shared memory manager) to manage the memory exchange between the other three blocks; the SCADA communication processand the simulation engine. Each one of those blocks is described in detail the next subsections.

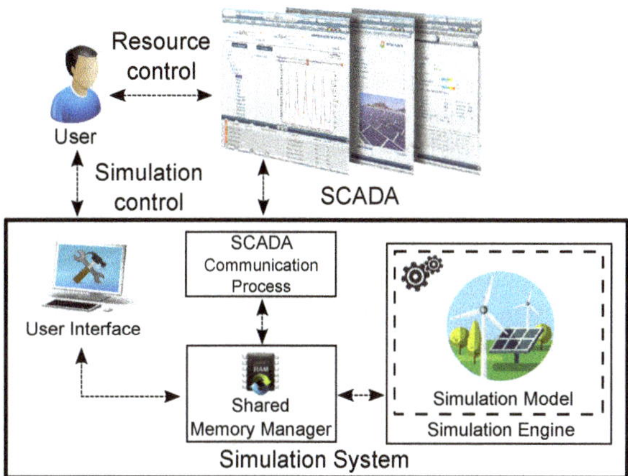

Figure 2. Elements of the simulator architecture.

The proposed simulation system was programmed as a standalone C++ application. Therefore, the shared memory manager, the SCADA Communication Process, and the simulation engine were programmed as external libraries that would be called by the simulation system's main process itself. In this sense, all of these blocks were written as a standardized C library interface that made it possible to change or modify each one independently. Specifically, the proposed architecture allowed advanced users to change the communication method with SCADA (through the SCADA communication process) or even to change the script language for the simulation engine, but maintaining the same simulation system.

Thus, thanks to this architecture, the user could interact with the system in two ways: through the SCADA or using the configuration interface. Besides, this architecture also made it possible to combine two control scenarios for the SCADA simultaneously: with simulated and real facilities, but allowing the user to adapt the simulation model for each specific use case. The different blocks that made up the architecture of the proposed solution are described in the next sections.

3.1. Shared Memory Manager

As can be seen in Figure 2, the simulation engine, the SCADA communication process, and the configuration interface share data by means of a shared memory area. Specifically, these three processes need to be read or written in the common memory zone.

The shared memory manager controls the common memory block. This manager is responsible for maintaining the consistency of the data, even when having multiple accesses over the same data points from different processes at the same time. Specifically, this block is responsible for:

- Creating the configuration and simulation shared variables when a new simulation starts.
- Saving or providing configuration parameter values to/from the configuration interface and the simulation engine.

- Saving or providing simulation signal values to/from the SCADA communication process, the configuration interface, and the simulation engine.

In order to do that, the shared memory manager has a memory synchronization mechanism, so it ensures that the data points are consistent for the simulated models and for SCADA. This manager guarantees that one process does not change the data points of other processes in the middle of an execution step. This manager work as follows:

- If a simulator model modifies the value of a data point, this new value is immediately available for this model, but it is not immediately written into the common memory block. Thus, for SCADA and the rest of the modules, the data value has not been modified yet.
- If the configuration interface or the SCADA communication process requests a write operation, this new value is stored in a temporary location, but not in the common memory block, until all the modules are synchronized at the end of the current simulation step. For example, if SCADA sends a run operation to a DER, no block will see this order, until the next simulation step is complete. This procedure guarantees data consistency, avoiding errors coming from a data change at the middle of an execution step.
- When an execution step is finished, the new data values are synchronized in the common memory block. In the same way, before executing a new simulation step, the simulation engine copies the data values into a temporary location in order to hold the new data values coming from SCADA and the configuration interface.
- If a data point is written more than once from SCADA and/or from the configuration interface, the shared memory manager will only consider the last one, discarding the previous values.

Every data point has an identification name (ID), so it allows other processes to retrieve the data values from the common memory area. Thus, this ID must be unique.

This memory model is especially suitable for the development of Finite State Machine (FSM) models. This procedure and memory model allow the simulator to follow a strategy typically used in automation equipment, such as the Programmable Logic Controllers (PLCs), which have proven its robustness over the years.

3.2. SCADA Communication Process

This block is designed as a data gateway between SCADA and the simulation tool. In this way, there exist many alternatives of communication interfaces for SCADAs. In renewable applications, some typical standards are Modbus [39], Distributed Network Protocol 3 (DNP3) [40], IEC 60870-5 [41], and IEC 62541 (Object linking and embedding for Process Control-Unified Architecture, OPC-UA) [42], as stated in [43,44]. A special mention should be given to OPC-UA, an interesting secure [44] platform for Industry 4.0 [45].

In this sense, it is difficult to define a general exchange mechanism that covers all current SCADA solutions. Thus, as an example, a communication based on an OPC-UA interface was implemented for the proposed solution.

Moreover, the proposed modular architecture allows the system an easy adaptation to other technology. Therefore, only changing the SCADA communication library makes it possible to interact with other SCADAs or applications where a OPC-UA interface is not available.

As the rest of the simulation tool, this interface is designed to be simple and easy to maintain and is based on the next methods:

- Constructor: This method can be used to inform to SCADAs that they need to start to exchange information with the simulator, instead of the real system. It is called every time a new simulation starts.
- Destructor: This method is called at the end of the simulation. It can be used to reconfigure the SCADA communication to the real deployment.

- Connector to the memory manager: This method allows SCADA to add, read, or write data in the common memory area (as was described in Section 3.1).

3.3. Simulation Engine

The simulation engine is responsible for executing the simulation process. This behavior is depicted in Algorithm 1, where *Event* represents the external interaction with the applications, *State* the internal current state of the application, and *State.log* the standard outputs of the application.

Algorithm 1 Simulation execution cycle.

Input: *model*: JS model to execute, *Event*: event generated {*Load_Model, Launch_Model, Stop_Model, Timer_Event*}, *State*: current state {*Error, Loaded, Running*}, t: current timestamp, Δt: time step

switch *Event* **do**

  case *Load_Model* **do**

    if *jsEngine.test*(*model*) $=$ *FALSE* **then**

      State.log.error $\leftarrow$ "Model not valid" *State* $\leftarrow$ *Error*

    else

      $\Delta t \leftarrow$ *jsEngine.execute*(*model.init*()) *State* $\leftarrow$ *Loaded*

  case *Launch_Model* **do**

    switch *State* **do**

      case *Error* **do**

        State.log.error $\leftarrow$ "Error, model cannot be executed, or no model is loaded"

      case *Running* **do**

        State.log.info $\leftarrow$ "simulation is running"

      case *Loaded* **do**

        Event.Timer_Event.value $\leftarrow t + \Delta t$ *State* $\leftarrow$ *Running*

  case *Timer_Event* **do**

    if *jsEngine.execute*(*model.step*(*Event.Timer_Event.value*)) $=$ *FALSE* **then**

      State.log.error $\leftarrow$ "Execution error" *State* $\leftarrow$ *Error*

    else if ($t \geq$ *Event.Timer_Event.value* $+ \Delta t$) **then**

      State.log.error $\leftarrow$ "Execution timeout" *State* $\leftarrow$ *Error*

    else

      Event.Timer_Event $\leftarrow$ *Event.Timer_Event.value* $+ \Delta t$

  case *Stop_Model* **do**

    if *State*=*Running* **then**

      Event.Timer_Event $\leftarrow$ *NULL* *State* $\leftarrow$ *Loaded*

As can be seen, the simulation engine was designed as an event-based architecture. Therefore, the events could be triggered by the user or by the simulator itself. The possible events that could trigger an action were:

- Load_Model. This event is triggered when a new model is uploaded to the simulator. When receiving this signal, and after checking if the simulation model if valid, the simulator calls the jsinit() function (see Section 3.3.2). At this point, the model is ready to be launched.
- Launch_Model. When the user wants to start or resume the simulation, this event is triggered. If the uploaded model is valid, the simulator starts a timer that will trigger a signal periodically based on the simulation step time.
- Timer_Event. During a simulation, this event is called periodically based on the simulation step time, consisting of the main simulation cycle. This event cyclically calls the function $jsjsEngine.execute(model.step(Event.Timer_Event.value), which is responsible for the execution of the following sequence$

1. Read the inputs (signals and parameters) from the shared memory manager. If there is no new data for a required value in the simulation, it keeps the last one.
2. Run the simulation step, executing the simulation model based on the input data. This model can be made up by an aggregation of different individual submodels. Thus, in each simulation step, the following tasks will be executed:

   - The different individual submodels are simulated in a certain order. The order may be important, since the output of some models may be the inputs of others.
   - In the specific case that a final aggregation is needed, once all the submodels are simulated, the overall output data are obtained by executing the aggregation submodel. This submodel is always the last one to run in a simulation step.

3. Update the tables of input and output data on the shared memory manager with the obtained results.

- Stop_Model: When a simulation is stopped, this event needs to be called. This event stops the timer, and therefore, the simulation is stopped. In order to resume the simulation, the Launch_Model signal needs to be called again.

To do these tasks, the simulation engine was designed to execute Discrete-Event Simulation (DES) co-simulations [46], based on individual submodels defined by a Finite State Machine (FSM) [47]. The advantage of this structures is that it allows the system to know if its execution time specification is suitable for the simulation in course. Therefore, if the model execution time is greater than the defined simulation step, the simulation stops immediately, generating an error message.

Additionally, E. Byon et al. [48] implemented a DES model for a wind turbine; they stated that DES provides "hierarchical and modular model construction and an object-oriented substrate supporting repository reuse", which is an advantage in the definition of DER models and reduces the development time.

As was described before, the simulation engine block was implemented as an external library, similarly to the SCADA communication process. Therefore, only changing the library, the proposed simulator may allow advanced users to change between different languages to define the simulation. Some examples of languages could be Network Simulator 3 (NS-3) [49], C++ [50], or Modelica [51–53], each one with its advantages and disadvantages. However, most of these languages have a complex learning curve and are not traditionally used by SCADA users.

In this sense, Wagner exposed in [54] that "although JavaScript cannot compete with strongly typed compiled languages (such as C++, Java and C#) on speed, it provides sufficient performance for many types of simulations and outperforms its competitors on ease of use and developer productivity". However, according to some comparisons [55] [56], the difference in performance between a native C/C++ code and a JS implementation is in the order of 35-45%, which is similar to Java performance. Due to this and following this approach, ECMAScript 2009 [57] was chosen as an appropriate solution for the scripting language for the model simulation engine.

ECMAScript is a standardized and platform independent version of JavaScript (JS). Nevertheless, as was described in [58], modern JS engines are very efficient. Many of the performance issues are due to inefficient usage of the provided Application Programming Interfaces (APIs). In this sense, to solve this problem, a reduced extra instruction set over JS (see Table 1) was provided for the proposed simulation tool.

Table 1. List of the extra instruction set defined for the simulation scripts.

Name	Description	Function Syntax
include	Add an external script file to the current simulation.	jsinclude(*<file>*js);
fsRead	Read a file, and return its content as a JavaScript string.	*<text>*js=fsRead(*<file>*js);
fsWrite	Write the content of a JavaScript text in a file. By default, any existing file is overwritten. Alternatively, it is possible to append the content at the end of an existing file by adding an active flag (with a "true" value) as the third parameter in the call.	jsfsWrite(*<file>*js,*<text>*js); or jsfsWrite(*<file>*js,*<text>*js,*<flag>*js);
logInfo	Send an information message to the general log. This message does not stop the script execution.	jslogInfo(*<message>*js);
logWarning	Send an alarm message to the general log. This message does not stop the script execution.	jslogWarning(*<message>*js);
logError	Send an error message to the general log. This message stops the simulation immediately.	jslogError(*<message>*js);
addInt	Create an integer-type variable in the shared memory.	jsaddInt(*<name>*js,*<intValue>*js);
addReal	Create a real-type variable in the shared memory.	jsaddReal(*<name>*js,*<realValue>*js);
addText	Create a text-type variable in the shared memory.	jsaddText(*<name>*js,*<textValue>*js);
getInt	Read the value of an integer-type variable from the shared memory.	*<intValue>*js=getInt(*<name>*js);
getReal	Read the value of a real-type variable from the shared memory.	*<realValue>*js=getReal(*<name>*js);
getText	Read the value of a text-type variable from the shared memory.	*<textValue>*js=getText(*<name>*js);
setInt	Update the value of an integer-type variable in the shared memory.	jssetInt(*<name>*js,*<intValue>*js);
setReal	Update the value of a real-type variable in the shared memory.	jssetReal(*<name>*js,*<realValue>*js);
setText	Update the value of a text-type variable in the shared memory.	jssetText(*<name>*js,*<textValue>*js);

This instruction set makes up a simple API, specifically designed for simulation purposes, reducing the complexity in the definition of simulation models. This approach improves the usability of the proposed simulation tool, making the simulation models easily modifiable by SCADA administrators, not requiring other more complex JS environments such as nodeJS [50,59,60].

However, JS still has some efficiency weaknesses. For example, it is not possible to simulate complex transients with the simulator, but its performance is enough to execute common models of DER systems, where the execution time is typically in an order of magnitude of seconds. This is especially true for training purposes, where the accuracy of the model in power generation is not the most important task.

Moreover, as was explained before, in the case of need, the JS module could be substituted by another one, based on any other more efficient language, only changing the corresponding shared library of the simulator, but maintaining its interface. Because, as described before, the simulator was designed in a modular way, each block could be substituted.

3.3.1. Simulation Model Definition

The simulation engine was designed to implement FSM. Specifically, every model should have the architecture depicted in Figure 3. As can be seen in this figure, every model should have two types of inputs (signals and parameters). Moreover, the model results are stored as outputs at the simulation end. Summing up, the description of the input/output data is the following.

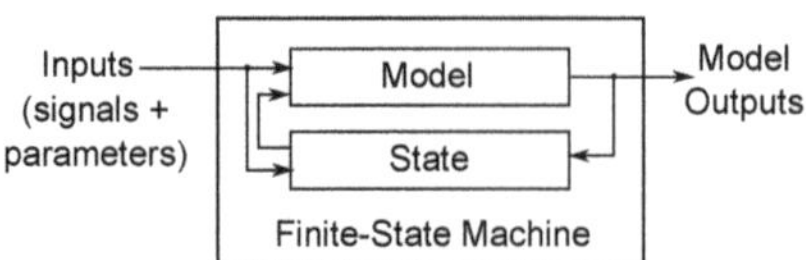

Figure 3. Elements of a DER simulation model.

- Signal inputs: This data represent the information related to the model operation. They come from SCADA or from the other submodels.
- Parameter inputs: These data represent the characteristics that define the behavior of the models (e.g., settings, threshold, limits, etc.) and the simulation scenario. They come from the configuration interface.
- Model outputs: These data represent the information obtained as a result of the execution of the different elements of the model. They will be sent to the shared memory manager (after every simulation step is finished), where they could be used by SCADA or by other models.

This notwithstanding, it is only a functional separation; all of the signals or parameters are actually stored in the same common memory area and are handled by the shared memory manager.

Taking this philosophy into account, one model could behave differently depending on its input signals and its configuration parameters. For example, it can be stopped, running or generating energy with a certain curve of generation depending on the inputs, with an anomalous behavior in which the generated energy is unusual, in a fail state, etc. In this sense, the same model can be used multiple times, with different behaviors, by calling different instances, but with different inputs.

In the proposed simulation engine, every model must implement two different parts: an initialization and the behavior for the simulation step. The initialization is responsible for creating the model interface and defining its initial behavior according to some received static parameters, as for example performance, generation curve, initial state, etc. Later, a behavior could be defined as an FSM for its execution in each simulation step. This part of the model will be responsible for obtaining the outputs according to the combination of the inputs and the current state of the model.

3.3.2. Description Model Language

As previously mentioned, the models are defined by means of scripts written in JS. In order to be executed in the simulation engine, two main functions have to be defined: the jsinit() function and the jsstep() function. The structure of the scrips of the models are depicted in Algorithm 2.

Algorithm 2 Structure of a model.

Data: $States : internal FSM states\{S_1, S_2, \cdots, S_n\}$
```
/* Include submodels' files  */
```
$include("submodel") \cdots$
```
    /* create state and simulation parameters  */
```
$state \leftarrow States.S_1 \quad param1 \leftarrow 0 \cdots$
```
    /* instance submodels  */
```
$submodel_i \leftarrow Submodel(i) \cdots$
Function *init()* **is**
```
    /* Do initialization  */
    ...
```
    return $\leftarrow \Delta t$
Function *step(timestep)* **is**
```
    /* Execute submodels' steps */
```
$submodel_i.step(timeStamp) \cdots$
```
        /* Execute internal FSM */
```
    switch *State* **do**
       ...
       case s_i **do**
```
            /* state S₁ code  */
            ...
```
```
                /* Test if state changes  */
```
           if *event$_j$* **then**
              $nextState \leftarrow States.S_j$
           ...
       ...
    $state \leftarrow nextState \quad$ **return** $\leftarrow 1$

Firstly, the jsinit() function is responsible for initializing the model. It is executed using the jsEngineAPI, designed to call JS codes from a C++ Application. In jsinit(), the signals shared with SCADA or other submodels (through the shared memory manager) and the local variables used by the model must be created and initialized. Additionally, this function must return the required simulation time step to the simulation engine. If the model needs to use other submodels, they should be initialized in this function.

Once the simulation starts, the jsstep() function is called periodically with the simulation time step returned by the jsinit() function. The jsstep() function is responsible for executing all the actions associated with each simulation step: read the inputs, obtain the next state of the simulation model, and obtain the outputs. Additionally, this function receives a parameter that allows the model to know how long the model has been active since it was started; this is especially useful for, at some point, simulating specific scenarios (errors, overgeneration, etc.) with which the operators could train.

As a summary, when a simulation is running, the system is executing the flow depicted in Figure 4.

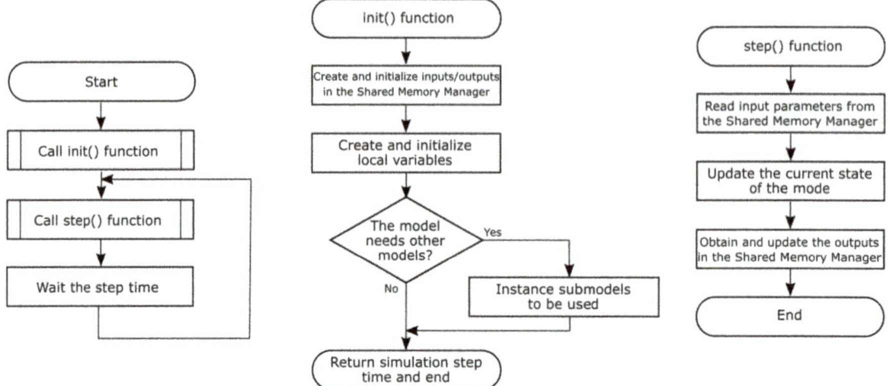

Figure 4. Simulation model script flow.

3.4. User Configuration Interface

The simulator includes a configuration interfacein order to set the model parameters and to manage the simulation engine. This interface is a web-based application that can be used with a standard web navigator. This web interface has three main interfaces: simulation, parameters, and logs(see Figure 5). All the described functions can be managed from these three interfaces.

- Simulation interface: Using the Simulation interface, the user can load different scripts of DER models in the simulator. When a new script is loaded into the simulator, it checks that the script is correct and able to be executed in the simulation engine. At this moment, the user can run the script. Scripts can also be stopped and restarted using this interface.
- Parameters interface: This interface shows the available inputs and their values in the common memory area. These values can be monitored and modified using this interface, even if the simulation engine is running. For example, this interface could be used to define the initial values of a model, before it is running. Data values can be written in single mode or in group mode, where all parameters of the group will be modified simultaneously.
- Logs interface: This interface contains logs of the current simulation. These logs come from the different modules of the simulator and from the different model scripts under simulation, using the extra instruction set designed for this purpose (see Table 1). Therefore, this interface is useful for the monitoring of the whole system and its evolution. This interface is also important for debugging purposes. For example, if a model has a syntax error, it is summarized in the log, next to its line number. Moreover, if the model has an execution error (as if it cannot guarantee the time step), this information is also added.

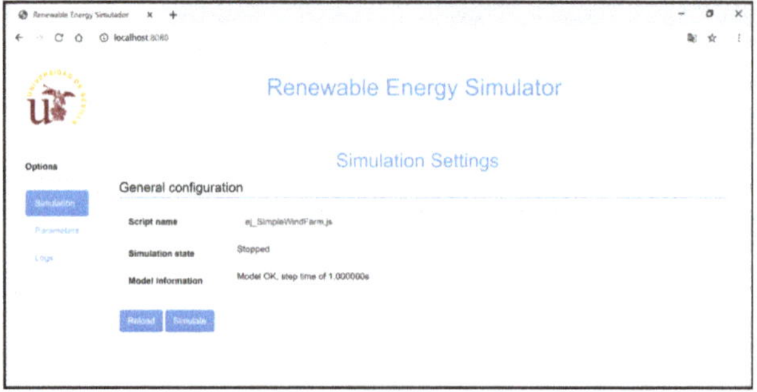

Figure 5. User web interface of the simulator.

4. Simulation Tool Abilities

As was described in previous sections, the main application of the proposed simulation tool is to train operators of renewable DER systems. Due to this, and as will be seen in the case of a real application (Section 5), the proposed JS simulation models are simple to understand, allowing the users to develop and tune their own models. However, beyond the application of this tool in training stages, its functionality can be extended to cover a wide set of scenarios.

In this sense, with the intention of showing the versatility of the proposed tool and where it can be used, this section briefly describes some applications and models that can be easily implemented over the proposed tool, mainly for training purposes, but also for other uses. Specifically, two sets can be distinguished: data sources and renewable DER models, where the first set characterizes the typical inputs of the models of the second set.

4.1. Data Sources

In renewable DER systems, a renewable energy source (i.e., solar irradiance in PVs or wind in turbine generators) is obviously necessary. Therefore, in many application, it is important to model their availability. One alternative to define these models is the use of AutoRegressive Moving Average (ARMA) time series, which is a common approach to generate synthetic information in both wind speed [61] or solar irradiance [62] scenarios. *ARMA(p,q)* refers to a model that adds p autoregressive terms with q moving-average terms, according to the expression of Equation (1).

$$X_t = \sum_{i=1}^{p} \phi_i X_{t-i} + \sum_{j=1}^{q} \theta_j \epsilon_{t-j} + \epsilon_t \tag{1}$$

where X_t is the actual wind speed at a certain time, ϕ_i are the p autoregressive coefficients, θ_j are the q moving average coefficients, and ϵ_t is a white noise that can be generated using a JS function as for example jsMath.random(). The main shortcoming of the *ARMA* model to generate renewable data sources is that this model considers a stationary time series. Due to this, to deal with seasonal trends, some authors proposed the use of seasonal *ARMA*, adding trend functions based on a probability distribution such as Weibull [63] or using the AutoRegressive Integrated Moving Average (*ARIMA*) [64], a more computationally intensive algorithm.

Another common method relies on the use of Artificial Neural Networks (ANNs), for both irradiance [65] and wind speed [66]. An ANN is composed of a number of highly interconnected processing elements $Y_{i,j}$, called neurons, where each output follows Equation (2):

$$Y_{i,j} = g \left(b_{i,j} + \sum_{k=1}^{n} \omega_{i,j,k} \cdot Y_{i,j-1} \right) \tag{2}$$

where i and j represent the rows and columns of the network and $g(x)$ the activation function, ReLU (Equation 3) currently being the most used one [67].

$$g(x) = max(0, x) \tag{3}$$

This method can offer a very good accuracy, but it has an elevated computational cost, especially for large networks as current deep learning models propose (e.g., [68]). Therefore, this approach could not be adequate for applications with fast simulation with a reduced time step. Nevertheless, it can be implemented in JS, using libraries such as Machine Learning in the Browser (MLitB) [69], which even supports, in some cases, GPU acceleration.

However, all those methods lack an important thing. They are focused on generating information in normal operational conditions, but not under anomalies. For example, in training applications, the proposed tool could be used to repeat specific patterns that represent anomalies (i.e., high temperature, wind issues, untwisting, etc.), which could be directly obtained from historical incidences of a real installation or manually generated to force specific conditions.

In this sense, the provided tool allows reading these values from external files, for example the Comma Separated Value (CSV) format. Moreover, it is important to consider that using external CSV does not reduce the capacities of the solution, due to it not being incompatible with the use of synthetic data, which can be obtained, for example, from tools such as ClimGen [70].

4.2. DER Models

Once the external data (not defined by the user) are available, the next step is to apply them to the DER production models. Thus, these inputs are not the only ones in the submodel. Additionally, the user control signals should also be considered. Thanks to this architecture, depending on the model chosen, the operators could obtain different behaviors for the same simulation scenarios based on their actions. In this sense, to understand the proposed simulator's abilities, this subsection identifies some models that could be easily implemented with the proposed tool and that might be able to simulate different configurations of these resources. Specifically, this study was carried out for the two main sources of renewable energy that are currently growing: solar systems and wind power systems. Thus, as will be seen below, these models may not have a complex implementation, so they can be easily applied in the proposed simulation tool.

4.2.1. Wind Power Systems

Roughly, a wind farm could be seen as an aggregation of a set of individual wind turbines. Therefore, the overall system could be characterized by the different functions of each one separately. Traditionally, this behavior is characterized by the power curve, a relationship between the wind speed and the active power generation. In this sense, manufacturers usually provide this curve in their technical data. However, this is usually generic for the turbine model and does not fully conform to reality. Due to this, statistical methods to fit this curve are a common solution [71,72]. Some authors proposed models based on polynomial regression [73]. Specifically, approaches with local approximations for the power curve, like the B-spline curves, have shown good results [74]. However, recent studies like [75] proposed monotone smooth regression to improve this approximation.

In this sense, it is important to note that this improvement model only complicates the characterization process, establishing specific continuity restrictions. Besides, its execution does not change; it is based on a third degree polynomial function (see Equation (4)), defined for each interval as:

$$aP\left(w(t)\right) = \begin{bmatrix} p_{i,1} & p_{i,2} & p_{i,3} & p_{i,4} \end{bmatrix} \begin{bmatrix} w(t)^3 \\ w(t)^2 \\ w(t) \\ 1 \end{bmatrix} \quad if \quad \begin{cases} i \in \mathbb{N} = [1, n] \\ i \ : \ WS_{min\_i} \leq w(t) < WS_{max\_i} \end{cases} \tag{4}$$

where n is the number of uniform intervals (defined by knots) in which the spline has been divided, each of them being bounded between a minimum (W_{min_i}) and maximum WS_{max_i} wind speed value. Thus, $[p_{i,1} \ p_{i,2} \ p_{i,3} \ p_{i,4}]$ is a vector with the coefficients a, b, c, and d of the polynomial, obtained as the row i from the weight matrix p.

Further, other more complex methods as Hidden Markov Models (HHMs), ANNs, or Support Vector Machines (SVMs) have also been proposed. In this sense, the work in [76] made an extensive study in this way. However, these approaches demand a higher computational cost. Therefore, in this work, the approach outlined above will be appropriate in most cases.

4.2.2. PV systems

Like wind generation, solar energy is another renewable energy that is increasing its share in the last few years. In this sense, the work in [76] showed an interesting review of methods (based on Random Forest (RF), k-Nearest Neighbors (k-NN), etc.) to estimate the solar production. However, these approaches only model the system. They do not allow the necessary flexibility to be used for a simulation in tools like the proposal.

As an alternative, solar farms could also be seen as some aggregation of photovoltaic subsystems (PV panel + inverter), whose generation is directly related to the solar irradiation that they receive. In this sense, Kazem and Khatib [77] showed a parametrical model of this set. PV panel generation is characterized by Equation (5). Specifically, its output power increases linearly with the solar radiation $G(t)$, but decreases with the ambient temperature $T_{amb}(t)$.

$$P_{PVpanel}[G(t), T_{amb}(t)] = P_{Peak}\left(\frac{G(t)}{G_{std}}\right) - \left\{\alpha_T\left[\frac{NOCT}{800}G(t) + T_{amb}(t) - T_{std}\right]\right\} \tag{5}$$

where G_{std} and T_{std} are the standard PV test conditions, α_T is the temperature coefficient, and P_{Peak} is the panel rated power. Thus, $NOCT$ is the nominal operating cell temperature (experimental data of the panel measured for 800 W/m^2, at $T_{amb} = 20\,°C$, and under a wind speed of 1.5 m/s). These data were defined in the manufacturer's datasheet.

Thus, similarly to the PV panel, the inverter could be also characterized, in this case with the addition of a performance parameter η_{INV}, defined by Equation (6):

$$\eta_{INV} = \frac{P_{in} - P_{loss}}{P_{in}} \tag{6}$$

η_{INV} makes it possible to estimate the total generation of the systems P_{PV}, following Equation (7):

$$P_{PV} = \eta_{INV} \cdot P_{PVpanel}[G(t), T_{amb}(t)] \tag{7}$$

where P_{in} is the input power and P_{loss} represents the energy lost in the conversion process. Thus, the performance of an inverter is supplied by the manufacturer (typically around 95%), but it usually has lower values if it is below 30% of its capacity.

4.2.3. Energy Storage Systems

Other crucial elements within DR systems are the Energy Storage Systems (ESS). Thus, the detailed modeling of these elements can be complex. However, for an electrical system, they can be seen as point energy stores, whose availability is usually represented through the State-of-Charge (SoC) of Equation (8), where $SoC(t)$ is the SoC (in p.u.) of the system in each moment, $SoC(t-1)$ is the SoC at the previous simulation step, P_{out} is the power delivered by the ESS along the period Δt, and C_{ESS} is the energy capacity (in KWh).

$$SoC(t) = SoC(t-1) - \frac{P_{out}}{C_{ESS}} \cdot \Delta t \tag{8}$$

Thanks to the simulator architecture and its ability to associate memory with each model, it is possible to implement this kind of resource in an easy way.

5. Study Case

One proof of the usefulness of the proposed tool could be found in the application where it was developed. This simulation tool was developed as a part of the SEAPEMproject, which is focused on developing a system for offshore wind farm support. Specifically, the proposed simulation tool was developed for its integration with the Bluence$^{®}$ suite by Isotrol$^{®}$ [78],A a complete Information and

Communications Technology (ICT) solution that makes it possible to manage large amounts of data in renewable energy plants. Bluence$^{®}$ is the core of the Bluence$^{®}$ Control Center (see Figure 6), a 24/7 support center for renewable energy power plants.

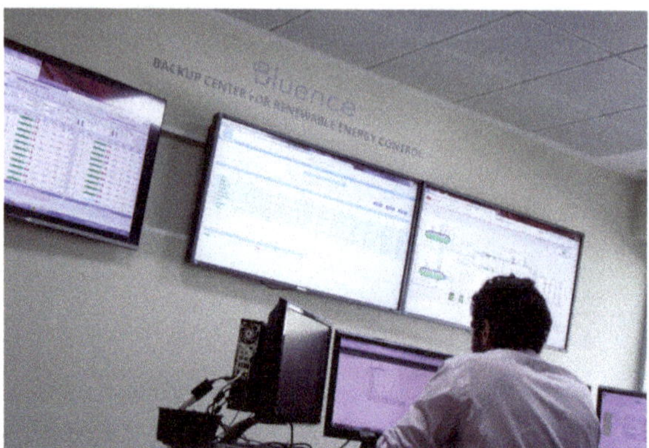

Figure 6. Bluence$^{®}$ Control Center by Isotrol$^{®}$.

Bluence$^{®}$ Control Center has been designed to support tasks such as redundancy and security, the dispatching center, incident management, and 24/7 remote support. It is in this scenario where Isotrol$^{®}$ detects the lack of simulation tools specifically designed for operator training, or for the start up of new SCADAs, scenarios where the proposed simulator becomes an interesting utility for planning, operation, and testing tasks.

Moreover, to understand the usefulness of the proposed simulation tool really, it is interesting to analyze some practical application examples. Therefore, this section describes in detail three functional models that make it possible to understand its potential integration for SCADAs with renewable energy management in training operator tasks. In this sense, one of the most relevant technologies of this kind of energy source is wind generation. Due to this, the proposed examples focus on this technology. However, it is interesting to highlight the possibilities of the simulator, allowing different levels of model accuracy. Besides, as described before, it can model plants of other technologies such as: PV, Combined Heat and Power (CHP) plants, etc.

Specifically, three models are described in next subsections. The first one describes a wind farm of ten generators using a simple turbine model. Later, the second applies an empirical model of a wind turbine, using monotonic splines. Finally, the last one proposes a more complex turbine model focused on training SCADA users over different scenarios.

5.1. Wind Farm with a Turbine, Simple Model, Based on the Manufacturer's Specifications

The capability of simulating the behavior of an asset is very interesting for facilities setting up and training their users. Obviously, the model actually depends on the specific use and the required precision. In this sense, the first example proposes an implementation of a simple wind farm model. This model proposes a hierarchical implementation (as can be seen in Figure 7), following the model philosophy described in previous sections, and using three submodels: one for the load wind historical dataset (see the code in Appendix A.1), another to implement a simple turbine model, which is directly based on the manufacturer's power curve (see the code in Appendix A.3), and the last submodel to aggregate the overall generation of the wind turbine set (see the code in Appendix A.2).

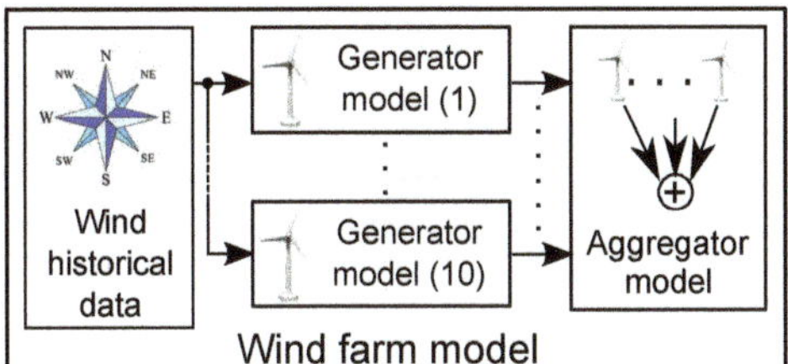

Figure 7. Elements of the wind farm model.

```javascript
// ej_Simple_Wind_farm.js: Simulation example for aggragation of 10 simple wind turbines.
include('LibSubmodels/Submodel_SourceFile.js');
include('LibSubmodels/Submodel_WindTurbine.js');
include('LibSubmodels/Submodel_Aggregator.js');

// Simulation parameters.
var aggregatorId = 0; var startSrcId = 11; var windSrcId = 12;
var turbineId = [1,2,3,4,5,6,7,8,9,10];
var model = "ECOTECNIA 74/1670"; var stopTime = 60;
var windTable = [0,2,3,4,5,6,7,8,9,10,11,12,13,25];
var powerTable = [0,0,7,50,118,226,378,580,840,1138,1463,1640,1670,1670];
var initDate = "01-nov-2017"; var initTime = "12:00:00.0000";
var windFile = "Signals/wind_"+initDate+".csv";
var startFile = "Signals/start_"+initDate+".csv";

// Sumodel instances.
var startSrc = new Submodel_SourceFile(startSrcId,"Start Source File");
var windSrc = new Submodel_SourceFile(windSrcId,"Wind Source File");
var turbines = new Array(turbineId.length);
for (i=0 ; i<turbineId.length ; i++)
  turbines[i] = new Submodel_WindTurbine(turbineId[i],model,windTable,powerTable,stopTime);
var aggregator = new Submodel_Aggregator(aggregatorId);

// Init function definition.
function init() { var timeStep = 1;
  startSrc.init(timeStep,initDate+" "+initTime,startFile);
  windSrc.init(timeStep,initDate+" "+initTime,windFile);
  for (i=0 ; i<turbineId.length ; i++)
    turbines[i].init(timeStep);
  aggregator.init(timeStep,turbineId,"activePower","totalPower");
  return timeStep;
};

// Step function definition.
function step(timeStamp) {
  startSrc.step(timeStamp); windSrc.step(timeStamp);
  for (i=0 ; i<turbineId.length ; i++)
    turbines[i].step(timeStamp,getReal("srcValue"+startSrcId),getReal("srcValue"+windSrcId));
  aggregator.step(timeStamp);
  return 1;
};
```

As can be seen in the code of the proposed model for training SCADA users, this implementation loads the different submodel files from a repository, creates the common signals between them, and defines the required main jsinit() and jsstep() functions. In this case, these two main functions are made up as a sequence of calls of the jsinit() and jsstep() functions of each submodule.

Specifically, as can be seen in the previous script, the implementation of the wind farm model was made up by ten turbines, fed with the same wind source. These operations were characterized by their power curve (the relation between the wind speed and the generated power) of the turbines, Model 74/1670 by the ECOTECNIA manufacturer for this case (see Table 2). To do that, after the jsinit() function's definition, this model instantiates a simple wind turbine submodel ten times (see the code in Appendix A.3), setting the specific configuration in the definition of each one.

Table 2. ECOTECNIA 74/1670 behavior *.

Wind Speed (m/s)	Power (kW)	Power Coefficient CP (-)	Thrust Coefficient CT (-)	Wind Speed (m/s)	Power (kW)	Power Coefficient CP (-)	Thrust Coefficient CT (-)
1	0	-	-	14	1670	0.23	0.30
2	0	-	-	15	1670	0.19	0.24
3	7	0.10	1.04	16	1670	0.15	0.20
4	50	0.30	0.88	17	1670	0.13	0.16
5	118	0.36	0.77	18	1670	0.11	0.14
6	226	0.40	0.78	19	1670	0.09	0.12
7	378	0.42	0.79	20	1670	0.08	0.10
8	580	0.43	0.79	21	1670	0.07	0.09
9	840	0.44	0.79	22	1670	0.06	0.08
10	1138	0.43	0.73	23	1670	0.05	0.07
11	1463	0.42	0.67	24	1670	0.05	0.06
12	1640	0.36	0.53	25	1670	0.04	0.05
13	1670	0.29	0.39				

* Note: Measurement conditions: Density = 1.225 kg/m^3, height = 0 m, and temperature = 15 °C.

This turbine submodel characterizes an FSM whose interface and behavior (state diagram) are shown in Figure 8. Specifically, this submodel has three states: the *PRODUCTION* state, which represents when the turbine is in normal generation mode and following the power generation curve, the *STOPPED* state, which represents when the turbine is completely stopped, and the *STOPPING* state, which characterizes the stopping behavior when an stop order is received.

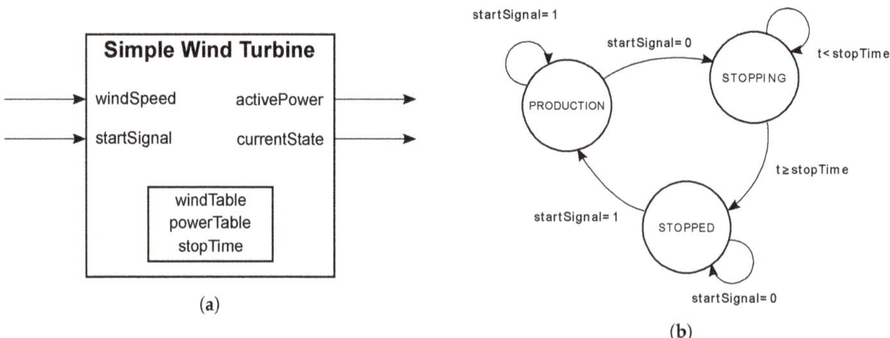

Figure 8. Simple wind turbine model. (**a**) I/O interface. (**b**) State diagram.

In standard operation, the active power, $aP\left(w(t)\right)$, generated by the turbine will be obtained by a piecewise linear interpolation, through Equation (9), where parameter $w(t)$ represents actual wind speed input and *WS* and *P* are respectively the wind speed and active power reference data, obtained form Table 2, both in the $\mathbb{R}^{1 \times n}$ range.

$$aP\left(w(t)\right) = \begin{cases} P[1] & if & w(t) < WS[1] \\ \dfrac{P[i+1] - P[i]}{WS[i+1] - WS[i]} \cdot (w(t) - WS[i]) + P[i] & if & \begin{cases} i \in [1, n-1] \\ i \; : \; WS[i] \leq w(t) \leq WS[i+1] \\ w(t) > WS[n] \end{cases} \\ P[n] & if \end{cases} \tag{9}$$

Thus, the model output *activePower* is calculated depending on the current state, as can be seen in Table 3, and taking into account that *t* is the current time, t_0 is the instant when starting the stopping action, and *stopTime* the time to stop.

Table 3. Wind turbine *activePower* generation depending on the current state.

STATE	PRODUCTION	STOPPING	STOPPED
activePower	$aP\,[w(t)]$	$\left(1 - \dfrac{t - t_0}{stopTime}\right) \cdot aP\,[w(t)]$	0

As Section 4.1 describes, we propose to obtain the wind profile from from external files in CSV format. As described before, training users on security assets requires generating specific patterns during the training, as happens in abnormal scenarios, such as high winds or gusts. Common models to forecast or generate synthetic winds usually do not allow this, it being easy to obtain this information from incidences of the historical database of the real system.

Using the proposed model for this study case and exciting it with a wind profile based on historical signals, it was easy to simulate the behavior of the overall wind farm with different time steps (as can be seen in Figures 9–11).

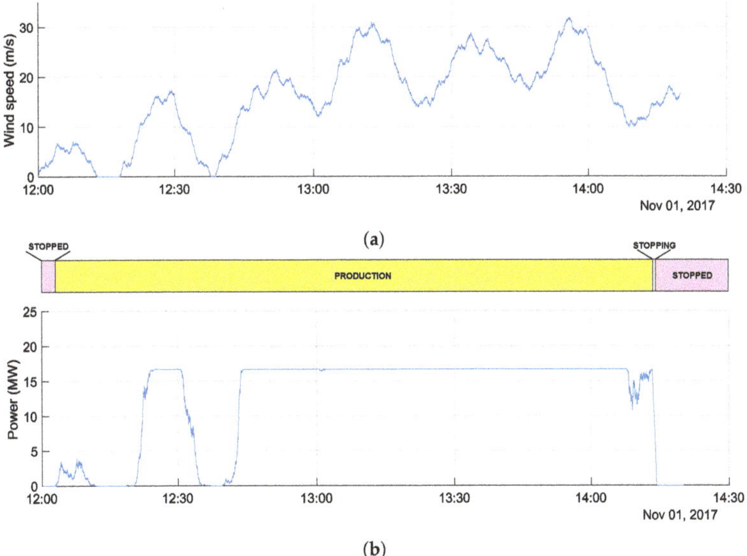

Figure 9. Simulation of a 10 turbine park (time step of 1 s). (**a**) Real wind speed. (**b**) Generated power.

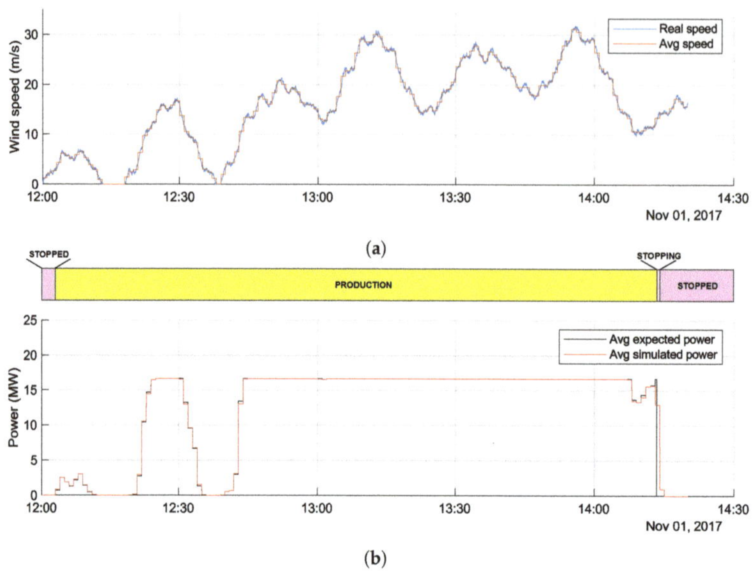

Figure 10. Simulation of a 10 turbine park (time step of 1 min). (**a**) Real and average wind speed (1 min aggregation). (**b**) Generated power.

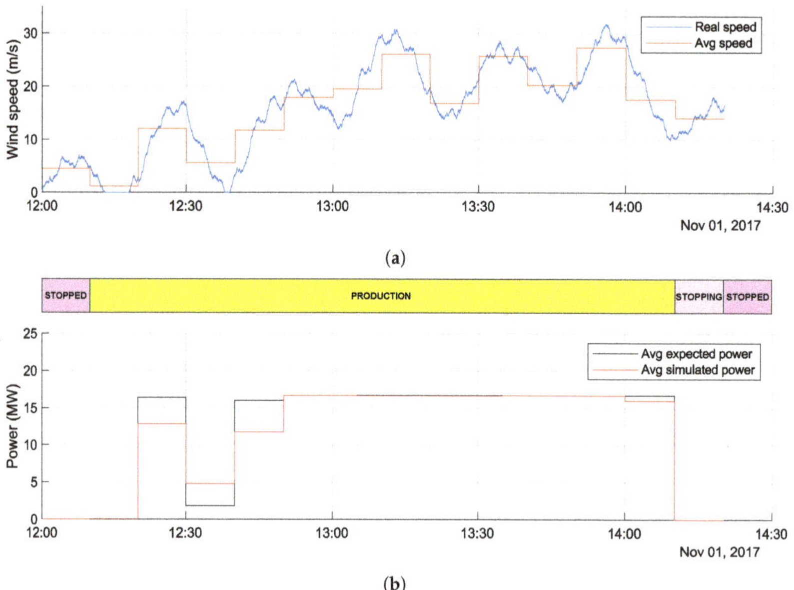

Figure 11. Simulation of a 10 turbine park (time step of 10 min). (**a**) Real and average wind speed (10 min of aggregation). (**b**) Generated power.

In these figures, three cases were simulated for the previous wind farm with the same wind profile for two hours and twenty minutes. In this case, in order to simplify the model, all wind generators were fed with the same wind profile, but different ones could be used as input for each one if desired. Figure 9 is the most accurate simulation, with a simulation step of one second (1 s); Figure 10 has a step of one minute (1 min); and Figure 11 has a simulation step of ten minutes (10 min). Thus, Figure 9 was used as a reference for Figures 10 and 11 to understand with an example the strong relationship that can exist between the simulated power error and the simulation time step.

Specifically, with the same wind profile, using a 1 min simulation step, the absolute error in the generated power simulation was 1.26%. However, using a 10 min simulation step, this absolute error increased up to 8.99%.

Thus, once the simulation scenario was defined and it was verified that the simulation was possible, the next step should be the comparison between the real and the simulated generated data. However, this analysis was only done in intervals of 10 min, because the available dataset from a real wind farm facility (see Table 4 with an example of turbine generation data details) only contained records of wind speed and generated energy at this frequency. Thus, taking into account that the power curve of wind turbines is a non-linear function (as can be seen in Figure 12b), the simulation for these average wind speed values did not completely match the real generation, but it was not a big issue in the training applications. This error strongly depended on the distribution of wind speed along time.

Table 4. Historical dataset details.

Description	Value
Total time	9480 h (13 months)
Total time in production state	4645.3 h
Average power real (in production state)	763.45 kW
Average power expected (in production state)	617.50 kW
Power error	21.90%

This fact is depicted in Figure 12 where the difference between the real and simulated generation is plotted as a function of the wind speed. The error represented was the signed difference between the real power generator during every 10 min period and the estimated power. Specifically, the estimated power was obtained assuming the average wind speed and the manufacturer curve during the same period. This figure was made up from the available dataset using only one turbine record.

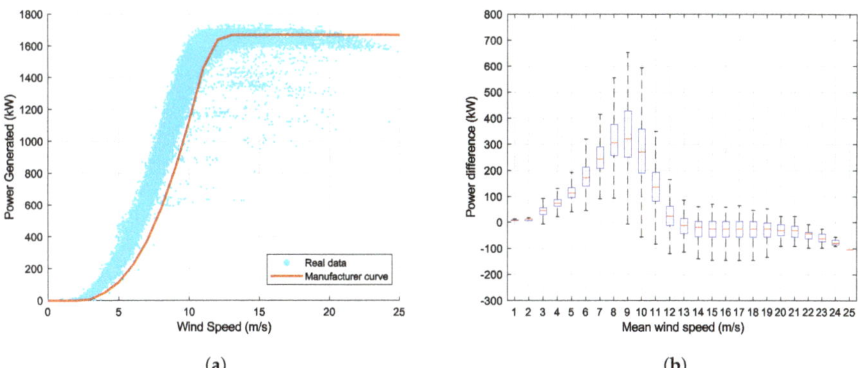

Figure 12. Simple wind turbine model. (**a**) Manufacturer's power curve vs. real data. (**b**) Distribution error between real data and manufacturer's curve (10 min of aggregation).

A curious effect appears in Figure 12b, which represents its distribution error (error between the real data and the manufacturer's power curve). On the one hand, from 3 m/s (the start of the production) until about 13 m/s, the distribution error was mainly positive (the turbine generated a power greater than that corresponding to the mean of the wind speed, assuming the power curve). Of course, in this interval, the second derivative of power curve was positive, and this explained why the difference was positive. On the other hand, from 13 m/s until 25 m/s, the second derivative was negative or zero. Of course, this caused the distribution error of the power difference to be negative due to the distribution of wind speeds. This was why from wind speeds of 13 m/s and higher, the experimental power values were less than the expected according to the manufacturer curve. Therefore, in this zone of the graph, the mean errors resulted in being negative.

To sum up, those differences in the real data were normal and understandable. Therefore, the simulated data recorded each ten minutes were not enough to obtain accurate results to reflect the dynamics of a wind turbine faithfully. Nevertheless, it is important to note that the goal of this paper is not to propose an accurate wind turbine model, for which, as previously shown, it would require a dataset with a higher temporal resolution. As was discussed throughout this paper, this is focused on the development of an operational simulation tool designed to be integrable in a SCADA, which allows training or testing by operators in a safe way for real facilities. In this sense, the power curve from the manufacturer was a valid approximation for a learning application. Using this simple model, an operator could observe the relation between both variables and check if the power was at the adequate limits. A power generation beyond those limits would suppose a problem in the turbine, which should be managed and solved by the operator.

With all of the above, this does not mean that the proposed tool did not allow the simulation of precise models. However, the proposed models were more focused on a correct representation of the operational evolution of the system than on a faithful representation of the behavior of the turbines. Nevertheless, if the main goal were implementing a more precise model, this improvement was compatible with the proposed tool, only writing a script that would define its behavior.

Furthermore, to show the simulation capacity of the proposed system, some performance results were obtained with a PC with an Intel® Core™ i5-6400T CPU with 8GB of RAM. Using the proposed simulator and the model of this subsection, which consisted of the aggregation of a series of wind generators, this model was executed using different numbers of generators, obtaining as a result a mean value of 10.93 microseconds of execution time per generator in every step.

Based on this result, it was possible to obtain the maximum number of generator instances that could be used for a fixed time step. For example, using one second as a time step in the aforementioned PC, the maximum number of generator instances was 91,743 for a valid on-line simulation. It is important to remember that, as described before, in order to be a valid simulation, the time step needed to be higher than the execution time required by the models.

5.2. Wind Farm with a Turbine Model, Obtained from Empirical Data

This example describes another application scenario. In this case, a wind turbine empirical model was based on a a monotone smooth regression spline [75], previously discussed in Section 4.2.1. This example had two objectives: (a) demonstrate how a previous model could be easily adapted or modified by an operator; (B) show how to implement in the simulator the models depicted in Section 4.

The empirical information used to model the wind turbine was obtained from a historical dataset from a wind farm based on ECOTECNIA 74/1670 wind generators, the same depicted in Table 4. Using this dataset and following Mehrjoo et al.'s procedure [75], the empirical model depicted in Figure 13a was obtained.

In this case, the monotonic spline was divided into 25 uniform intervals (defined by 26 knots), whose coefficients a, b, c, and d conformed to the matrix of coefficients p, which are used in Equation (4).

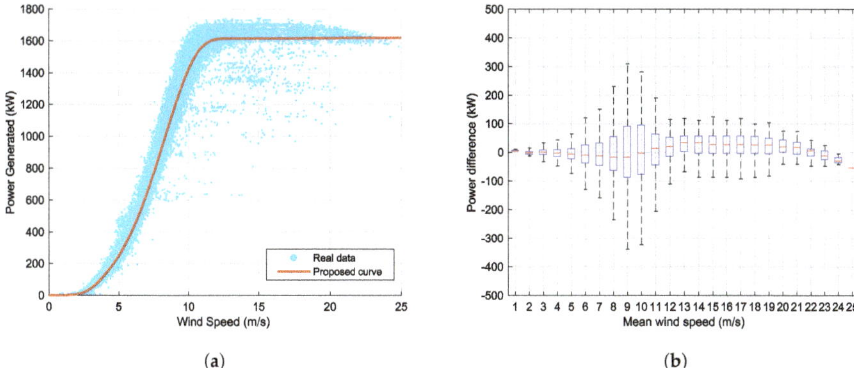

Figure 13. Simple wind turbine model. (**a**) Estimated monotonic spline for the power curve vs. real data. (**b**) Distribution error between real data and the monotonic spline model (10 min of aggregation).

The use of 25 intervals and considering that the wind speed was limited in the operational range (speeds out of this range caused the turbine to stop), the power from wind speed and using this model could be obtained using the Algorithm 3, as can also be seen in the JS code below.

Algorithm 3 Power estimation with monotonic spline.

Data: p: $[a, b, c, d]$ parameters' matrix, size 25×4.
Input: w: current wind value, in $[0, 25]$ range
Output: aP: power obtained
begin
$\quad \lfloor \; i \leftarrow \lfloor w \rfloor \quad aP \leftarrow p[i,1] \cdot w^3 + p[i,2] \cdot w^2 + \cdot p[i,3]w + p[i,4]$

```
const p = [[-2.3382,4.0083,0,1.0287e-07],[6.851,-23.559,27.568,-9.1892],
    [2.9074,0.10221,-19.755,22.359],[-3.3688,56.588,-189.21,191.82],
    [3.5426,-26.348,142.53,-250.51],[-1.3627,47.23,-225.36,362.64],
    [1.6336,-6.7022,98.235,-284.55],[-6.8934,172.36,-1155.2,2640.2],
    [-8.3494,207.31,-1434.8,3385.7],[-11.558,293.95,-2214.6,5725],
    [-0.0096583,-52.51,1250,-5823.7],[11.195,-422.28,5317.5,-20738],
    [6.4142,-250.15,3252,-12476],[-4.9439e-10,1.2205e-08,-2.6495e-09,1616.3],
    [2.2112e-08,-9.3881e-07,1.3333e-05,1616.3],[9.1158e-09,-3.5556e-07,4.6082e-06,1616.3],
    [-4.9996e-08,2.4802e-06,-4.0739e-05,1616.3],[0.31814,-16.225,275.83,53.303],
    [-0.7423,41.038,-754.92,6237.8],[0.84817,-49.618,967.56,-4671.2],
    [-0.74202,45.793,-940.67,8050.2],[0.318,-20.988,461.74,-1766.6],
    [-1.2137e-06,8.2095e-05,-0.0018491,1619.5],[1.7214e-06,-0.00012043,0.0028089,1619.5],
    [-3.3594e-06,0.00024539,-0.0059707,1619.6]];

function aP(w) { var i;
  if (w < 0) i = 0;
  else if (w > p.length) i = p.length-1;
  else  i = Math.floor(w);
  return p[i][0]*w^3 + p[i][1]*w^2 + p[i][2]*w + p[i][3];
};
```

As can be seen in Figure 13, this model offered better accuracy than the previous one. However, it is important to remark that accuracy in power estimation is not the main goal of a tool designed for training SCADA users. Simplicity to adapt to new scenarios and to have a correct behavior that can be simulated in real time are more important factors. However, this class of models demonstrated the abilities of the simulator to be used in different applications.

5.3. Wind Turbine Complex Model

In order to improve the skills of the operators, it is interesting to have more internal operational states than defined in the previous simple turbine models. In this sense, the additional states should include information not only about the wind turbine itself, but also about the state of its control system.

Specifically, this new model was developed using data extracted from a real turbine control system, which contained information about states, errors, signals, and power measurements. Analyzing this dataset, it was possible to identify the states depicted in Table 5.

Table 5. Table of states of the control system.

State	Short Name	Description
PRODUCTION	P	The normal state for power production.
UNTWISTING	Ut	The nacelle is performing the untwisting of the cable. It is rotating, and the production is stopped.
UNTWISTING ANOMALY	UtA	The untwisting process is taking an excessive time. Activated after 25 min if the angle advance has been very low during that time.
EXCESSIVE WIND	WE	The mean speed of the wind is too high (greater than 25 m/s). The turbine is stopped, as the manufacturer recommends.
MAXIMUM GUST	WM	High wind gust detected (even if the average speed is not too high). Production stopped to avoid problems in the turbine.
STOP-MANUAL	SM	Manual stop of the production.
STOP-POWER REGULATION	SPR	Production stopped due to an order of the Balance Responsible Parties (BRP) or Independent System Operator (ISO).
STOP-BIRDS	SB	Production stops due to birds traveling near the turbine.
CHECKING	Chk	The system is checking if there is some problem, or it is reconnecting after a communication error.
ERROR	Err	Some error.
COM ERROR	CErr	Communication error.
MAINTENANCE-CORRECTIVE	MC	Corrective maintenance to correct some problem.
MAINTENANCE-PREVENTIVE	MPre	Preventive maintenance.
MAINTENANCE-PROGRAMMED	MPro	Programmed maintenance.
ALARM	Alarm	It is not a state itself, but it represent extra information about the system.

In summary, these states were the normal production state (called *PRODUCTION* here), manual stop ordered by the operator (called *STOP-MANUAL* here), and others. In this model, the state *PRODUCTION* had the same behavior as the *PRODUCTION* state of the previous model (in this case, according to the requisites of the company, it was defined by Equation (9), but it could also use the model defined in Section 5.2). In the wind turbine complex model, when the system was in a state different from *PRODUCTION*, the generation was zero. This notwithstanding, the transition between *PRODUCTION* and any other state followed the behavior described by the *STOPPING* of the simple model, by the equation of Table 3. Every state required a different attention and/or operation by the SCADA users, so this information was valuable to help them in their learning process. Thus, as an example, some interesting cases of the above ones were selected to understand these operational states and their relationship with the operation of SCADA.

The first one (Figure 14) corresponded to the untwisting process anomaly. At 4:01:06, the turbine started the untwisting process, but as can be seen, the nacelle direction did not change. This was due to an anomaly in the rotation process. Therefore, after 25 min, the system entered the state *UNTWISTING ANOMALY* (because the angle practically did not change in that period). After some minutes, the system finally achieved rotating, and the system returned to the state *PRODUCTION*.

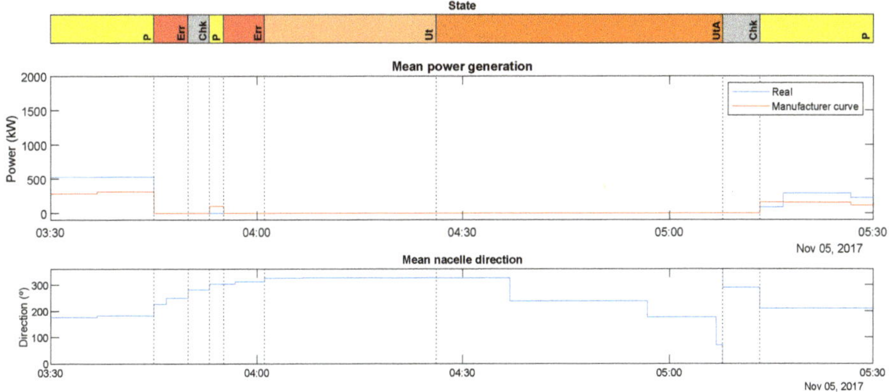

Figure 14. Untwisting anomaly.

When a untwisting anomaly happens, the SCADA operator has the task of finding the cause and solving it. In this case, the operator should check the registered alarms and errors and should choose a corrective maintenance to perform. This process can be appreciated in Figure 15.

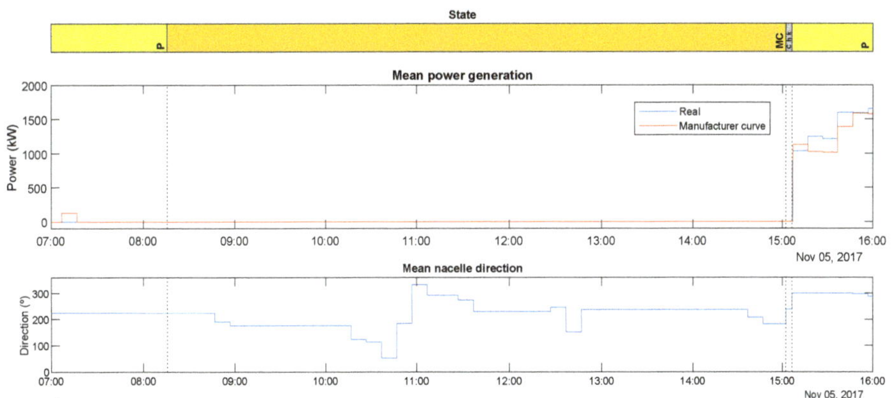

Figure 15. Corrective maintenance.

Regarding the stopping of the turbine, there are some reasons for this to occur. Specifically, the three main reasons are: the stop due to birds passing (state *STOP-BIRDS*), the stop order by the Balance Responsible Parties (BRP) or Independent System Operator (ISO) due to power regulation (state *STOP-POWER REGULATION*), and the manual stop directly performed by the operator (state *MANUAL STOP*).

In this sense, Figure 16 shows an event of stopping for birds passing through the zone. Besides, Figure 17 shows a stop case for power regulation from a BRP. The production during the stop time was equal to zero, even with adequate wind speed for power production.

Regarding the normal operation of the wind turbines, the automation system was programmed to avoid damages caused by excessive wind. This was done thanks to the states *EXCESSIVE WIND* (this happened when the wind speed was more than 25 m/s during a few seconds) and *MAXIMUM GUST* (when very fast wind gusts were detected). As can be seen in Figure 18, both states could occur during a windy day, and they were important for training purposes.

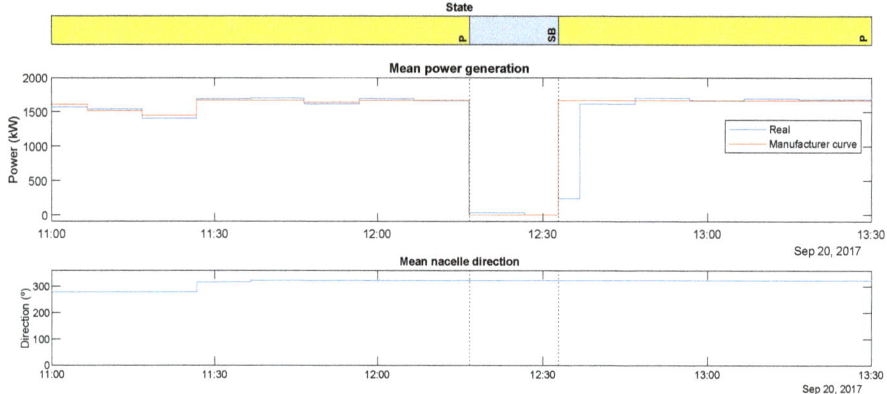

Figure 16. Stop due to birds passing.

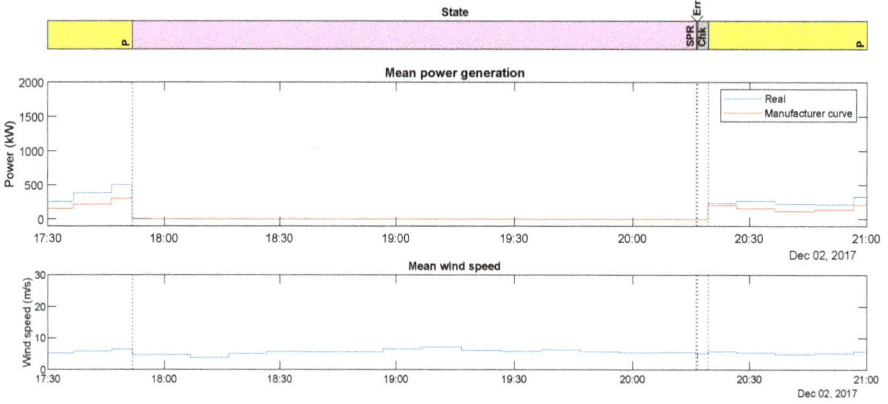

Figure 17. Stop due to power regulation performed by the BRP.

Additionally, some examples of common errors that could cause the error state are included in Table 6.

Table 6. List of possible control system errors.

Error Type
Inverter failure
Inverter communication failure
Pitch failure
Pitch communication error
Blade 1 error
Blade 2 error
Blade 3 error

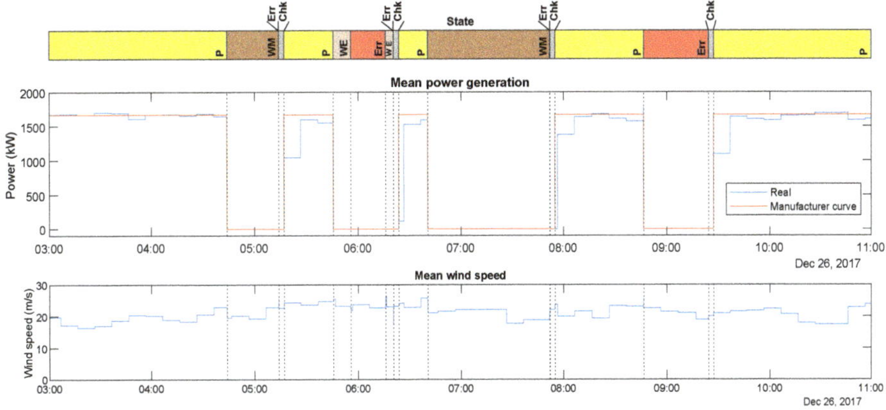

Figure 18. Stop due to too fast wind.

Of course, the movement from one state to another required a few signals (some of them depended on the element under control itself, and others depended on the operator actions, being therefore important in training scenarios). They are summarized in Table 7.

Table 7. List of possible control system signals.

Signal Type	Signal Name	From
Failure	Error Detected	Internal Error
Stop	Stop-Birds	Operator
Stop	Stop-Manual	Operator
Stop	Stop-Power Regulation	Independent System Operator
Maintenance	Maintenance-Corrective	Operator
Maintenance	Maintenance-Preventive	Operator
Maintenance	Maintenance-Programmed	Operator

Considering all of the above, the proposed model structure for the control system is shown in Figure 19. As can be seen, it used an extension of the turbine model described above, adding the nacelle control and its associated states.

Once all the signals and connections were defined, the states of the control system were implemented in a model for the proposed simulator. This model was done using a double structure: mode and state. The modes (also called procedures) of operation were "global" groups of states, while the states described more specifically how the control system managed the turbine (as listed in Table 5), offering a clear vision of the model and simplifying the model implementation. The summary for this structure is depicted in Figure 20.

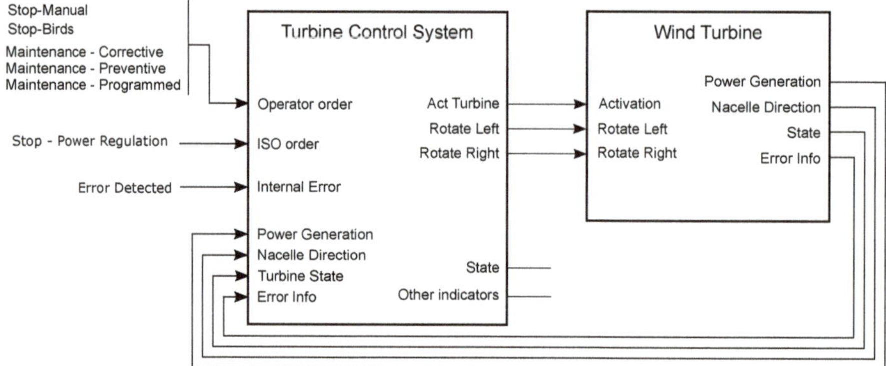

Figure 19. Turbine control system blocks.

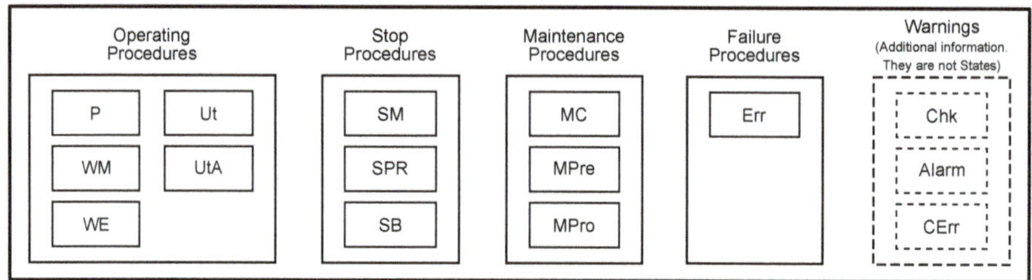

Figure 20. Turbine control system: procedures and states.

The relationship between procedures and the signals that produce an evolution between them are shown in the DES model of Figure 21.

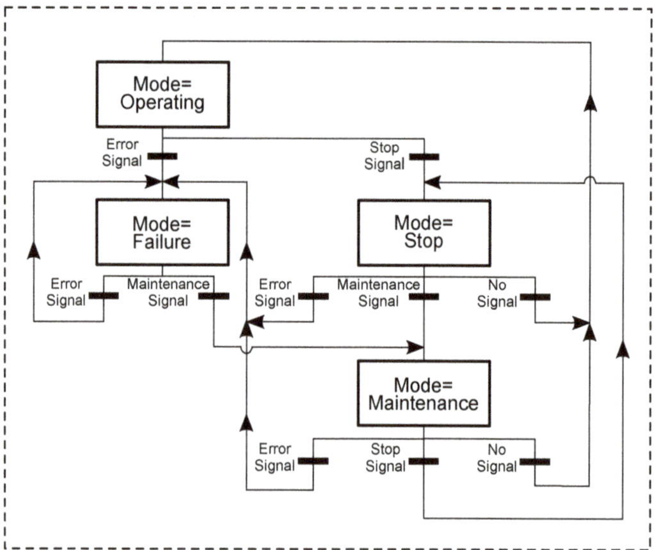

Figure 21. DES diagram of the global procedure modes in the proposed model.

Thus, in each simulation cycle, after deciding the current mode, the DES would be evaluated (see Figure 22). Specifically, the result of this evaluation would be the current state, which would characterize its specific behavior (described above in Table 5).

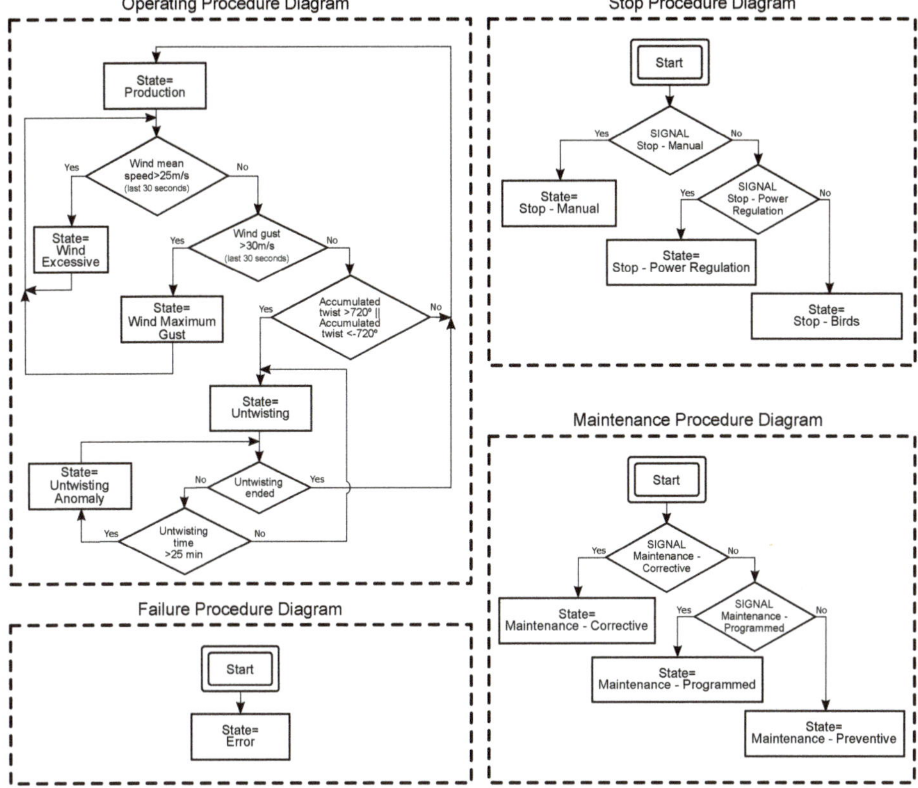

Figure 22. Procedure diagrams.

This proposed model was particularly interesting for the operator training process. Especially, some of the most important states were the stop states, the untwisting anomaly, and errors, which required an operator intervention to check the system and/or a quick response from them, in order to restore the normal turbine operation. In summary, as was explained previously, the actions that a SCADA operator could practice in a simulated scenario using this model were:

- Corrective maintenance
- Manual stop
- Programmed maintenance
- Birds stop
- Predictive maintenance

Thus, based on the proposed model, the information observed by the operator would be the generation power, the list of errors (if any), and the information about the environment (wind speed and direction).

6. Conclusions

As described before, the presence of renewable generation systems within the energy mix is increasing. However, due to the enormous variability and uncertainty associated with this type of resource, they require sophisticated planning tools in order to manage these resources in a proper way.

Throughout this paper, different examples of the simulation use in the field of the renewable energy were exposed. Nevertheless, these solutions are traditionally focused on the simulation of the generation process or on the facilities' design, leaving the operation of these systems in the background. Conversely, the need for simulation tools that help to operator in the improvement training and setting task was clearly identified.

In this sense, this paper proposed the integral design of a simulation tool, which could be easily connected to a real SCADA, making it possible to perform simulation under different scenarios. Unlike other simulation tools, which did not have this feature, the proposed simulator could interact directly with real SCADAs similar to a real installation, the interaction not being appreciable by the users, making training more realistic.

Currently, the simulator supports OPC-UA as a communication interface with SCADAs. Furthermore, the communication mechanisms of the simulation tool were flexible and could be configured depending on each SCADA thanks to the modular architecture in which the simulator was designed.

The simulation was defined based on scripts models, making the simplest programming interface possible. This development was carried out taking into account the skills of typical SCADA administrators and operators and establishing JavaScript as the script programming language, easier to learn and being closest to their profile.

As a proof of the usefulness of the proposed tool, a series of possible simulated input data sources for renewable energy simulations was briefly described. In addition, renewable DER models (wind power, PV, and storage systems) were also presented, showing the potential of the proposed simulator.

Additionally, a study case was depicted. Specifically, this study case contained three examples of a wind farm simulation model. A simple one was based on the generation curve given by the manufacturer which only had three states: stop, production, and stopping. Another one was based on empirical data using a recent proposed modeling technique: monotonic splines. A third one had different states and control signals, emulating a more realistic wind turbine system, being more useful in the operator training process.

As can be seen from the script examples, the proposed programming language and the provided API were simple. Thanks to this, plant administrators could create new simulation scenarios, without needing advanced programming skills. Thus, these models would allow them to sort out that which is lacking as typically detected in the support of renewable SCADAs, such as: setting up facilities or the training of novel operators.

Author Contributions: Conceptualization: D.F.L., E.P., and A.G.; methodology: D.F.L.; investigation: D.F.L. and E.P.; software: D.F.L. and E.P.; validation: E.P., S.G., and A.P.; formal analysis: E.P. and A.P.; writing, original draft preparation: E.P., D.F.L., S.G., and A.P.; writing, review and editing: A.G. and C.L.; supervision: A.G. and C.L.; project administration: C.L. All authors read and agreed to the published version of the manuscript.

Funding: This research was funded by "Centro para el Desarrollo Tecnológico Industrial" (CDTI)Grant Number ITC-20131053 under the project "SEA-PEM: Sistema Estratégico de Apoyo Aparques Eólicos Marinos" and the "Ministerio de Ciencia, Innovación y Universidades", Government of Spain, under the project "Bigdata Analitycs e Instrumentación Cyberfísica para Soporte de Operaciones de Distribución en la Smart Grid", Number RTI2018-094917-B-I00. This last project funded Sebastián García. Moreover, Antonio Parejo is supported by the scholarship "Formación de Profesorado Universitario (FPU)", Grant Number FPU16/03522 from the "Ministerio de Educación y Formación Profesional", Government of Spain.

Acknowledgments: The authors would like to thank Isotrol Company, especially José García Franquelo, for their help and support. Additionally, the authors would also like to thank CDTI and the Ministerio de Ciencia, Innovación y Universidades, Government of Spain, for their funding and support.

Conflicts of Interest: The authors declare no conflict of interest.

Appendix A. JavaScript Definitions of the Submodel Libraries

This Appendix shows a set of submodel examples that were used in the study cases, making it possible to understand the usefulness of the proposed simulation tool.

Appendix A.1. Load Historical Data Submodel

This JS code implements the decodification of historical files. These files contain two columns: timestamp and value. The model finds the initial record and returns the corresponding value in each interaction.

```javascript
function Submodel_SourceFile(elemId,type) {
  // Submodel Id and characteristics
  this.id = elemId; this.type = type; this.model = "FILE";
  // Submodel constants.
  this.initDate;    this.source;  this.f_index = 0;
  // Init function definition.
  this.init = function(timeStep,initDate,source) {
    this.source = source;
    this.initDate = new Date(initDate);
    var filetext = fsRead(source);
    this.lines = filetext.split('\n');
    // Source submodel outputs.
    addReal("srcValue"+this.id, 0.0);
    return timeStep;
  };
  // Step function definition.
  this.step = function(timeStamp) {
    var data;
    if (this.f_index < this.lines.length-1) {
      var relativetimeStamp = this.initDate.getTime()+1000*timeStamp;
      var flag = true;
      do{
        var textline = this.lines[this.f_index+1];
        data = textline.split(',');
        var texttimeStamp = new Date(data[0].replace(/^\s*|\s*$/g, ""));
        if (texttimeStamp.getTime()<=relativetimeStamp)
          this.f_index++;
        else
          flag = false;
      } while(flag)
    }
    else {
      var textline = this.lines[this.lines.length];
        data = textline.split(',');
        var texttimeStamp = new Date(data[0].replace(/^\s*|\s*$/g, ""));
        if (texttimeStamp.getTime()<=relativetimeStamp)
          this.f_index = this.lines.length;
    }
    var textline = this.lines[this.f_index];
    data = textline.split(',');
    setReal("srcValue"+this.id, data[1].replace(/^\s*|\s*$/g,""));
    return 1;
  };
};
```

Appendix A.2. Power Aggregator Submodel

This JS code implements the power aggregation, summing up in each step the element of a vector that groups the individual powers of each element.

```javascript
function Submodel_Aggregator(elemId) {
  // Submodel Id and characteristics
  this.id = elemId; this.type = "Aggregator"; this.model = "ADD";
  // Aggregator submodel constants.
  this.elementsIds = new Array(0); this.inNamePattern = ""; this.outNamePattern = "";
  // Init function definition.
  this.init = function(timeStep,elementsIds,inNamePattern,outNamePattern) {
    // Aggregator submodel outputs.
    addReal(outNamePattern+this.id, 0.0);
    // Aggregator submodel constants.
    this.elementsIds = elementsIds;
    this.inNamePattern = inNamePattern; this.outNamePattern = outNamePattern;
    return timeStep;
  };
  // Step function definition.
  this.step = function(timeStamp) {
    var totalPower = 0;
    for (i=0 ; i<this.elementsIds.length ; i++)
      totalPower += getReal(this.inNamePattern+this.elementsIds[i]);
    setReal(this.outNamePattern+this.id, totalPower);
    return 1;
  };
};
```

Appendix A.3. Wind Turbine Simple Submodel

This JS code implements a simple submodel of a wind turbine that characterizes its behavior through its manufacture's power curve (wind speed-power generation relationship). In this submodel, dynamic behavior and other operating states are neglected.

```javascript
function Submodel_WindTurbine(elemId,model,windTable,powerTable,stopTime) {
  // Submodel Id and characteristics
  this.id = elemId; this.type = "Wind Generator"; this.model = model;
  // Submodel constants.
  this.wTable = windTable; this.pTable = powerTable;
  this.stopTime = stopTime;
  // Init function definition.
  this.init = function(timeStep) {
    // Wind turbine submodel inputs.
    addInt("startSignal"+this.id, 0);
    addReal("windSpeed"+this.id, 0);
    // Wind turbine submodel outputs.
    addReal("activePower"+this.id, 0.0);
    addText("currentState"+this.id, "STOPPED");
    // Wind turbine submodel constants.
    this.nextState = getText("currentState"+this.id);
    this.stopTS;
    return timeStep;
  };
```

```javascript
// Step function definition.
this.step = function(tStamp,startSignal,windSpeed) {
  setText("currentState"+this.id, this.nextState);
  setInt("startSignal"+this.id, startSignal);
  setReal("windSpeed"+this.id, windSpeed);
  switch (getText("currentState"+this.id)) {
    case "STOPPED":
      setReal("activePower"+this.id, 0.0);
      if (getInt("startSignal"+this.id) == 1)
        this.nextState = "PRODUCTION";
      break;
    case "RUNNING":
      setReal("activePower"+this.id, linInterp(this.wTable,this.pTable,windSpeed));
      if (getInt("startSignal"+this.id) == 0) {
        this.nextState = "STOPPING"; this.stopTS = tStamp+stopTime;
      }
      break;
    case "STOPPING":
      factor = (this.stopTS-tStamp)/stopTime; if(factor<0){factor = 0.0;}
      setReal("activePower"+this.id,factor*linInterp(this.wTable,this.pTable,windSpeed));
      if (tStamp >= this.stopTS)
        this.nextState = "STOPPED";
      break;
  }
  return 1;
};
// Auxiliary function.
function linInterp(Vx,Vy,x) {
  var index = 0; var flag = true; var y;
  if (x<=Vx[0]) { y = Vy[0]; }
  else if (x>=Vx[Vx.length-1]) { y = Vy[Vy.length-1]; }
  else {
    while ((flag)&&(index<Vx.length-1)) {
      if ((x>=Vx[index])&&(x<Vx[index+1])) {
        y = (x-Vx[index])*((Vy[index+1]-Vy[index])/(Vx[index+1]-Vx[index]))+Vy[index];
        flag = false;
      }
      index++;
    }
  }
  return y;
};
};
```

References

1. Ackermann, T.; Andersson, G.; Söder, L. Distributed generation: A definition1. *Electr. Power Syst. Res.* **2001**, *57*, 195–204. [CrossRef]
2. Farhangi, H. The path of the smart grid. *IEEE Power Energy Mag.* **2010**, *8*, 18–28. [CrossRef]
3. Kezunovic, M.; McCalley, J.D.; Overbye, T.J. Smart Grids and Beyond: Achieving the Full Potential of Electricity Systems. *Proc. IEEE* **2012**, *100*, 1329–1341. [CrossRef]
4. Ardito, L.; Procaccianti, G.; Menga, G.; Morisio, M. Smart Grid Technologies in Europe: An Overview. *Energies* **2013**, *6*, 251–281. [CrossRef]

5. Personal, E.; Guerrero, J.I.; Garcia, A.; Peña, M.; Leon, C. Key performance indicators: A useful tool to assess Smart Grid goals. *Energy* **2014**, *76*, 976–988. [CrossRef]

6. European Commission. *Directive (EU) 2019/944 of the European Parliament and of the Council of 5 June 2019 on Common Rules for the Internal Market for Electricity and amending Directive 2012/27/EU*; The Publications Office of the European Union (Publications Office): Luxembourg, 2019.

7. Wu, J.; Liu, M.; Lu, W. Measurement-based online distributed optimization of networked distributed energy resources. *Int. J. Electr. Power Energy Syst.* **2020**, *117*, 105703. [CrossRef]

8. England, B.S.; Alouani, A.T. Multiple Loads-Single Smart Meter for Measurement and Control of Smart Grid. In Proceedings of the 2019 IEEE Innovative Smart Grid Technologies, Chengdu, China, 21–24 May 2019; doi:10.1109/ISGT-Asia.2019.8881529. [CrossRef]

9. Parejo, A.; Personal, E.; Larios, D.F.; Guerrero, J.I.; García, A.; León, C. Monitoring and Fault Location Sensor Network for Underground Distribution Lines. *Sensors* **2019**, *19*, 576. [CrossRef]

10. Guerrero, J.; Personal, E.; García, A.; Parejo, A.; Pérez, F.; León, C. Distributed Charging Prioritization Methodology Based on Evolutionary Computation and Virtual Power Plants to Integrate Electric Vehicle Fleets on Smart Grids. *Energies* **2019**, *12*, 2402. [CrossRef]

11. Glaessgen, E.; Stargel, D. The Digital Twin Paradigm for Future NASA and U.S. Air Force Vehicles. In Proceedings of the 53rd AIAA/ASME/ASCE/AHS/ASC Structures, Structural Dynamics and Materials Conference & 20th AIAA/ASME/AHS Adaptive Structures Conference & 14th AIAA, Honolulu, HI, USA, 23–26 April 2012; American Institute of Aeronautics and Astronautics: Reston, VA, USA, 2012. [CrossRef]

12. lombardi, m.; Cammarota, A.; Refoyo Mayoral, J. Development, applications and benefits of the network digital twin. In Proceedings of the 25th International Conference on Electricity Distribution (CIRED2019), Madrid, Spain, 3–6 June 2019.

13. Jain, P.; Poon, J.; Singh, J.P.; Spanos, C.; Sanders, S.R.; Panda, S.K. A Digital Twin Approach for Fault Diagnosis in Distributed Photovoltaic Systems. *IEEE Trans. Power Electron.* **2020**, *35*, 940–956. [CrossRef]

14. Marten, D.; Lennie, M.; Pechlivanoglou, G.; Nayeri, C.; Paschereit, C. Implementation, Optimization, and Validation of a Nonlinear Lifting Line-Free Vortex Wake Module Within the Wind Turbine Simulation Code QBLADE. *J. Eng. Gas Turbines Power* **2016**, *138*. doi:10.1115/1.4031872. [CrossRef]

15. Akour, S.; Al-Heymari, M.; Ahmed, T.; Khalil, K. Experimental and theoretical investigation of micro wind turbine for low wind speed regions. *Renew. Energy* **2018**, *116*, 215–223. doi:10.1016/j.renene.2017.09.076. [CrossRef]

16. Obeng, M.; Gyamfi, S.; Derkyi, N.S.; Kabo-bah, A.T.; Peprah, F. Technical and economic feasibility of a 50 MW grid-connected solar PV at UENR Nsoatre Campus. *J. Clean. Prod.* **2020**, *247*, 119159. [CrossRef]

17. Kut, P.; Nowak, K. Design of photovoltaic systems using computer software. *J. Ecol. Eng.* **2019**, *20*, 72–78. doi:10.12911/22998993/112907. [CrossRef]

18. Honrubia-Escribano, A.; Jiménez-Buendía, F.; Sosa-Avendaño, J.L.; Gartmann, P.; Frahm, S.; Fortmann, J.; Sørensen, P.E.; Gómez-Lázaro, E. Fault-Ride Trough Validation of IEC 61400-27-1 Type 3 and Type 4 Models of Different Wind Turbine Manufacturers. *Energies* **2019**, *12*, 3039. [CrossRef]

19. IEC. *IEC 61400-27-1:2015. Wind Turbines—Part 27-1: Electrical Simulation Models—Wind Turbines*; IEC Central Office: Geneva, Switzerland, 2015.

20. Jia, K.; Gu, C.; Xuan, Z.; Li, L.; Lin, Y. Fault Characteristics Analysis and Line Protection Design Within a Large-Scale Photovoltaic Power Plant. *IEEE Trans. Smart Grid* **2018**, *9*, 4099–4108. [CrossRef]

21. Varma, R.K.; Siavashi, E.M. PV-STATCOM: A New Smart Inverter for Voltage Control in Distribution Systems. *IEEE Trans. Sustain. Energy* **2018**, *9*, 1681–1691. [CrossRef]

22. Parejo, A.; Sanchez-Squella, A.; Barraza, R.; Yanine, F.; Barrueto-Guzman, A.; Leon, C. Design and Simulation of an Energy Homeostaticity System for Electric and Thermal Power Management in a Building with Smart Microgrid. *Energies* **2019**, *12*, 1806. [CrossRef]

23. Saleh, M.; Esa, Y.; Mohamed, A. Applications of Complex Network Analysis in Electric Power Systems. *Energies* **2018**, *11*, 1381. [CrossRef]

24. Schneider, K.P.; Mather, B.A.; Pal, B.C.; Ten, C.W.; Shirek, G.J.; Zhu, H.; Fuller, J.C.; Pereira, J.L.R.; Ochoa, L.F.; de Araujo, L.R.; et al. Analytic Considerations and Design Basis for the IEEE Distribution Test Feeders. *IEEE Trans. Power Syst.* **2018**, *33*, 3181–3188. [CrossRef]

25. Viana, M.S.; Manassero, G.; Udaeta, M.E. Analysis of demand response and photovoltaic distributed generation as resources for power utility planning. *Appl. Energy* **2018**, *217*, 456–466. [CrossRef]

26. Jha, R.R.; Dubey, A.; Liu, C.C.; Schneider, K.P. Bi-Level Volt-VAR Optimization to Coordinate Smart Inverters With Voltage Control Devices. *IEEE Trans. Power Syst.* **2019**, *34*, 1801–1813. [CrossRef]

27. Camilo, F.M.; Pires, V.F.; Castro, R.; Almeida, M. The impact of harmonics compensation ancillary services of photovoltaic microgeneration in low voltage distribution networks. *Sustain. Cities Soc.* **2018**, *39*, 449–458. [CrossRef]

28. EPRI. *Distribution Resource Integration and Value Estimation (DRIVE)*, Version 2.0; EPRI: Palo Alto, CA, USA, 2018.

29. Ismael, S.M.; Aleem, S.H.A.; Abdelaziz, A.Y.; Zobaa, A.F. State-of-the-art of hosting capacity in modern power systems with distributed generation. *Renew. Energy* **2019**, *130*, 1002–1020. [CrossRef]

30. Smith, J.; Rylander, M.; Rogers, L.; Dugan, R. It's All in the Plans: Maximizing the Benefits and Minimizing the Impacts of DERs in an Integrated Grid. *IEEE Power Energy Mag.* **2015**, *13*, 20–29. [CrossRef]

31. Steinbrink, C.; Blank-Babazadeh, M.; El-Ama, A.; Holly, S.; Lüers, B.; Nebel-Wenner, M.; Ramírez Acosta, R.P.; Raub, T.; Schwarz, J.S.; Stark, S.; et al. CPES Testing with mosaik: Co-Simulation Planning, Execution and Analysis. *Appl. Sci.* **2019**, *9*, 923. [CrossRef]

32. OFFIS. Mosaik: A Flexible Smart Grid Co-Simulation Framework. 2019. Available online: https://mosaik.offis.de/ (accessed on 19 January 2020).

33. Schwarz, J.S.; Witt, T.; Nieße, A.; Geldermann, J.; Lehnhoff, S.; Sonnenschein, M. Towards an Integrated Development and Sustainability Evaluation of Energy Scenarios Assisted by Automated Information Exchange. In *Smart Cities, Green Technologies, and Intelligent Transport Systems*; Donnellan, B., Klein, C., Helfert, M., Gusikhin, O., Pascoal, A., Eds.; Springer International Publishing: Cham, Switzerland, 2019; pp. 3–26.

34. Sahnoun, M.; Baudry, D.; Mustafee, N.; Louis, A.; Smart, P.A.; Godsiff, P.; Mazari, B. Modelling and simulation of operation and maintenance strategy for offshore wind farms based on multi-agent system. *J. Intell. Manuf.* **2019**, *30*, 2981–2997. [CrossRef]

35. Zhang, P.; Li, F.; Bhatt, N. Next-Generation Monitoring, Analysis, and Control for the Future Smart Control Center. *IEEE Trans. Smart Grid* **2010**, *1*, 186–192. [CrossRef]

36. Lin, K.S., New Cost-Consequence FMEA Model for Information Risk Management of Safe And Secure SCADA Systems. In *Software Engineering, Artificial Intelligence, Networking and Parallel/Distributed Computing*; Springer International Publishing: Cham, Switzerland, 2020; pp. 33–51._3. [CrossRef]

37. Liu, Y.; Cheng, H.; Kong, X.; Wang, Q.; Cui, H. Intelligent wind turbine blade icing detection using supervisory control and data acquisition data and ensemble deep learning. *Energy Sci. Eng.* **2019**, *7*, 2633–2645. [CrossRef]

38. Schlechtingen, M.; Santos, I.F.; Achiche, S. Wind turbine condition monitoring based on SCADA data using normal behavior models. Part 1: System description. *Appl. Soft Comput.* **2013**, *13*, 259 – 270. [CrossRef]

39. Modbus Organization, I. Modbus Application Protocol Specification V1.1b3. 2012. Available online: http://www.modbus.org/docs/Modbus_Application_Protocol_V1_1b3.pdf (accessed on 21 January 2020).

40. IEEE. *1815-2012—IEEE Standard for Electric Power Systems Communications-Distributed Network Protocol (DNP3)*, IEEE Standards Association: Piscataway, NJ, USA, 2012.

41. IEC. *IEC 60870-5:2020 SER Series. Telecontrol Equipment and Systems—Part 5: Transmission Protocols—All Parts*; IEC Central Office: Geneva, Switzerland, 2020.

42. IEC. *IEC TR 62541-1:2016. OPC Unified Architecture—Part 1: Overview and Concepts*; IEC: Geneva, Switzerland, 2016.

43. IEEE. *IEEE Standard for Secure SCADA Communications Protocol (SSCP)*; IEEE Std 1711.2-2019; IEEE Standards Association: Piscataway, NJ, USA, 2020, pp. 1–37. [CrossRef]

44. Figueroa-Lorenzo, S.; Añorga, J.; Arrizabalaga, S. A Role-Based Access Control Model in Modbus SCADA Systems. A Centralized Model Approach. *Sensors* **2019**, *19*, 4455. [CrossRef]

45. Rinaldi, J. *OPC UA Unified Architecture: The Everyman's Guide to the Most Important Information Technology in Industrial Automation*; CreateSpace Independent Publishing Platform: Scotts Valley, CA, USA, 2016.

46. Gomes, C.; Thule, C.; Broman, D.; Larsen, P.G.; Vangheluwe, H.L. Co-Simulation: A Survey. *ACM Comput. Surv. (CSUR)* **2018**, *51*. [CrossRef]

47. Wagner, F.; Schmuki, R.; Wagner, T.; Wolstenholme, P. *Modeling Software with Finite State Machines: A Practical Approach*; Auerbach Publications—Taylor & Francis Group: Abingdon, UK, 2006.

48. Byon, E.; Pérez, E.; Ding, Y.; Ntaimo, L. Simulation of wind farm operations and maintenance using discrete event system specification. *Simulation* **2011**, *87*, 1093–1117. [CrossRef]

49. Zhang, J.; Hasandka, A.; ; Alam, S.M.S.; Elgindy, T.; Florita, A.R.; Hodge, B.-M. Analysis of Hybrid Smart Grid Communication Network Designs for Distributed Energy Resources Coordination. In Proceedings of the 2019 IEEE Power & Energy Society Innovative Smart Grid Technologies Conference (ISGT), Washington, DC, USA, 18–21 February 2019; doi:10.1109/ISGT.2019.8791581. [CrossRef]

50. Cordasco, G.; D'Auria, M.; Spagnuolo, C.; Scarano, V. Heterogeneous Scalable Multi-languages Optimization via Simulation. *Commun. Comput. Inf. Sci.* **2018**, *946*, 151–167. doi:10.1007/978-981-13-2853-4_13. [CrossRef]

51. Yoo, Y.; Lee, S.; Yoon, J.; Lee, J. Modelica-based dynamic analysis and design of lift-generating disk-type wind blade using computational fluid dynamics and wind tunnel test data. *Mechatronics* **2018**, *55*, 1–12. doi:10.1016/j.mechatronics.2018.08.003. [CrossRef]

52. Ding, H.; Zhang, Y.; Ye, K.; Hong, G. Development of a model for thermal-hydraulic analysis of helically coiled tube once-through steam generator based on Modelica. *Ann. Nucl. Energy* **2020**, *137*. doi:10.1016/j.anucene.2019.107069. [CrossRef]

53. Beiron, J.; Montañés, R.; Normann, F.; Johnsson, F. Dynamic modeling for assessment of steam cycle operation in waste-fired combined heat and power plants. *Energy Convers. Manag.* **2019**, *198*. doi:10.1016/j.enconman.2019.111926. [CrossRef]

54. Wagner, G. Introduction to simulation using JavaScript. In Proceedings of the 2016 Winter Simulation Conference (WSC), Arlington, VA, USA, 11–14 December 2016; pp. 148–162. [CrossRef]

55. Zakai, A. Fast Physics on the Web Using C++, JavaScript, and Emscripten. *Comput. Sci. Eng.* **2018**, *20*, 11–19. [CrossRef]

56. Taheri, S.; Vedienbaum, A.; Nicolau, A.; Hu, N.; Haghighat, M.R. OpenCV.js. In Proceedings of the 9th ACM Multimedia Systems Conference on MMSys '18, Amsterdam, The Netherlands, 12–15 June 2018. ACM Press: New York, NY, USA, 2018. [CrossRef]

57. Ecma International. ECMA-262 Standard—ECMAScript Language Specification. 2011. Available online: http://www.ecma-international.org/ecma-262/5.1/ECMA-262.pdf (accessed on 24 January 2020).

58. Selakovic, M.; Pradel, M. Performance issues and optimizations in Java script: An empirical study. In Proceedings of the International Conference on Software Engineering, Austin, TX, USA, 14–22 May 2016; pp. 61–72. [CrossRef]

59. Rahimly, P.; Rahimly, O.; Poveshchenko, Y.; Podryga, V.; Gasilova, I. Application Software for the Simulation of Fluid Dynamics and Transphase Processes in Collectors with Gas-Hydrate Depositions. *Math. Model. Comput. Simulations* **2019**, *11*, 789–798. doi:10.1134/S2070048219050168. [CrossRef]

60. Luo, Y.; Chhabda, J. Hybrid Real/Virtual Simulation in an Engineering Laboratory Course. In Proceedings of the ASCE International Workshop on Computing in Civil Engineering 2017, Seattle, WA, USA, 25–27 June 2017.

61. Chen, J.; Rabiti, C. Synthetic wind speed scenarios generation for probabilistic analysis of hybrid energy systems. *Energy* **2017**, *120*, 507–517. [CrossRef]

62. David, M.; Ramahatana, F.; Trombe, P.; Lauret, P. Probabilistic forecasting of the solar irradiance with recursive ARMA and GARCH models. *Sol. Energy* **2016**, *133*, 55–72. [CrossRef]

63. Nayak, A.K.; Mohanty, K.B. Analysis of Wind Characteristics using ARMA & Weibull Distribution. In Proceedings of the 2018 National Power Engineering Conference (NPEC), Madurai, India, 9–10 March 2018; IEEE: Piscataway, NJ, USA, 2018; doi:10.1109/npec.2018.8476717. [CrossRef]

64. Das, U.K.; Tey, K.S.; Seyedmahmoudian, M.; Mekhilef, S.; Idris, M.Y.I.; Deventer, W.V.; Horan, B.; Stojcevski, A. Forecasting of photovoltaic power generation and model optimization: A review. *Renew. Sustain. Energy Rev.* **2018**, *81*, 912–928. [CrossRef]

65. Pazikadin, A.R.; Rifai, D.; Ali, K.; Malik, M.Z.; Abdalla, A.N.; Faraj, M.A. Solar irradiance measurement instrumentation and power solar generation forecasting based on Artificial Neural Networks (ANN): A review of five years research trend. *Sci. Total Environ.* **2020**, *715*, 136848. [CrossRef]

66. Filik, Ü.B.; Filik, T. Wind Speed Prediction Using Artificial Neural Networks Based on Multiple Local Measurements in Eskisehir. *Energy Procedia* **2017**, *107*, 264–269. [CrossRef]

67. Zhao, G.; Zhang, Z.; Guan, H.; Tang, P.; Wang, J. Rethinking ReLU to Train Better CNNs. In Proceedings of the 2018 24th International Conference on Pattern Recognition (ICPR), Beijing, China, 20–24 August 2018; IEEE: Piscataway, NJ, USA, 2018. [CrossRef]

68. Son, J.; Park, Y.; Lee, J.; Kim, H. Sensorless PV Power Forecasting in Grid-Connected Buildings through Deep Learning. *Sensors* **2018**, *18*, 2529. [CrossRef]
69. Meeds, E.; Hendriks, R.; Faraby, S.A.; Bruntink, M.; Welling, M. MLitB: Machine learning in the browser. *PeerJ Comput. Sci.* **2015**, *1*, e11. [CrossRef]
70. Duveiller, G.; Donatelli, M.; Fumagalli, D.; Zucchini, A.; Nelson, R.; Baruth, B. A dataset of future daily weather data for crop modelling over Europe derived from climate change scenarios. *Theor. Appl. Climatol.* **2015**, *127*, 573–585. [CrossRef]
71. Liu, X. An Improved Interpolation Method for Wind Power Curves. *IEEE Trans. Sustain. Energy* **2012**, *3*, 528–534. [CrossRef]
72. Jung, J.; Broadwater, R.P. Current status and future advances for wind speed and power forecasting. *Renew. Sustain. Energy Rev.* **2014**, *31*, 762–777. [CrossRef]
73. Lydia, M.; Selvakumar, A.I.; Kumar, S.S.; Kumar, G.E.P. Advanced Algorithms for Wind Turbine Power Curve Modeling. *IEEE Trans. Sustain. Energy* **2013**, *4*, 827–835. [CrossRef]
74. Shokrzadeh, S.; Jozani, M.J.; Bibeau, E. Wind Turbine Power Curve Modeling Using Advanced Parametric and Nonparametric Methods. *IEEE Trans. Sustain. Energy* **2014**, *5*, 1262–1269. [CrossRef]
75. Mehrjoo, M.; Jozani, M.J.; Pawlak, M. Wind turbine power curve modeling for reliable power prediction using monotonic regression. *Renew. Energy* **2020**, *147*, 214–222. [CrossRef]
76. Theo, W.L.; Lim, J.S.; Ho, W.S.; Hashim, H.; Lee, C.T. Review of distributed generation (DG) system planning and optimisation techniques: Comparison of numerical and mathematical modelling methods. *Renew. Sustain. Energy Rev.* **2017**, *67*, 531–573. [CrossRef]
77. Kazem, H.A.; Khatib, T. Techno-economical assessment of grid connected photovoltaic power systems productivity in Sohar, Oman. *Sustain. Energy Technol. Assessments* **2013**, *3*, 61–65. [CrossRef]
78. ISOTROL. WebPage. Available online: https://www.isotrol.com/web/ (accessed on 27 January 2020).

Article

Control of Hybrid Diesel/PV/Battery/Ultra-Capacitor Systems for Future Shipboard Microgrids

Muhammad Umair Mutarraf [1,*], Yacine Terriche [1], Kamran Ali Khan Niazi [1], Fawad Khan [2], Juan C. Vasquez [1] and Josep M. Guerrero [1]

[1] Department of Energy Technology, Aalborg University, 9220 Aalborg, Denmark; yte@et.aau.dk (Y.T.); kkn@et.aau.dk (K.A.K.N.); juq@et.aau.dk (J.C.V.); joz@et.aau.dk (J.M.G.)
[2] Department of Information Security, National University of Sciences and Technology, H-12, Islamabad 44000, Pakistan; fawad.khan.xdu@gmail.com
* Correspondence: mmu@et.aau.dk; Tel.: +45-91778118

Received: 29 June 2019; Accepted: 04 September 2019; Published: 7 September 2019

Abstract: In recent times, concerns over fossil fuel consumption and severe environmental pollution have grabbed attention in marine vessels. The fast development in solar technology and the significant reduction in cost over the past decade have allowed the integration of solar technology in marine vessels. However, the highly intermittent nature of photovoltaic (PV) modules might cause instability in shipboard microgrids. Moreover, the penetration is much more in the case of utilizing PV panels on ships due to the continuous movement. This paper, therefore, presents a frequency sharing approach to smooth the effect of the highly intermittent nature of PV panels integrated with the shipboard microgrids. A hybrid system based on an ultra-capacitor and a lithium-ion battery is developed such that high power and short term fluctuations are catered by an ultra-capacitor, whereas long duration and high energy density fluctuations are catered by the lithium-ion battery. Further, in order to cater for the fluctuations caused by weather or variation in sea states, a battery energy storage system (BESS) is utilized in parallel to the dc-link capacitor using a buck-boost converter. Hence, to verify the dynamic behavior of the proposed approach, the model is designed in MATLAB/SIMULINK. The simulation results illustrate that the proposed model helps to smooth the fluctuations and to stabilize the DC bus voltage.

Keywords: shipboard microgrids; photovoltaic (PV) systems; energy storage technologies; microgrids; hybrid energy storage systems (HESS)

---

1. Introduction

Electrification in marine vessels has been a trend to improve efficiency and minimize emissions [1–4]. There are severe restrictions imposed by the International Convention for the Prevention of Pollution (MARPOL) from ships [5]. This international convention covers the prevention of pollution of the marine environment by marine vessels. The Kyoto Protocol and Paris Agreement further discuss the reduction of these emissions. The Kyoto protocol was the first to propose limitations on greenhouse emissions and to provide a schedule to prevent from global warming. It was endorsed in 1997 at the Kyoto conference. Further, the Paris agreement was made on the 4th of November 2016. The main motivation of this agreement was to keep the temperature below 2 °C and to put further efforts to narrow it down upto 1.5 °C [6,7].

In 2012, the International Marine Organization (IMO) announced that global NO_x and SO_x emissions from world shipping exhibits approximately 15% and 13%, respectively [8]. Further, it illustrated that for entire international shipping, the CO_2 emissions are found to be 2.2% of the global CO_2 emissions whereas the CO_2 emissions from the entire world shipping is noted to be 2.6%. IMO predicts that by 2050, CO_2 emissions from entire international shipping might raise by 50–250%. Figure 1 illustrates CO_2 emissions produced by international shipping from 1990 to 2016.

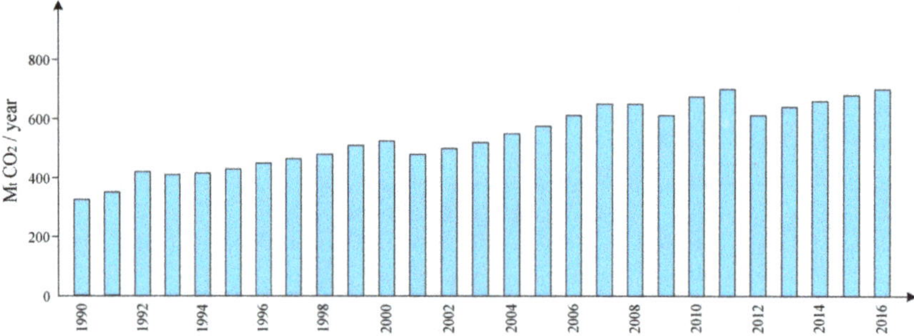

Figure 1. Fossil CO_2 emissions by International Shipping [9].

The existing topologies mostly utilized in ships comprise of Battery Energy Storage Systems (BESS) and Hybrid energy storage systems (HESS) typically battery and flywheel. H. Lan et al. used a hybrid model consisting of PV modules, a diesel engine, and flywheel energy storage system (FESS) to improve the power quality and to smooth the fluctuations that are caused by PV systems [9]. The study in [10] proposes a method to determine the optimal sizing of a diesel generator, a PV generation system, and BESS for the shipboard power system to save the fuel and minimize the investment cost and greenhouse emissions. The prototype in [11] consists of a PV module, battery energy storage system, and diesel generation system that can be operated in island mode and grid-connected mode as well. The aim of the hybrid model was to minimize the cost and support the grid in grid-connected mode.

S. Wen et al. propose a hybrid model that comprises of PV generation system, diesel engines, and HESS. The HESS utilized in the study consist of an ultra-capacitor, lead-acid and lithium-ion battery to cater for fluctuation caused by the rolling of ships [12]. In this study, a mathematical model of a PV system is proposed considering the rolling effect of ships. As a PV system on the ship has more fluctuations, HESS is proposed by employing a discrete fourier transform to break down the needed balancing power into several time varying components. Further, a particle swarm-based algorithm is utilized to perform the cost analysis and to optimize the capacity and size of various types of energy storage systems (ESS) used. The study in [13] uses a load profile sharing approach between Li-ion battery and diesel generator to minimize the impact of fluctuations for diesel generator. The high frequency components are allocated to a Li-ion battery whereas low-frequency components are are assigned to diesel generator.

H. Liu et al. propose a mathematical model for PV system considering the integrated motion (basic movement and movement caused by rocking) of a ship and sea conditions [14]. The proposed methodology employed an ultra-capacitor to minimize the fluctuations; hence, ensuring system stability, minimal usage of battery, and an increase in the lifetime of a battery. The study in [15] utilizes a Proportional Integral (PI)-based energy management system to control the charging and discharging of the battery and ultra-capacitor. A low-pass-based filter control is developed to separate the storage reference between ultra-capacitor and battery. This reference power is then sent to a dual active bridge to control the exchange of power between the medium voltage dc system and HESS. J. Tang et al. [16] proposed a HESS comprising a battery and ultra-capacitor with an improved Maximum power tracking (MPPT) algorithm to minimize the intermittency of the PV system. Frequency Hierarchical-based control algorithm is utilized, which assigns low-frequency fluctuations to a battery whereas high-frequency fluctuations are assigned to the ultra-capacitor. Khooban et al. [17] designed an optimal fuzzy based on PD+I load frequency controller for shipboard microgrids. The tuning of the coefficients of the controller is done by a black-hole optimization algorithm to limit the frequency fluctuations. The study in [18] integrated an ultra-capacitor with a PV system to smooth the fluctuations and boost the LVRT capability. The control strategy is designed to attain the scheduling between ultra-capacitor and PV system to minimize the dc-link over-voltage problem. The study in [19] proposed a hybrid system (battery and ultra-capacitor) for shipboard microgrids to support pulse loads and peak demand by using an energy management system based on fuzzy logic. Table 1 shows the list of marine vessels that utilize PV modules.

Table 1. Summary of marine vessels utilizing PV modules.

Ship's Name	Solar Power	Battery Type	Power of Battery	Power Generation Sources	Reference
e-Boat (AIDE Energy)	105 kW	Li-ion (LiFePO$_4$)	48 kWh	PV + Battery	[20]
Greenline 33	1.8 kW	Li-ion (LiPO)	11.5 kW	Battery + DG + PV	[21]
Greenline 39	1.2 kW	Li-ion (LiPO)	3 kW	Battery + DG + PV	[22]
Greenline 40	1.8 kW	Li-ion (LiPO)	3 kW	Battery + DG + PV	[23]
Duffy London	–	Li-ion	2380–3400 kWh	PV + Battery	[24]
Silent 80	26 kWp	–	240 kW	Battery + DG + PV	[25]
Silent 64	15 kWp	–	120 kWh	Battery + DG + PV	[26]
Silent 55 Ferry	10 kWp	–	120 kWh	Battery + DG + PV	[27]
MS Türanor Planet Solar	93.5 kW	Li-ion	1130 kWh	PV + Battery	[28]
Aquarius Eco Ship	–	Lead acid	–	PV + Battery	[29]
M/V Auriga Leader	–	Nickel-hydrogen	–	PV + Battery + Diesel	[30,31]

In order to decrease greenhouse emission and save fuel, numerous solutions have been proposed, for example, wind, solar, waste heat recovery systems, alternative fuels, and hybrid propulsions are among the mostly applied techniques to acquire environmental limitations set by IMO. Although the aforementioned technologies help to reduce emissions and improve fuel efficiency, their intermittent nature or/and slower response requisite the use of ESS such as batteries, flywheel, ultra-capacitor etc. As higher energy and power density cannot be attained by a single energy storage device, novel approaches, for instance, HESS should be investigated for future shipboard microgrids.

In this paper, a PV system is integrated with the shipboard microgrids to minimize the use of fossil fuels and to decrease the greenhouse emission. In order to cater for the highly intermittent nature of PV systems, an ultra-capacitor is hybridized with a Li-ion based battery. The ultra-capacitor energy storage system is utilized as it has higher power density and efficiency as compared to FESS. Among batteries, Li-ion is considered to have high energy and power density. Moreover, a low pass filter and PI-based energy management system is proposed such that high fluctuations will be catered for by the ultra-capacitor whereas low frequency fluctuations are smoothed by the battery. Further, BESS is used in parallel with the dc-link of variable frequency drive (VFD) utilizing a buck-boost converter to cater for the fluctuations caused by the propulsion motor; these fluctuations usually occur due to the variation in the sea conditions and weather. As PV characteristics of a Li-ion battery show that the voltage remains constant, most of the time around 80% in nominal region, the use of a buck-boost converter for this application can be ignored.

The rest of this study is formulated as follows. In Section 2, a comparison of ESS is conducted. The mathematical models are presented in Section 3, followed by modeling of the PV system, battery and ultra-capacitor. Section 4 presents the control strategy for the hybrid PV/UC/battery power system. Section 5 shows the simulation results and verification of the proposed model. Finally, a conclusion is drawn from the study and presented in Section 6.

2. Energy Storage Technologies for Shipboard Microgrids

The energy storage system consists of power an electronics-based converter, energy storage device, and the control mechanism. The popular ESS currently in use for several applications include secondary batteries, Superconducting magnetic energy storage (SMES), ultra-capacitors, Pumped hydro, and Fuel cells (FC). The storage devices aforementioned differ from one another in terms of life-cycle, energy and power density, efficiency, charge, and discharge rate, etc.

2.1. Comparison of Different Energy Storage Technologies

The use of a lead–acid battery was started in the 1890s and is still extensively utilized in those applications in which cost is considered a major problem [32]. It is used in applications such as stand-alone PV system, Uninterruptible Power Supplies, and as a starter in vehicles. These batteries are inexpensive as compared to other types of batteries and have quite low self-discharge (typically 0.3% a day). Moreover, lead can be recyclable and these batteries are available at cheap prices. The main drawback of this sort of battery is its lower energy density, short life-cycle, and hazardousness of lead. On the other hand, Lithium-ion (Li-ion)-based batteries have high nominal voltage of around 3.7 V for each cell as compared to 2 V in case of lead acid batteries [33]. Li-ion-based batteries are more suitable for the application of shipboard microgrids as these batteries have high efficiency, high power and density, and long cycle life as compared to other sort of batteries. The main obstacle is its higher cost and the safety concerns, as metal oxides electrodes are typically unbalanced and can decompose with raise in temperature. In order to cater for this problem, the battery units are prepared with a monetizing unit to evade overcharging and discharging. Table 2 illustrates the technical features of different types of energy storage technologies.

Table 2. Technical features of ESS [33–42].

System	Energy Density	Power Density	Efficiency	Life Time	Response Time	Self-Discharge per Day
	(kWh/kg)	(kW/kg)	(years)	(%)		%
Lead Acid	30–50	75–300	3–12	65–80	ms	0.1–0.3
Li-ion	75–200	150–315	5–15	90–97	ms–s	0.1–0.3
Flywheel	10–30	400–1500	20–30	90–95	ms–s	100
Ultra-capacitor	20+	100,000+	10–20	85–98	ms	20–40

The applications in which high-power density is required for smaller duration, ultra-capacitors and flywheel can be utilized. The main characteristics of ultra-capacitors are their long life-cycle, high power density, faster charging and discharging because of lower internal resistance. The higher cost, sensitivity to over-voltage, and lower energy density limit the use of ultra-capacitor to only those applications that are required with high power for a shorter duration. The pros and cons of ultra-capacitor and Li-ion battery are depicted in Table 3.

Table 3. Comparison between Li-ion and Ultra-capacitor.

ESS	Li-Ion	Ultra-Capacitor
Advantages	High efficiency (90–97%) [36] Low self-discharge (<5% per month) [36] Lower maintenance [37] Fast response time (<5 ms) [36] High reliability [37]	High efficiency (85–98%) [36] Fast response time (<5 ms) [36] Enhanced life cycle [39] Broader operating temperature (−40 °C to 65 °C) [43] More than 500,000 duty cycles [43]
Disadvantages	High cost [37] Life-cycle shorten by deep discharge [38] Require special overcharging and discharging protection circuit [37]	High cost [39] Short charge and discharge time [36] Low energy density [36] High rate of self-discharge (14% per month) [36]

2.2. Hybridization of Battery and Ultra-Capacitor

In the literature, HESS consisting of battery and ultra-capacitor has been extensively studied for several applications such as in transportation [42], stand-alone PV [43], and terrestrial microgrids [44]. As high power and high energy density can not be achieved by a single ESS, there is an utmost need to hybridize a higher power density device with a high energy density device. The life-cycle of a battery is found out to be quite low usually in the range of 1500–4500 cycles, and low power density (below 1 kW/kg) whereas an ultra-capacitor has high power density (above 10 kWh/kg) and long life-cycle. Hence, these energy storage technologies can be hybridized. There are two possibilities to hybridize them either internally or externally. In the internal hybridization, the devices are hybridized at the electrode level as depicted in Figure 2a. Ultra-battery is an example of internal hybridization in which lead-acid battery is hybridized with ultra-capacitor [45]. In this study, a hardwire connection is considered as shown in Figure 2b.

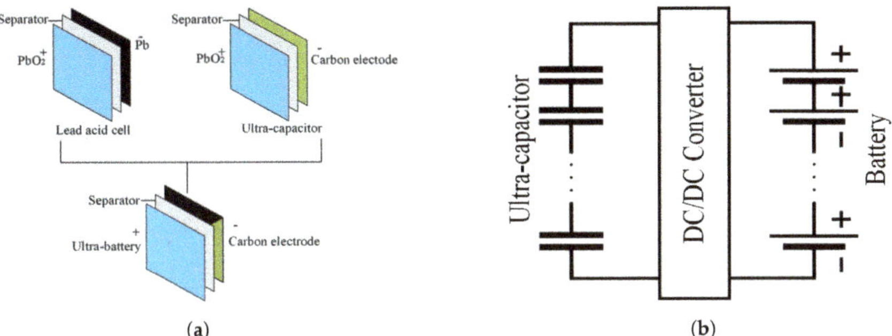

(**a**)　　　　　(**b**)

Figure 2. Possible hybridization (**a**) Ultra-battery (**b**) External Hybridization.

3. Model of the Shipboard Microgrid

The proposed model depicted in Figure 3 consists of the PV generation system, diesel engine, BESS, ultra-capacitor, boost converter, buck-boost converters, active front end converter, and an inverter. The ultra-capacitor and battery are connected using buck-boost converter in parallel with the dc-link capacitor so it enhances the stability of the shipboard microgrid. The total power flow in the proposed model is depicted in the Equation (1).

$$P_{total} = P_{PV} + P_{DG} + P_{ESS} - P_{Load} \tag{1}$$

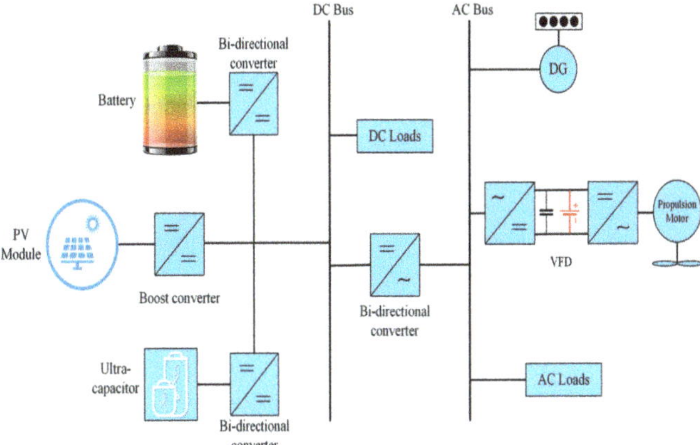

Figure 3. Hybrid Shipboard Microgrid.

The detailed parameters of the synchronous generator, PV system, ultra-capacitor, and Li-ion battery utilized in this paper are given in the Table 4.

Table 4. Parameters of devices utilized in the study.

Device	Parameters	Values
Synchronous Generator	Nominal Power	80 kW
	Line to line voltage	400 V
	Frequency	50 Hz
PV panel–Single panel (Waaree Energies WU-120) [46]	Rated Power	120.7 W
	Open circuit voltage (V_{oc})	21 V
	Short circuit current (V_{sh})	8 A
	Voltage at maximum power (V_{mp})	17 V
	Current at maximum power (I_{mp})	7.1 A
	Number of cells in parallel	30
	Number of cells in series	15
	Temperature	25 °C
Ultra-capacitor (PC 2500) [47]	Rated capacity	2517.54 F
	Rated voltage	385 V
	Initial voltage	385 V
	Operating temperature	25 °C
	Number of series connected Ultra-capacitors	154
	Number of parallel connected Ultra-capacitor	1
	Capacity of parallel connected Ultra-capacitor	2500 F
	Equivalent dc series resistance	0.001
Li-ion battery (Panasonic CGR18650AF) [48]	Nominal Voltage	288 V
	Rated Capacity	168 Ah
	Maximum Capacity	182 Ah
	Fully charge voltage	339 V
	No of cells in series	87
	No of cells in parallel	82

3.1. Diesel Engine Modeling

The diesel generator comprises a speed governor, prime mover, compound excitation transformer, and a synchronous generator [49]. The block diagram of a diesel generator set is shown in Figure 4 [50]. The engine's speed is usually maintained at a fixed speed by using speed governor typically 1500 or 1800 rpm. [51]. The reason for utilizing diesel gen-sets are their relative ease of starting, reliability, compact portability and power density such that their usage is widely spread. In order to control the current and/or voltage, the diesel engine model requires some control mechanisms. The output of the synchronous machine can be controlled by controlling the field current. The excitation system is responsible to provide field voltage E_{fd} for a synchronous machine and is utilized to provide an initial magnetic field to start the synchronous machine as well.

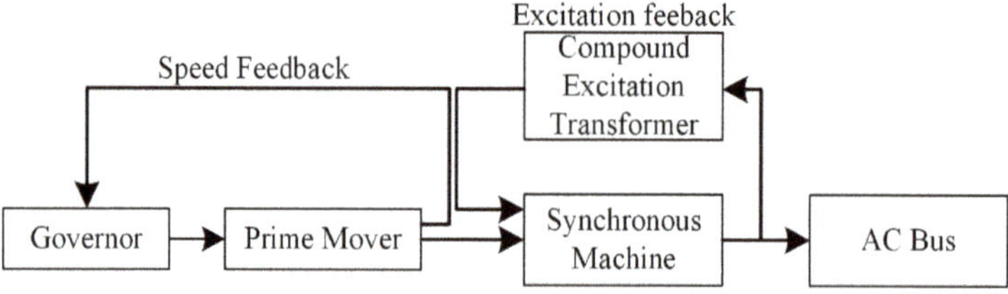

Figure 4. Diesel engine block diagram.

Figure 5 illustrates the speed and excitation feedback control system. The output current and terminal voltage are detected by the compound excitation transformer of excitation feedback system and the speed of the generator is detected by speed feedback. In the past, mechanical governors were utilized but due to errors, such as large inherent droop, they were replaced with electric governors. All blocks in an electronic governor are electrical excluding the actuator block. Hence, higher accuracy can be achieved and a droop of 0.5% is observed in contrast to 4% in mechanical governors.

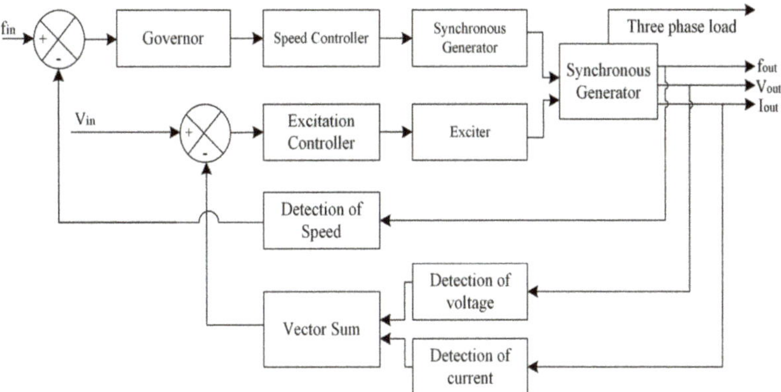

Figure 5. Feedback control system of a generator.

3.2. PV System Modeling

The PV model in the Simulink library is utilized in this paper; the PV source comprises of several strings of PV modules linked in parallel whereas each string contains the number of series PV modules as well. The equivalent circuit of a PV module based on an electrical system is illustrated in Figure 6 [52]. The mathematical model of a PV module can be expressed by the Equations (2)–(7).

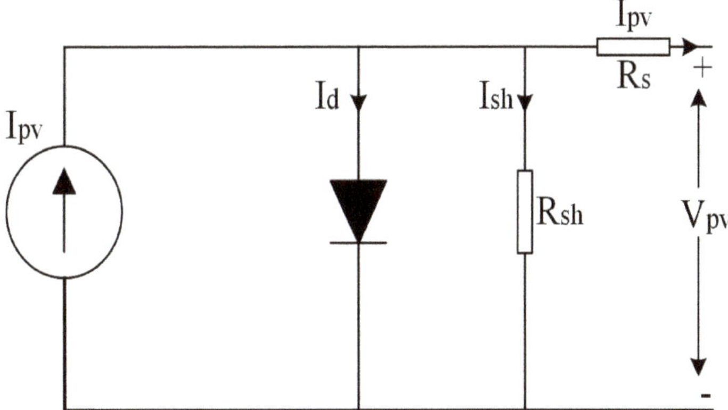

Figure 6. Equivalent circuit of a PV module.

$$I_{pv} = \frac{I_r}{100}[I_{sc} + K_i(T - 298)] \tag{2}$$

where I_{pv} is at the nominal condition (25 °C and 1000 W/m^2 (A)).

$$I_{Rs} = \frac{I_{sc}}{exp(qVoc/N_sTnk) - 1} \tag{3}$$

where q = 1.6 $*$ 10^{-19} C and k = 1.3805 $*$ 10^{-23} J/K.

$$I_o = I_{Rs}[\frac{T}{T_r}]^3 exp[\frac{qE_{go}}{k*n}(\frac{1}{T} - \frac{1}{T_r})] \tag{4}$$

where $T_r = 298$ K and $E_{go} = 1.1$ eV.

$$V^t = \frac{Tk}{q} \qquad (5)$$

$$I_{sh} = \frac{VN_p/N_s + IR_s}{R_{sh}} \qquad (6)$$

$$I = N_p I_{ph} - N_p I_o [exp(\frac{V/N_s + IR_s/N_p}{V^t n)} - 1] - I_{sh} \qquad (7)$$

The PV characteristic curve of PV module with specified temperature (25 °C) and varying irradiance is shown in Figure 7 where Figure 7a illustrates the I-V characteristic and Figure 7b shows the P-V characteristics of solar array.

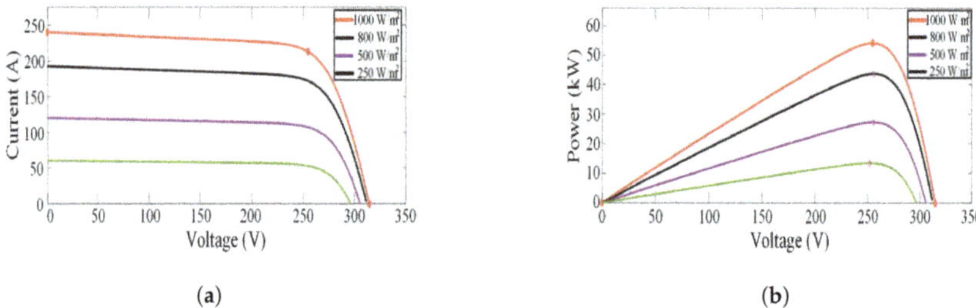

(**a**)

(**b**)

Figure 7. Characteristic curves of PV cell at 25 °C and varying irradiance (**a**) I-V (**b**) P-V.

The characteristic curve at varying temperature and specified irradiance (1000 W/m²) is depicted in Figure 8 where Figure 8a illustrates the I-V characteristic and Figure 8b shows the P-V characteristics of solar array.

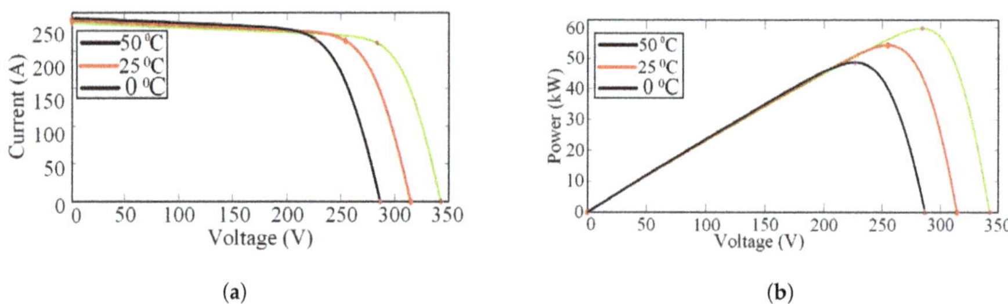

(**a**)

(**b**)

Figure 8. Characteristic curves of PV cell at varying temperature and specified irradiance (1000 W/m²) (**a**) I-V (**b**) P-V.

The efficiency of a PV module can be maximized using maximum power pint (MPPT)-based algorithms. In this paper, a Perturb and observe (P & O)-based algorithm is utilized due to the simplicity of the method and its accuracy of tracking. The P & O algorithm tracks the MPPT by increasing or decreasing the voltage repeatedly at the MPPT of the PV system. The inputs to the MPPT algorithm are the output current (I_{pv}) and voltage (V_{pv}) of PV module. The flow chart of P & O algorithm is illustrated in Figure 9. In order to implement this methodology, the voltage and current of PV modules must be initially measured [53]. The output of the P & O algorithm is the estimated MPPT voltage (V_{mp}), which is utilized to control DC–DC converter in order to attain continuously PV source open-circuit voltage.

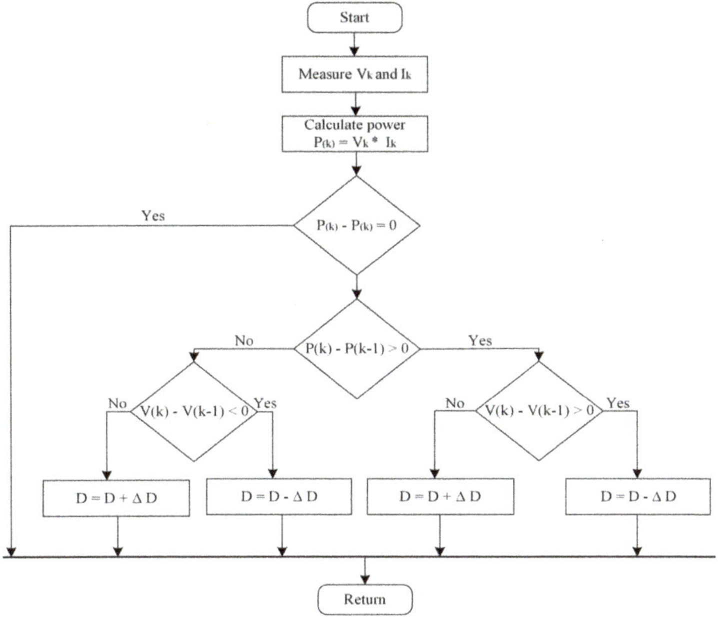

Figure 9. Flow chart of P & O algorithm.

3.3. Battery Modeling

The battery models are categorized mainly as an electric circuit model and electrochemical model, the rest of the models are usually derived from these; for instance: a dynamic battery model that is derived from a combination of an electrochemical and an electrical model. Some researchers further categorized this as an analytical and mathematical model [54]. The former one is based on the empirical data such as Peukert's equation, which involves integrating current to improve the model of a battery. On the other hand, the latter is a simplified electrochemical model like, for instance, Stepherd's equation. The reason for utilizing a mathematical model in this paper is the high efficiency and cost-effective tools such as MATLAB/SIMULINK [55].

Physics-based, empirical, and semi-empirical models have been proposed in the literature to diagnose the aging effect and its impact on the life time of a battery [56]. This paper did not take into consideration the temperature effect, aging effect, and degradation of a Li-ion battery. The calendar aging is an important factor especially in transportation applications and is considered for future work. The reason for utilizing a Li-ion battery in this paper is because of its low maintenance cost, lower self-discharge, high energy and power density in comparison with other batteries. The charge and discharge model of a Li-ion battery is illustrated in the Equations (8) and (9) respectively.

3.3.1. Charge Model (i$_l$) < 0

For Li-ion battery the charging model is shown in following equation:

$$f_1(i_t, i_l, i) = E_o - K\frac{Q}{0.1Q + i_t}i_l - K\frac{Q}{Q - i_t}i_t + Aexp(-Bi_t) \tag{8}$$

3.3.2. Discharge Model (i$_l$) > 0

For Li-ion battery the discharging model is shown in following equation:

$$f_2(i_t, i_l, i) = E_o - K\frac{Q}{Q - i_t}i_l - K\frac{Q}{Q - i_t}i_t + Aexp(-Bi_t) \tag{9}$$

The SOC and efficiency of the battery in terms of charge and discharge can be defined as:

$$SOC(t)_{ch/dis} = \begin{cases} SOC(t-1)_{ch} + \frac{1}{C_e} \int_{t-1}^{t} i_{bat,ch}(t)dt \\ SOC(t-1)_{dis} - \frac{1}{C_e} \int_{t-1}^{t} i_{bat,dis}(t)dt \end{cases}$$

$$E_{bat,ch/dis} = \begin{cases} E(t-1)_{ch} + (P_{bat,ch}(t) \cdot \eta_{ch})\Delta T \\ E(t-1)_{dis} + (\frac{P_{bat,dis}(t)}{\eta_{dis}})\Delta T \end{cases}$$

The discharge characteristics of a Li-ion-based battery is shown in Figure 10. This type of battery has a flat voltage curve in the usable discharge range. The discharge characteristics are divided into three parts: nominal region, exponential region, and discharge curve. In the exponential region, voltage drops in a very fast manner whereas the mid point voltage is the nominal voltage of the battery during charging and discharging and if this curve is flatter and shows that there will be less variation in voltage. The fully charged voltage is higher than nominal voltage whereas when it reaches end of life, the voltage will be less than mid point voltage.

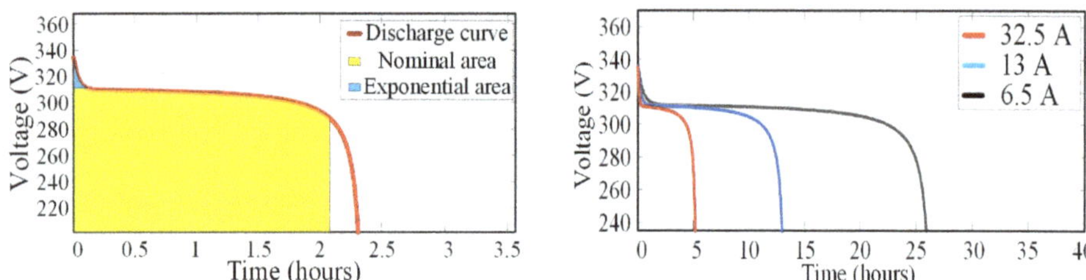

Figure 10. Nominal Current Discharge Characteristic of Li-ion battery.

3.4. Ultra-Capacitor Modeling

The configuration of the ultra-capacitor utilized in this study is depicted in Figure 11 which comprises an ultra-capacitor model and a DC–DC converter. This model is the combination of equivalent internal resistance (R_{IR}), equivalent parallel resistance (R_{PR}), and capacitance of an ultra-capacitor. The configuration acts as a boost converter in discharging mode whereas the buck converter in the charging mode. As the response time of the ultra-capacitor is very fast and has the capability to smooth the fluctuations produced by the highly intermittent nature of the PV modules, especially when it is integrated with the shipboard microgrids as compared to terrestrial microgrids.

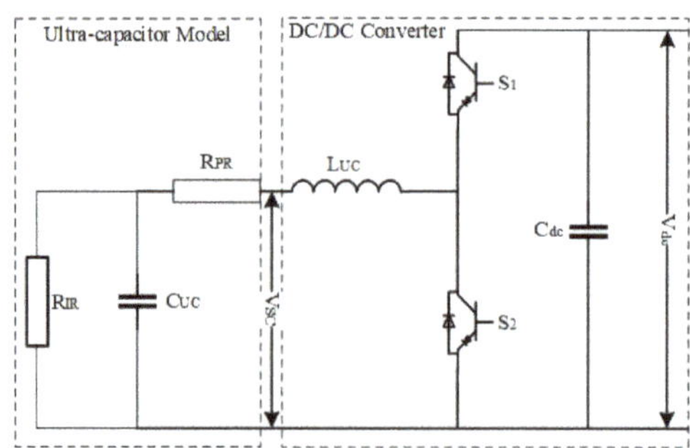

Figure 11. Configuration of the Ultra-capacitor.

The output voltage of an ultra-capacitor can be expressed in terms of Stern's equation as expressed in Equation (10).

$$V_{UC} = \frac{N_s d Q_T}{N_e N_p \varepsilon \varepsilon_o A_i} + \frac{2 N_e N_s R T}{F} sinh^{-1} \frac{Q_T}{N_p N_e^2 A_i \sqrt{8 R T \varepsilon \varepsilon_o c}} - R_{UC} i_{UC} \tag{10}$$

4. Control Strategy of Hybrid Shipboard Microgrids

4.1. Active Front End Converter

A rectifier is usually a six-pulse-type as it consists of six diodes and is found to be a cost-effective approach to convert Alternating current (AC) power into Direct Current (DC). This approach produces harmonics into the shipboard power system, which results in an additional heat and losses. One of the possible approaches is to use multiple-rectifiers like 12-pulse, 18-pulse, etc. to minimize the harmonics but this approach increases the cost due to an increase in the number of diodes. Another approach is to add passive filters that inject low impedance into the power system to absorb harmonic frequencies, but the disadvantage of this approach is that a passive filter utilizes more energy and takes a large amount of space. An active front end converter is utilized in this study and it is an effective approach that minimizes the total harmonic distortion, flexible regulation of dc-link voltage, provides unity power factor, and relatively sinusoidal input currents [57]. The block diagram of the active front with the control mechanism is illustrated in Figure 12.

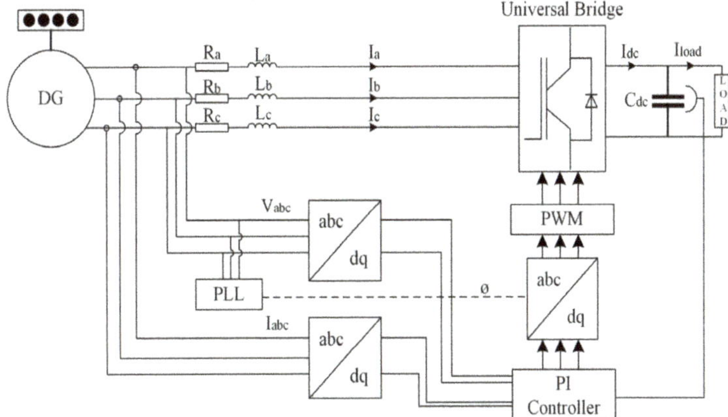

Figure 12. Control Strategy of Active Front End Converter.

In this technique, instead of utilizing diodes to convert AC into DC, insulated bipolar transistors (IGBTs) are used. The total harmonics in this approach are reduced to 3%, which in the conventional diode rectifier approach are around 25–30%. The schematic of the implemented control strategy of three-phase active front end converter is implemented in MATLAB/Simulink, and is depicted in Figure 13. At first, the dc-link voltage (V_{dc}) is compared with the reference value (V_{dc_ref}) and then the error is fed to the outer loop of PI controller. The output of the controller is I_{d_ref} which corresponds to the active power of the converter. The input reactive power of the active front end converter is adjusted by setting the current reference of q-axis (I_q) to 0 to achieve power factor equals to 1.

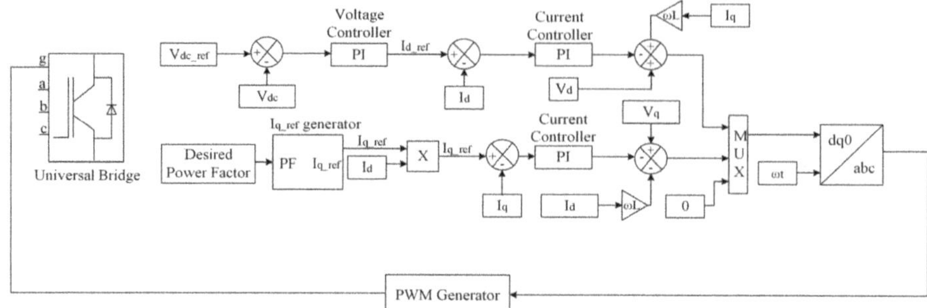

Figure 13. Control Strategy of Active Front End Converter.

4.2. Inverter Control

The control strategy applied for an inverter consists of two control loops, the outer loop which is pledged to control dc-link voltage whereas the inner loop is pledged to control grid current as illustrated in Figure 14. The current loops further solves power quality issues such as improved power factors and low THD. On the other hand, the voltage control loop is helpful in balancing the flow of power in the system. D-q control or synchronous reference frame-based control uses a reference frame transformation abc to dq (current park transformation and voltage park transformation), which transforms the voltage and current into a d-q frame. The transformed current is helpful in controlling the grid current whereas voltage detects frequency and phase of the grid. Therefore, the control variables are converted to dc values, hence, controlling and filtering become easier. (V_{d_ref}) and (V_{q_ref}) are the voltages in dq frame whereas to generate the gate signal for voltage source inverter, pulse width modulation (PWM) is utilized.

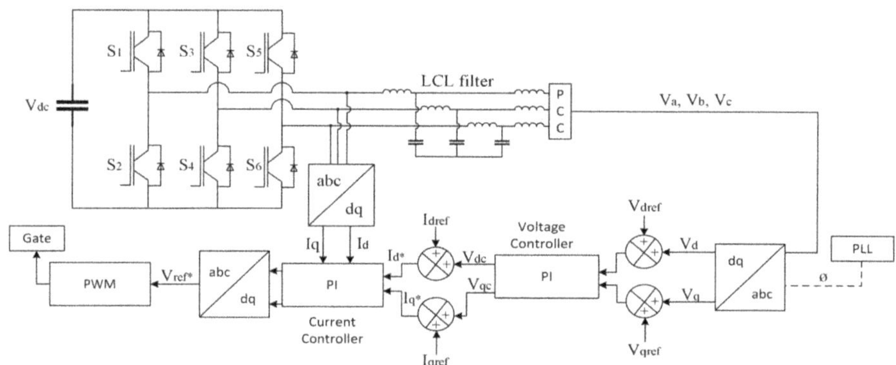

Figure 14. Schematic diagram of three phase inverter with PI controller.

4.3. Frequency Sharing Control

The proposed methodology is based on the frequency sharing approach in which load is shared between low and high-frequency components. The high-frequency components are catered by ultra-capacitor whereas low-frequency components are handled by the Li-ion battery. The block diagram of such an approach is illustrated in Figure 15.

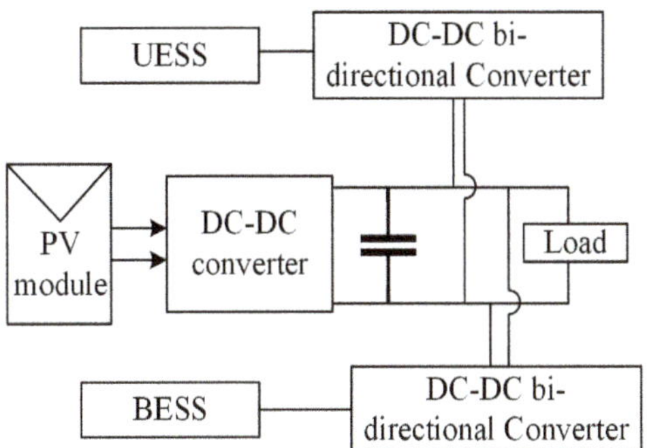

Figure 15. Block diagram of hybrid system (PV/battery/Ultra-capacitor).

The proposed energy management strategy relies on the frequency sharing approach. In this methodology, the load current is shared between high and low frequency components by the use of a low pass filter as expressed below:

$$\begin{cases} I_l = I_{scref} + I_{battref} \\ I_{scref} = HF \\ I_{battref} = LF \end{cases}$$

The Simulink model of the proposed approach is illustrated in Figure 16. The dc-link voltage of a PV module is compared with the reference dc voltage and is then fed to an outer loop voltage controller. The lower frequency components are then fed into an inner current loop and the signal is then fed to PWN which sends the gate signal to the dc-dc converter of a battery. On the other hand, ultra-capacitors are known by their higher power density and faster response. Therefore, higher frequency fluctuations will be catered by the ultra-capacitor. The inner current loop will send a signal by PWM to the gate of ultra-capacitor. The control strategy for such an approach can be expressed by Equations (11)–(13).

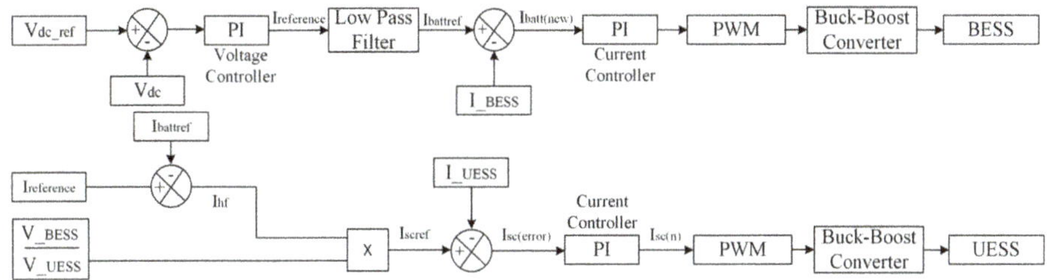

Figure 16. Schematic diagram of control strategy of a battery and ultra-capacitor energy storage system (UCESS).

$$I_{lf} = \frac{1}{1 + \tau s} \tag{11}$$

$$I_{reference} = (K_p + \frac{K_i}{s})(V_{dcref} - V_{dc}) \tag{12}$$

$$I_{batt(new)} = (\frac{1}{1 + \tau s})I_{reference} - I_{BESS}$$

$$I_{hf} = 1 - (\frac{1}{1 + \tau s}) \tag{13}$$

$$I_{uc(n)} = (I_{hf} \cdot \frac{V_{BESS}}{V_{UCESS}} - I_{UESS})(K_p + \frac{K_i}{s})$$

5. Simulation and Verification

The hybrid PV/Diesel/Battery/Ultra-capacitor model illustrated in Figure 3 is selected to test the proposed control approach. The approach adopted in this paper is categorized into two cases to show the effectiveness of the proposed approach. In the first case, due to the intermittent nature of PV systems, ultra-capacitor and Li-ion battery are hybridized to cater for the fluctuations. On the other hand, in the second case, BESS is used in parallel with the dc-link capacitor of VFD to mitigate the fluctuations caused by variation in sea conditions or due to weather.

5.1. Case 1

In the first case, as discussed before, the PV systems show more fluctuations in scenarios of integration with a shipboard microgrid due to the movement of ships and other factors such as moisture, weather, etc. The Li-ion battery and ultra-capacitor are hybridized and are integrated with the output of PV modules using dc-dc converter in parallel to the dc-link capacitor. In this study, at the start irradiation is set to be maximum and is varied at 0.8 s to check the effectiveness of the proposed approach. Figure 17a illustrates the variation in power, Figure 17b refers to the current, and Figure 17c shows the voltage with the variation in irradiance.

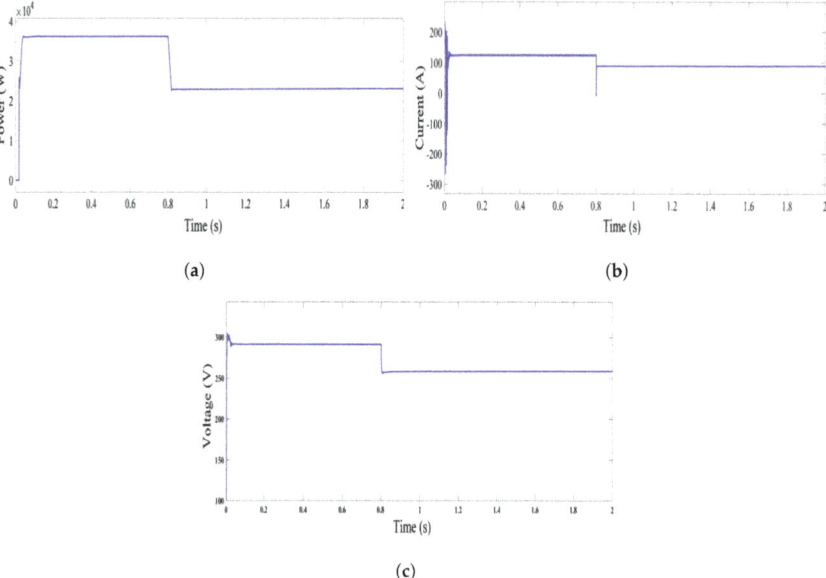

Figure 17. Simulation results of PV array (**a**) Power of solar module, (**b**) Current of solar module (**c**) Voltage of solar module.

A Li-ion-based battery is utilized in this study to cater for the low-frequency fluctuations. It can be inferred from Figure 18a–c, that at 0.8 s, there is a decrease in voltage, current, and power that show the variation in irradiance. Hence, the battery is providing the power to maintain the voltage at DC-link. It can be further observed from Figure 18d that the battery is getting charged at the start as the irradiance of the solar panel at that instant is considered to be maximum. At 0.8 s, due to variation in irradiance, the battery starts to support due to which the battery drains and ultimately the SOC of the battery decreases.

The ultra-capacitor is hybridized with the battery to cater for the higher frequency fluctuations. It can be inferred from Figure 19a–c, that at 0.8 s there is sudden decrease in voltage, current, and power that illustrates the variation in irradiance. In the start, SOC of the ultra-capacitor is constant as its fully charged. At 0.8 s due to sudden variation in irradiance, fluctuation is observed. The ultra-capacitor is responsible to cater for the high-frequency fluctuation. It can be seen from Figure 19d that SOC of the ultra-capacitor suddenly decreases and remain constant for the rest of the period as the battery is responsible to cater for rest of the duration.

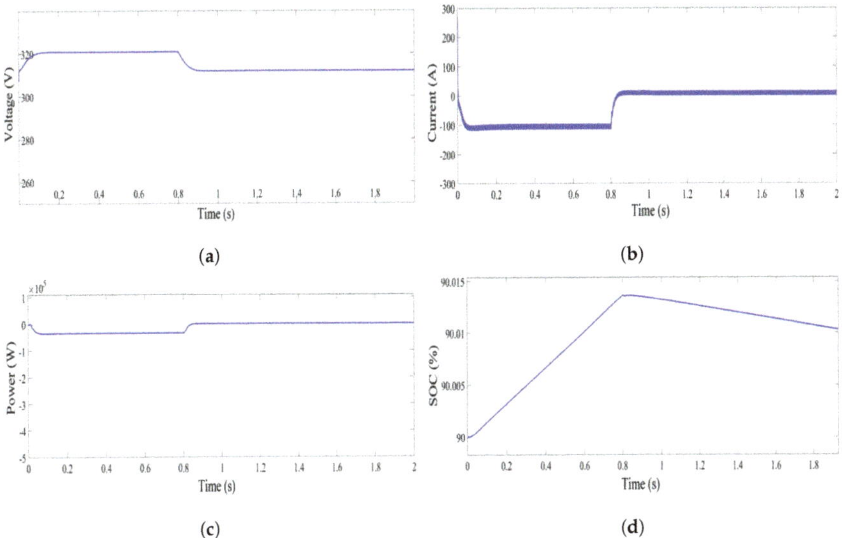

Figure 18. Simulation results of Li-ion battery (**a**) Voltage of Li-ion Battery, (**b**) Current of Li-ion Battery (**c**) Power of Li-ion Battery (**d**) State of charge (SOC) of Li-ion Battery.

Figure 20 illustrates the dc-link voltage, the variation in the generation of power from PV panel at 0.8 s is observed depicting the change in irradiance, which is catered by the hybrid energy storage system.

Figure 21 shows the comparison of stabilizing the dc-link voltage with varying irradiance by using proposed approach in comparison to earlier approach to stabilize dc-link voltage by using BESS only.

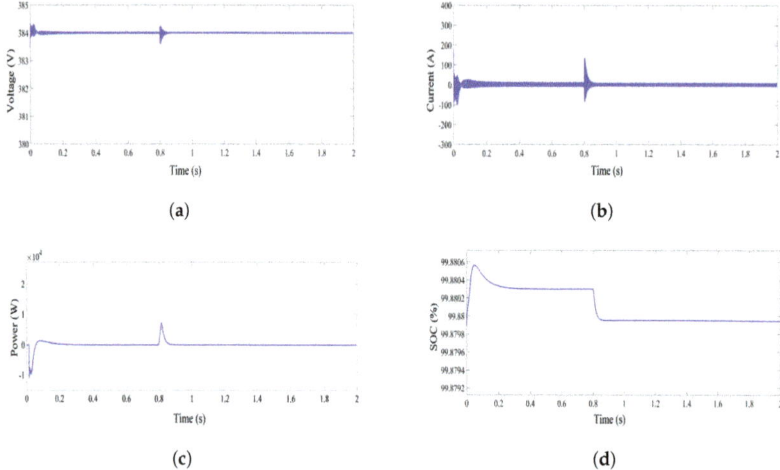

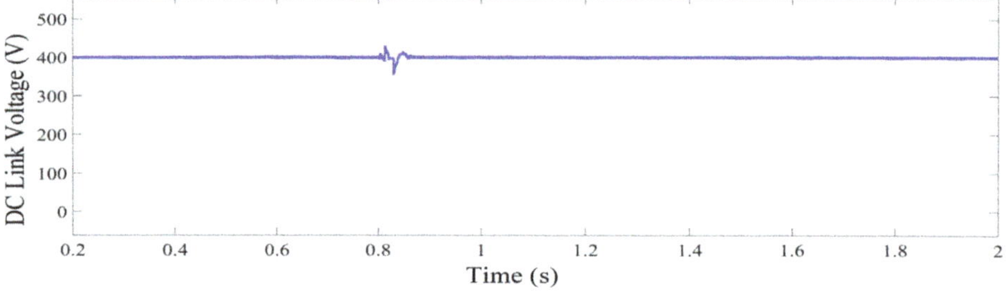

Figure 19. Simulation results of Ultra-capacitor (**a**) Voltage of Ultra-capacitor energy storage system, (**b**) Current of Ultra-capacitor energy storage system (**c**) Power of Ultra-capacitor energy storage system (**d**) SOC of Ultra-capacitor energy storage system.

Figure 20. DC link voltage with BESS and UC.

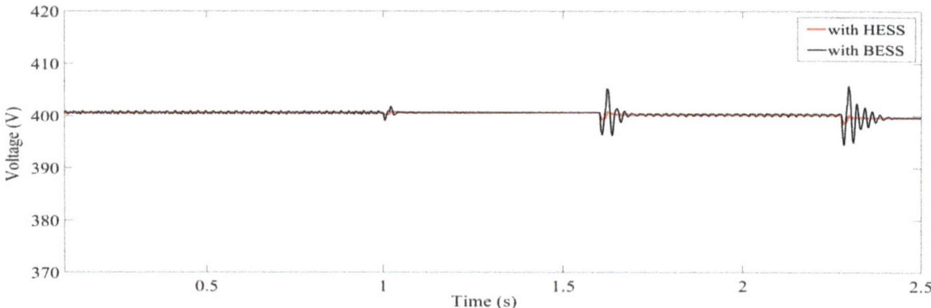

Figure 21. Comparison of using HESS for smoothing dc-link voltage with using only BESS.

5.2. Case 2

In the second case, to cater for the fluctuations caused by the propulsion motors due to variation in sea states which ultimately causes variation in power requirement, the battery is integrated in parallel with the dc-link capacitor using a buck-boost converter. It can be inferred from Figure 21a, that at 1.2 s the fluctuation is observed and it is catered by the BESS. On the other hand, Figure 22b illustrates that the SOC of BESS starts to decrease at 1.2 s.

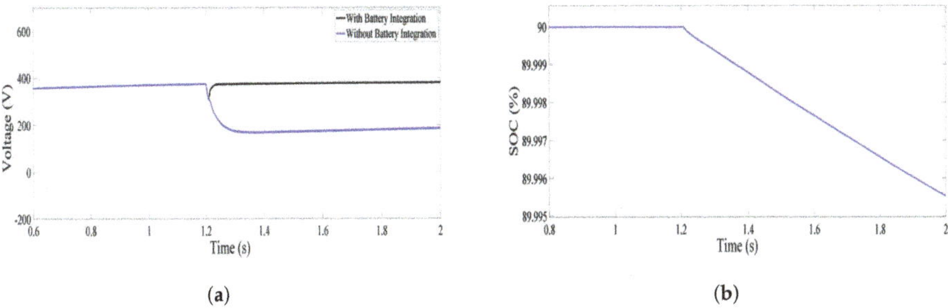

Figure 22. Simulation results with and without integration of battery (**a**) DC link voltage with and without BESS, (**b**) SOC of Battery.

Due to the impact of variation in power demand, there is variation in frequency observed and is depicted in Figure 23, it is observed that at 1.2 s the frequency starts to decrease due to the variation in load and is catered by BESS.

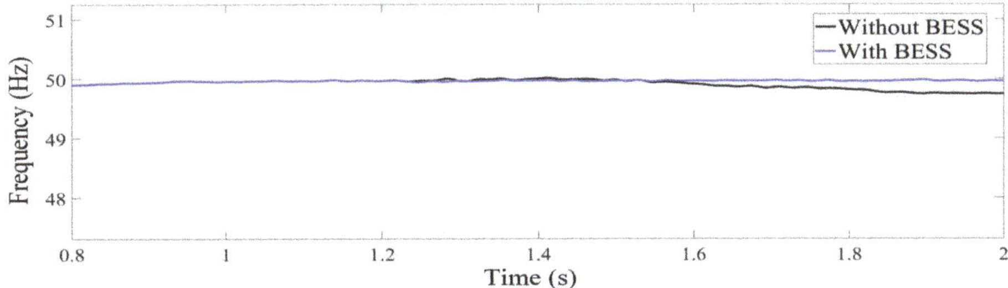

Figure 23. Frequency in case of variation with and without BESS.

The comparison between different approaches in the literature are presented in Table 5. Although in literature few researchers focused on frequency sharing approach for shipboard microrgids, some of them shared the frequency components between BESS and DG that ultimately adds burden on the DG. Few researches utilize sole energy storage such as flywheel or ultra-capacitor to minimize the fluctuations caused by PV systems, but these ESS devices can smooth or mitigate short term fluctuations only. Other techniques utilized fuzzy-based control and optimization algorithms for sizing several ESS, stabilizing dc-link voltage and reducing emissions, etc. The proposed approach in this study utilizes PI-based control for frequency sharing between BESS and an ultra-capacitor. The PI-based methodology employed in this work has less complexity and is often employed for industrial applications.

Table 5. Comparison between different input sources and design scenarios.

Author	Power Generation Sources	ESS	Technique	Scope	Reference
H. Lan et al.	DG + PV	FESS	PI Control	To smooth fluctuations	[9]
H. Lan et al.	DG + PV	BESS	MOPSO	Minimize CO_2 emissions	[10]
S. Wen et al.	DG + PV	BESS + Ultra-capacitor	PSO	Optimal sizing of several ESS	[12]
K. Bellache	DG	BESS	Polynomial control	Load sharing between DG & BESS	[13]
J. Tang	PV	BESS + Ultra-capacitor	Frequency hierarchical control	Smooth the fluctuations of PV	[16]
M. H. Khooban et al.	DG + PV	BESS + FESS	Fuzzy PD + I Control	Enhance performance of LFC	[17]
Y. Qiu et al.	DG + PV	Ultra-capacitor	PI control	Improve LVRT capability & smooth power	[18]
M. R. Banaei	DG + PV	FC + BESS	MPC	Stabilize dc-link voltage	[58]
R. Tang	DG + PV	BESS	MPC	Power flow scheduling	[59]
A. Dolatabadi	DG + PV	BESS	Non-linear programming	Minimize fuel, emissions and maintenance cost of DG	[60]
This study	DG + PV	BESS + Ultra-capacitor	PI Control	Frequency components sharing between BESS & UC	–

6. Conclusions

In this paper, a hybrid PV/Diesel/Battery/Ultra-capacitor is modeled and a control strategy based on a PI controller is developed. The PV panels behave differently when they are integrated with shipboard microgrids due to the movement and rolling of a ship. Therefore, there is an utmost need to integrate an ultra-capacitor to shipboard power systems to smooth the fluctuations that are caused by the highly intermittent nature of solar panels. In this study, the load current is shared on the basis of different frequency components to smooth the power. The simulation results show the effectiveness of the proposed hybrid model as they help to minimize the fluctuation. It can be inferred that an ultra-capacitor-based energy storage system is responsible to cater for the high-frequency fluctuations due to the faster response, whereas a battery is responsible to mitigate the low-frequency components because of its slow response. Moreover, battery life-time can be increased by handling short-term fluctuations through the ultra-capacitor. Further, in the second case, the integration of a battery in parallel with the dc-link capacitor helps to mitigate the fluctuations caused by either variation in sea conditions or due to weather. Future work will be based on a case study of a yacht by considering the impact of partial shading effect and to validate the proposed approach; it will be implemented on a hardware in loop setup.

Author Contributions: M.U.M. has written the original draft and performed the simulations; Y.T. contributed in terms of methodology and validation; , K.A.K.N. and F.K. were involved in Writing, reviewing & editing, J.C.V. and J.M.G. supervised this study and corrected the manuscript.

Funding: This research received no external funding.

Conflicts of Interest: The authors declare no conflict of interest.

Abbreviations

The following abbreviations are used in this manuscript:

PV	Photovoltaic
IMO	International marine organization
ESS	Energy storage system
HESS	Hybrid energy storage system
IPS	Integrated power system
VFD	Variable frequency drives
PI	Proportional Integral
BESS	Battery energy storage system
FESS	Flywheel energy storage system
UC	Ultra-capacitor
DG	Diesel generator
MOPSO	Multi-Objective Particle Swarm Optimization
PSO	Particle swarm optimization
MPC	Model predictive control
LFC	Load frequency control
I_{pv}	Solar generated current at the nominal condition (25 °C and 1000 W/m^2)
I_r	Solar radiation (W/m^2)
I_{sc}	Short circuit current (A)
K_i	Short circuit current of the cell
I_{Rs}	Reverse saturation current of PV module
q	charge of an electron
V_{oc}	Open circuit voltage (V)
N_s	Number of cells in series
n	The ideality factor of diode
k	Boltzmann's constant

I_o Saturation current of a PV module
T_r Nominal temperature
E_{go} Band gap energy of a semiconductor
V^t Thermal voltage of diode (V)
R_{sh} Shunt resistance (Ω)
N_s Number of modules connected in series
N_p Number of modules connected in parallel
I output current of PV module (A)
i_t Extracted capacity (Ah)
i battery current (A)
i_l Low frequency current dynamics (A)
E_o Constant voltage (V)
K Polarization constant (V/Ah)
Q Maximum battery capacity (Ah)
A Exponential voltage (V)
B Exponential capacity (Ah^{-1})
C_o Capacity of battery
η Efficiency of converter
A_i Interfacial area between electrolyte and electrode
Q_T Electric charge
ε Permittivity of material
ε_o Permittivity of free space
D Molecular radius
R Ideal gas constant
T Operating temperature
F Faraday constant
c molar concentration
R_{UC} Total resistance of UC
i_{UC} UC current

References

1. McCoy, T.J. Electric ships past, present, and future technology leaders. *IEEE Electrif. Mag.* **2015**, *3*, 4–11. [CrossRef]
2. Zahedi, B.; Norum, L.E.; Ludvigsen, K.B. Optimized efficiency of all-electric ships by dc hybrid power systems. *J. Power Sour.* **2014**, *255*, 341–354. [CrossRef]
3. Doerry, N. Naval Power Systems: Integrated power systems for the continuity of the electrical power supply. *IEEE Electrif. Mag.* **2015**, *3*, 12–21. [CrossRef]
4. Doerry, N.; Amy, J.; Krolick, C. History and the status of electric ship propulsion, integrated power systems, and future trends in the US Navy. *Proc. IEEE* **2015**, *103*, 2243–2251. [CrossRef]
5. The International Convention for the Prevention of Pollution from Ships. Available Online: http://www.imo.org/en/About/Conventions/ListOfConventions/Pages/International-Convention-for-the-Prevention-of-Pollution-from-Ships-%28MARPOL%29.aspx (accessed on 5 June 2019).
6. Rogelj, J.; Den Elzen, M.; Höhne, N.; Fransen, T.; Fekete, H.; Winkler, H.; Schaeffer, R.; Sha, F.; Riahi, K.; Meinshausen, M. Paris Agreement climate proposals need a boost to keep warming well below 2 °C. *Nature* **2016**, *534*, 631. [CrossRef]
7. EU Commission: The Paris Protocol—A Blueprint for Tackling Global Climate Change Beyond 2020. Available online: https://www.eesc.europa.eu/sites/default/files/resources/docs/15_362-ppaper_changement-clim_en.pdf (accessed on 27 August 2018).
8. International Maritime Organization: Prevention of Air Pollution from Ships. Available online: http://www.imo.org/en/OurWork/Environment/PollutionPrevention/AirPollution/Pages/Air-Pollution.aspx (accessed on 5 June 2019).
9. Lan, H.; Bai, Y.; Wen, S.; Yu, D.; Hong, Y.Y.; Dai, J.; Cheng, P. Modeling and stability analysis of hybrid pv/diesel/ess in ship power system. *Inventions* **2016**, *1*, 5. [CrossRef]

10. Lan, H.; Wen, S.; Hong, Y.Y.; David, C.Y.; Zhang, L. Optimal sizing of hybrid PV/diesel/battery in ship power system. *Appl. Energy* **2015**, *158*, 26–34. [CrossRef]

11. Lee, K. J.; Shin, D.; Yoo, D.W.; Choi, H.K.; Kim, H. J. Hybrid photovoltaic/diesel green ship operating in standalone and grid-connected mode–Experimental investigation. *Energy* **2013**, *49*, 475–483. [CrossRef]

12. Wen, S.; Lan, H.; Yu, D.C.; Fu, Q.; Hong, Y.Y.; Yu, L.; Yang, R. Optimal sizing of hybrid energy storage sub-systems in PV/diesel ship power system using frequency analysis. *Energy* **2017**, *140*, 198–208. [CrossRef]

13. Bellache, K.; Camara, M.B.; Dakyo, B. Hybrid Electric Boat based on variable speed Diesel Generator and lithium-battery-using frequency approach for energy management. In Proceedings of the 2015 International Aegean Conference on Electrical Machines & Power Electronics (ACEMP), 2015 International Conference on Optimization of Electrical & Electronic Equipment (OPTIM) & 2015 International Symposium on Advanced Electromechanical Motion Systems (ELECTROMOTION), Side, Turkey, 2–4 September 2015; pp. 744–749.

14. Liu, H.; Zhang, Q.; Qi, X.; Han, Y.; Lu, F. Estimation of PV output power in moving and rocking hybrid energy marine ships. *Appl. Energy* **2017**, *204*, 362–372. [CrossRef]

15. Khan, M.M.S.; Faruque, M.O. Management of hybrid energy storage systems for MVDC power system of all electric ship. In Proceedings of the 2016 North American Power Symposium (NAPS), Denver, CO, USA, 18–20 September 2016; pp. 1–6.

16. Tang, J.; Xiong, B.; Huang, Y.; Yuan, C.; Su, G. Optimal configuration of energy storage system based on frequency hierarchical control in ship power system with solar photovoltaic plant. *J. Eng.* **2017**, *13*, 1511–1514. [CrossRef]

17. Khooban, M.H.; Dragicevic, T.; Blaabjerg, F.; Delimar, M. Shipboard microgrids: A novel approach to load frequency control. *IEEE Trans. Sustain. Energy* **2017**, *9*, 843–852. [CrossRef]

18. Qiu, Y.; Yuan, C.; Tang, J. Integrating SCESS into a Ship-PV Power System to Mitigate Power Fluctuations and Improve LVRT Capability. *Arab. J. Sci. Eng.* **2019**, *44*, 1–13. [CrossRef]

19. Khan, M.M.S.; Faruque, M.O.; Newaz, A. Fuzzy logic based energy storage management system for MVDC power system of all electric ship. *IEEE Trans. Energy Convers.* **2017**, *32*, 798–809. [CrossRef]

20. Solar-Powered Super Yacht (E-Boat). Available online: http://www.aideenergy.com/index.php?option=module&lang=en&task=showlist&id=74 (accessed on 5 June 2019).

21. Greeline 33 Yacht. Available Online: https://inoffice.app.box.com/v/greenlineyachtsbrochure (accessed on 5 June 2019).

22. Grrenline 39. Available online: https://www.greenlinehybrid.si/yacht/greenline-39/ (accessed on 5 June 2019).

23. Greenline 40. Available online: https://www.greenlinehybrid.si/yacht/greenline-40/ (accessed on 5 June 2019).

24. Duffy London. Available online: https://duffylondon.com/product/archivenew/solaris-global-cruiser-2/ (accessed on 5 June 2019).

25. Silent 80 Yacht. Availble online: https://www.silent-yachts.com/silent80/ (accessed on 5 June 2019).

26. Silent 64 Yacht. Available online: https://www.silent-yachts.com/silent64/ (accessed on 5 June 2019).

27. Silent 55 Ferry. Available online: https://www.silent-yachts.com/silent55-ferry/ (accessed on 5 June 2019).

28. PlanetSolar, Solar-Powered Ship. Available online: https://www.ship-technology.com/projects/planetsolar/ (accessed on 5 June 2019).

29. Aquarius Eco Ship: Low Emission Green Ship Concept Design with Rigid Sails & Solar Power. Available online: https://www.ecomarinepower.com/en/aquarius-eco-ship (accessed on 5 June 2019).

30. M/V Auriga Leader. Available online: https://www.marineinsight.com/types-of-ships/auriga-leader-the-worlds-first-partially-propelled-cargo-ship/ (accessed on 5 June 2019).

31. World's First Solar-Power-Assisted Vessel Further Developed-Car Carrier Auriga Leader. Available online: https://www.nyk.com/english/news/2011/NE_110525.html (accessed on 5 June 2019).

32. Dufo-López, R.; Lujano-Rojas, J.M.; Bernal-Agustín, J.L. Comparison of different lead-acid battery lifetime prediction models for use in simulation of stand-alone photovoltaic systems. *Appl. Energy* **2014**, *115*, 242–253. [CrossRef]

33. Křivík, P.; Vaculík, S.; Bača, P.; Kazelle, J. Determination of state of charge of lead-acid battery by EIS. *J. Energy Storage* **2019**, *21*, 581–585. [CrossRef]

34. Mutarraf, M.; Terriche, Y.; Niazi, K.; Vasquez, J.; Guerrero, J. Energy Storage Systems for Shipboard Microgrids—A Review. *Energies* **2018**, *11*, 3492. [CrossRef]

35. Chen, H.; Cong, T.N.; Yang, W.; Tan, C.; Li, Y.; Ding, Y. Progress in electrical energy storage system: A critical review. *Prog. Nat. Sci.* **2009**, *19*, 291–312. [CrossRef]

36. Argyrou, M.C.; Christodoulides, P.; Kalogirou, S.A. Energy storage for electricity generation and related processes: Technologies appraisal and grid scale applications. *Renew. Sustain. Energy Rev.* **2018**, *94*, 804–821. [CrossRef]

37. Zubi, G.; Dufo-López, R.; Carvalho, M.; Pasaoglu, G. The lithium-ion battery: State of the art and future perspectives. *Renew. Sustain. Energy Rev.* **2018**, *89*, 292–308. [CrossRef]

38. Hoffart, F. Proper care extends Li-ion battery life. *Power Electron.* **2008**, 24–28.

39. Díaz-González, F.; Sumper, A.; Gomis-Bellmunt, O.; Villafáfila-Robles, R. A review of energy storage technologies for wind power applications. *Renew. Sustain. Energy Rev.* **2012**, *16*, 2154–2171. [CrossRef]

40. Zhao, H.; Wu, Q.; Hu, S.; Xu, H.; Rasmussen, C.N. Review of energy storage system for wind power integration support. *Appl. Energy* **2015**, *137*, 545–553. [CrossRef]

41. Chatzivasileiadi, A.; Ampatzi, E.; Knight, I. Characteristics of electrical energy storage technologies and their applications in buildings. *Renew. Sustain. Energy Rev.* **2013**, *25*, 814–830. [CrossRef]

42. Hemmati, R.; Saboori, H. Emergence of hybrid energy storage systems in renewable energy and transport applications—A review. *Renew. Sustain. Energy Rev.* **2016**, *65*, 11–23. [CrossRef]

43. Jing, W.; Lai, C.H.; Wong, W.S.; Wong, M.D. Dynamic power allocation of battery-supercapacitor hybrid energy storage for standalone PV microgrid applications. *Sustain. Energy Technol. Assess.* **2017**, *22*, 55–64. [CrossRef]

44. Zhou, H.; Bhattacharya, T.; Tran, D.; Siew, T.S.T.; Khambadkone, A.M. Composite energy storage system involving battery and ultracapacitor with dynamic energy management in microgrid applications. *IEEE Trans. Power Electron.* **2010**, *26*, 923–930. [CrossRef]

45. Cooper, A.; Furakawa, J.; Lam, L.; Kellaway, M. The UltraBattery—A new battery design for a new beginning in hybrid electric vehicle energy storage. *J. Power Sour.* **2009**, *188*, 642–649. [CrossRef]

46. Waaree Energies WU-120. Available online: https://db.photovoltaikforum.com/PvForum/api/download/1/65646/1/pdf/65644_WS_WU_120_160_EN_05_12 (accessed on 20 August 2019).

47. Datasheet PC 2500 Ultra-Capacitor. Available online: http://datasheet.iiic.cc/datasheets-1/maxwell_technologies/PC2500.pdf (accessed on 5 June 2019).

48. Panasonic CGR18650AF. Available online: http://www.houseofbatteries.com/documents/CGR18650AF.pdf (accessed on 20 August 2019).

49. Benhamed, S.; Ibrahim, H.; Belmokhtar, K.; Hosni, H.; Ilinca, A.; Rousse, D.; Ramdenee, D. Dynamic modeling of diesel generator based on electrical and mechanical aspects. In Proceedings of the 2016 IEEE Electrical Power and Energy Conference (EPEC), Ottawa, ON, Canada, 12–14 October 2016; pp. 1–6.

50. Hansen, J.F.; Adnanes, A.K.; Fossen, T.I. Mathematical modelling of diesel-electric propulsion systems for marine vessels. *Math. Comput. Model. Dyn. Syst.* **2001**, *7*, 323–355. [CrossRef]

51. Cooper, A.R.; Morrow, D.J.; Chambers, K.D.R. Development of a diesel generating set model for large voltage and frequency transients. In Proceedings of the IEEE PES General Meeting, Providence, RI, USA, 25–29 July 2010; pp. 1–7.

52. Patel, H.; Agarwal, V. MATLAB-based modeling to study the effects of partial shading on PV array characteristics. *IEEE Trans. Energy Convers.* **2008**, *23*, 302–310. [CrossRef]

53. Noman, A.M.; Addoweesh, K.E.; Mashaly, H.M. DSPACE real-time implementation of MPPT-based FLC method. *Int. J. Photoenergy*, **2013**. [CrossRef]

54. Arianto, S.; Yunaningsih, R.Y.; Astuti, E.T.; Hafiz, S. Development of single cell lithium ion battery model using Scilab/Xcos. *AIP Conf. Proc.* **2016**, *1711*, 060007.

55. Gomadam, P.M.; Weidner, J.W.; Dougal, R.A.; White, R.E. Mathematical modeling of lithium-ion and nickel battery systems. *J. Power Sour.* **2002**, *110*, 267–284. [CrossRef]

56. Abada, S.; Marlair, G.; Lecocq, A.; Petit, M.; Sauvant-Moynot, V.; Huet, F. Safety focused modeling of lithium-ion batteries: A review. *J. Power Sour.* **2016**, *306*, 178–192. [CrossRef]

57. Song, Z.; Tian, Y.; Chen, W.; Zou, Z.; Chen, Z. Predictive duty cycle control of three-phase active-front-end rectifiers. *IEEE Trans. Power Electron.* **2015**, *31*, 698–710. [CrossRef]

58. Banaei, M.R.; Alizadeh, R. Simulation-based modeling and power management of all-electric ships based on renewable energy generation using model predictive control strategy. *IEEE Intell. Transp. Syst. Mag.* **2016**, *8*, 90–103. [CrossRef]

59. Tang, R.; Wu, Z.; Li, X. Optimal operation of photovoltaic/battery/diesel/cold-ironing hybrid energy system for maritime application. *Energy* **2018**, *162*, 697–714. [CrossRef]

60. Dolatabadi, A.; Ebadi, R.; Mohammadi-Ivatloo, B. A two-stage stochastic programming model for the optimal sizing of hybrid PV/diesel/battery in hybrid electric ship system. *J. Operation Autom. Power Eng.* **2019**, *7*, 16–26.

 *energies*

Article

A Comprehensive Inverter-BESS Primary Control for AC Microgrids

Michele Fusero [1], Andrew Tuckey [2], Alessandro Rosini [3,*], Pietro Serra [1], Renato Procopio [3] and Andrea Bonfiglio [3]

[1] Grid Edge Solutions ABB S.p.A., 16145 Genova, Italy; michele.fusero@it.abb.com (M.F.); pietro.serra@it.abb.com (P.S.)
[2] Grid Edge Solutions ABB Australia Pty Ltd.; andrew.tuckey@au.abb.com
[3] Department of Electrical, Electronic, Telecommunication Engineering and Naval Architecture, University of Genoa, 16145 Genova, Italy; renato.procopio@unige.it (R.P.); a.bonfiglio@unige.it (A.B.)
*   Correspondence: alessandro.rosini@edu.unige.it

Received: 26 July 2019; Accepted: 1 October 2019; Published: 9 October 2019

Abstract: This paper proposes the design of a comprehensive inverter-BESS primary control capable of providing satisfactory performances both in grid-connected and islanded configurations as required by international standards and grid codes, such as IEEE Std. 1547. Such control guarantees smooth and fast dynamic behavior of the converter in islanded configuration as well as fast power control and voltage-frequency support in grid-connected mode. The performances of the proposed primary control are assessed by means of EMT (ElectroMagnetic Transients) simulations in the dedicated software DIgSILENT PowerFactory® (Germany, Gomaringen) . The simulation results show that the proposed BESS (Battery Energy Storage System) primary control is able to regulate frequency and voltage in Grid-Forming mode independently of the number of paralleled generators. This is achieved adopting a virtual generator technique which presents several advantages compared to the conventional one. Moreover, the proposed control can be switched to Grid-Support mode in order to provide fast control actions to allow frequency and voltage support as well as power control following the reference signals from the secondary level.

Keywords: primary control; BESS; microgrid; Grid Forming; Grid Support; Inverter Control; Grid Support; DIgSILENT PowerFactory; EMT simulations

---

1. Introduction

The need of pollutant and greenhouse gasses emission reduction and the promising increase of Renewable Energy Sources (RES) have made Microgrids (MG) one of the most interesting and challenging structure for industrial and academic researchers [1]. Several definitions have been provided for a MG but the most effective defines it as a group of interconnected loads and distributed energy resources with clearly defined electrical boundaries that acts as a single controllable entity with respect to the grid and can connect and disconnect from the grid to enable it to operate in both grid-connected or island modes [2]. It is easy to understand that one of the main topics in this field is the control system, and it is well known that it can be structured with a hierarchical structure [3]. Tertiary level control is an energy-production level that must be able to manage the power flow between the MG and the main grid [4]. Secondary level ensures that all the electrical variables into the MG are within the required values and it can include a synchronization control loop to seamlessly connect or disconnect the MG to or from the distribution system [5]. Primary control is typically a communication-less control layer; it is normally implemented in a decentralized manner in order to properly control voltage and frequency and/or active and reactive powers [6]. Finally, inner control loops are adopted to regulate the output voltage and to control the current while maintaining the system stable. Considering the normative aspect, IEEE Std. 1547 [7] clearly defines the inverter's primary control functionalities that can be essentially clustered in two operational configurations (i) Grid-Following Operation or Grid Support Mode (GSM) and (ii) Grid-Forming Operation. In GSM, the inverter is controlled as a current source in which the main control goals are to supply the load connect to the MG and to participate in frequency and voltage support [8,9]; for these reasons, GSM is the typical configuration in grid-connected state. Grid-Forming Operation is typically exploited in islanded mode where the inverter can be either the voltage and frequency master (stand-alone mode) or allow parallel operations with other Distributed Energy Resources (DERs). An obvious consequence of this high penetration of inverter-interfaced DERs is the reduction of total inertia and damping because most of the proposed control methods for Grid-Forming inverters, e.g., droop control methods [10], provides barely any inertia or damping support for the MG. For this reason, one of the most performing primary control technique for Grid-Forming inverters is the Virtual Synchronous Machine (VSM) or the Virtual Generator Mode (VGM) [11] where the control acts on the inverter in order to mimic the dynamical behavior of a traditional synchronous generator [12–15] virtually adding some inertia and frequency damping to the system and accordingly, improving MG stability and [16], since inertia response is the result of rotating heavy mass and it is proportional to the rotor speed, the VGM concept can also directly improve the frequency response [17]. In [14,18–20], the VGM primary control is developed using the complete model of the synchronous generator and this makes the algorithm complex and the controller tuning difficult. Simpler design models for the VGM control are proposed in [21–24], where only the inertial behavior of a synchronous generator is considered by imposing the swing equation in the primary controller.

In [25], a VGM control technique is proposed showing how it can theoretically provide all the required functionalities of Grid-Forming inverters (according to IEEE Std. 1547) and also presenting some practical applications of VGM control technique for Battery Energy Storage System (BESS) around the world.

Considering all these aspects, the aim of this paper is to present a new primary control for BESS able to guarantee good performances in grid-connected and islanded configurations providing:

- Regulation of frequency and voltage in Grid-Forming mode independently of the number of paralleled generators using the VGM technique in order to mimic the dynamic behavior of synchronous generators;
- Ability to guarantee black-start of the MG in Grid-Forming mode;
- Correct active and reactive power sharing in parallel with other DERs;
- Fast control actions in grid-connected mode to allow providing frequency and voltage support (GSM) as well as power control following the reference signals from the secondary level control;
- Synchronization and connection of the BESS to the external main grid or to other DERs in islanded mode with minimum transients;

The results will show that the new primary control proposed in this paper is also interesting not only from an academic point of view, but also from an industrial one for these aspects:

- The BESS converter is able to work both in VGM and in GSM guaranteeing the possibility to work in parallel with other DERs or an external main grid;
- The primary control can be switched from Grid-Forming mode to Grid-Support mode and vice versa without converter power interruption;
- Considering the Grid-Support mode, the proposed control is able to provide fast actions to the MG because this functionality is implemented in the primary level and not in the secondary one;
- When the support to the MG is not necessary, the control is able to use control signal coming from the secondary level control in order to satisfy other tasks reported in IEEE Std. 1547 such as State of Charge management, power smoothing, and compliance with power flow constraints imposed at the connection point with the external Main Grid (peak lopping);

- Considering the Grid-Forming operating mode, the proposed VGM technique is a PI-based one, which means that the tuning procedure can be easily manged by operators and not just by control engineers.

In summary, all these aspects and the level of detail in which they have been implemented and presented in the paper represent the main contribution of the work and a good starting point for actual implementation on industrial controller (which is the final aim of the ABB/University of Genoa's final goal).

The paper is organized as follows: in Section 2, a description of the BESS primary control is proposed in both the operating configurations. The test case MG implemented in the dedicated power system simulator DIgSILENT PowerFactory® is detailed in Section 3, while simulation results and comments are provided in Section 4. Some conclusive remarks are presented in Section 5.

2. Primary Control Method Description

The aim of this section is to provide an effective description of the proposed primary level BESS control: as stated before, it can provide a correct BESS converter regulation both in Grid-Forming and Grid-Support operating modes as it will be detailed in the following subsections. As one can see from Figure 1, the BESS converter has a R-L-C filter at its output (R_f, L_f and C_f) and the primary control needs some measurements from the field in order to guarantee optimal performances in the two operating modes. In detail, these measurements are active power P_{meas}^{BESS} and reactive power Q_{meas}^{BESS} at the output of the BESS converter, the RMS value of the controlled voltage V_{meas}, the controlled frequency f_{meas} and phase angle θ_{meas} coming from a Phase Locked Loop (PLL) control function synchronized at the connection bus to the MG. The output of the BESS primary control are the reference voltages v_{ref}^{α} and v_{ref}^{β} in the α-β stationary reference frame for the Grid-Forming operating mode and the d-q reference frame currents $i_{d,ref}^{GSM}$ and $i_{q,ref}^{GSM}$ for the Grid-Support operating mode. In order to generate the control signals for the BESS converter $v_{d,ref}^{inv}$ and $v_{q,ref}^{inv}$, voltage and current control loops are mandatory, more precisely when the Grid-Forming operating mode is required, voltage and current control loops are used in a cascade configuration, while in Grid-Support operating mode only the current control loop is required. In the next subsection, VGM and GSM techniques as well as voltage and current control loops are described in detail.

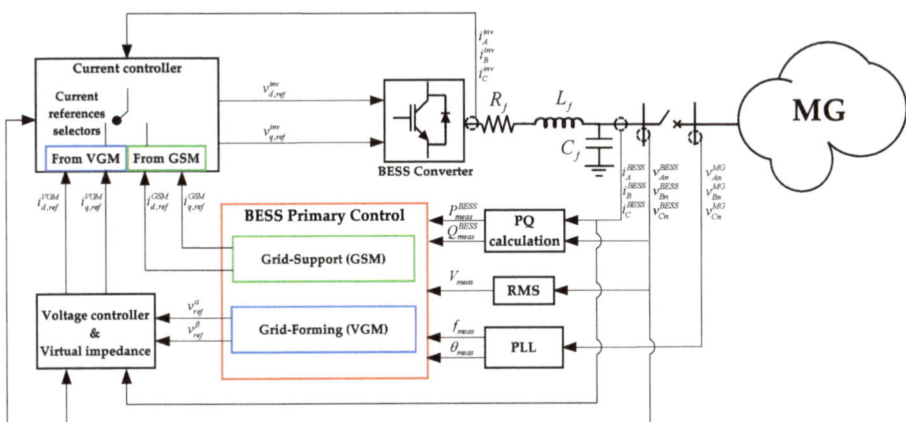

Figure 1. Proposed BESS (Battery Energy Storage System) primary control scheme.

2.1. Grid-Forming Operating Mode: VGM

Considering the Grid-Forming operating mode, the VGM technique presented in [25] is exploited and its control block diagram is reported in Figure 2 where AVR (Automatic Voltage Regulator) and Rotor Flux Model make the reactive power/voltage magnitude similar to that of a synchronous generator. More precisely, the Rotor Flux Model, modeled as an integrator with a gain K_ψ, responds to the reactive power Q_{meas}^{BESS} at the output of the inverter with an initial voltage variation of V_{ref}^{VGM} to model the machine flux variation through the following virtual electrical dynamic equation:

$$\frac{dV_{ref}^{VGM}}{dt} = K_\psi\left(Q_{AVR}^{BESS} - Q_{meas}^{BESS}\right) \tag{1}$$

where Q_{AVR}^{BESS} is the control action of the BESS converter AVR. Then, the AVR brings the RMS voltage at the output of the inverter V_{meas} back to its set point V_{set}. Similarly, the Inertia model and the Frequency Governor make the VGM similar to a synchronous generator active power/frequency dynamics. In particular, the Inertia model, described by an integrator and by two gains K_H for the inertia itself and K_d for the damping effect, reacts to the active power P_{meas}^{BESS} at the output of the inverter, drawing energy from the inertia and slowing the rotational speed of the virtual generator ω_{ref}^{VGM} using the following virtual mechanical dynamic equation:

$$\frac{d\omega_{ref}^{VGM}}{dt} = K_H\left(P_{GOV}^{BESS} - P_{meas}^{BESS} - K_d\omega_{ref}^{VGM}\right) \tag{2}$$

where P_{GOV}^{BESS} is the regulating action of the BESS converter Governor. Then, the Governor brings back the frequency f_{meas} to its set point f_{set}. To allow operation in parallel with other sources, the control includes droop factors m_{droop} and n_{droop} to achieve power sharing control in the steady state; while for stand-alone operation, the droop coefficients are usually set to zero. Then, in order to guarantee the black-start capability [26] of the MG using the BESS, two different selectors are implemented in the control diagram:

- If the measured RMS voltage V_{meas} is zero, the primary control is able to understand the necessity to provide a black-start procedure; so, imposing the logic signal named *Bl-St* equal to 1, the VGM channel is bypassed and the MG is energized using a ramp voltage reference $V_{ref}^{Bl\text{-}St}$ and the rated angular frequency ω_n.
- When the voltage reaches a specific percentage $k_\%$ of the rated voltage V_n, the VGM control channel is activated (reset of the control integrators) and the selectors switch to the VGM control actions V_{ref}^{VGM} and θ_{ref}^{VGM}.

Summarizing the above, voltage and frequency references logic for the black-start procedure is:

$$\begin{aligned} &\text{if}\quad 0 \leq V_{meas} \leq k_\% V_n \\ &\qquad Bl - St = 1 \Rightarrow V_{ref} = V_{ref}^{Bl\text{-}St} \ \& \ \theta_{ref} = \theta_{ref}^{Bl\text{-}St} \\ &\text{if}\quad V_{meas} \geq k_\% V_n \\ &\qquad Bl - St = 0 \Rightarrow V_{ref} = V_{ref}^{VGM} \ \& \ \theta_{ref} = \theta_{ref}^{VGM} \end{aligned} \tag{3}$$

Finally, the control actions are transformed in the α-β stationary reference frame.

As a final remark, it is worth pointing out that the tuning procedure of the VGM control is based on a "trial and error" strategy due to the intrinsic simplicity of the proposed primary controller. For example, the inertia parameter K_H of the virtual generator is set in order to have a precise frequency dynamic after an active power step. Then, the governor parameters are set in order to guarantee a desired time response.

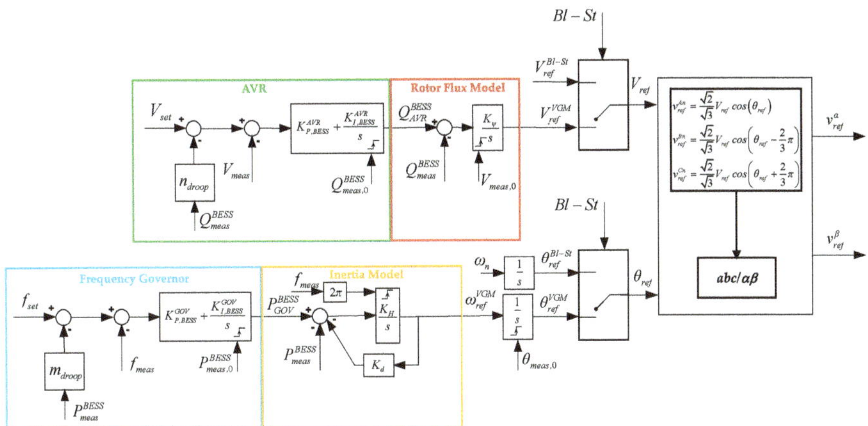

Figure 2. Proposed VGM (Virtual Generator Mode) control model for Grid-Forming operating mode.

2.2. Grid-Support Operating Mode: GSM

Considering now the grid-connected operating mode, the proposed control must provide fast control actions to allow frequency and voltage support (GSM-*fV*) as well as power control following the reference signals from the secondary level control (GSM-*PQ*). To meet these control requests, the block diagram in grid-connected mode is depicted in Figure 3.

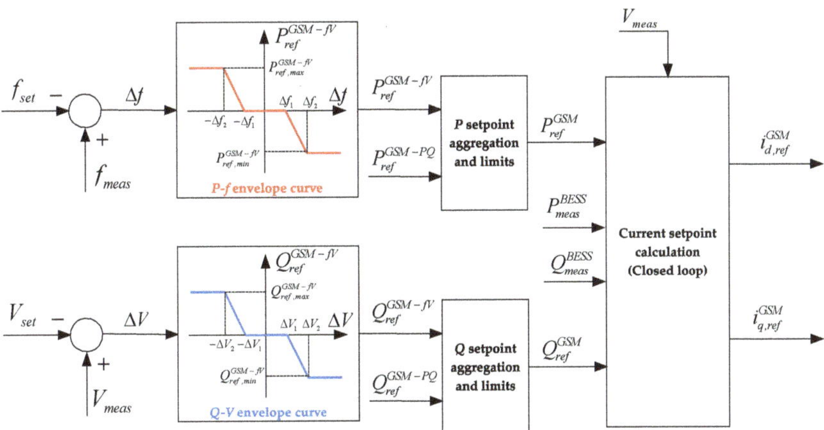

Figure 3. Proposed GSM (Grid Support Mode) control diagram in grid-connected operating mode.

In grid-connected mode, the BESS inverter is controlled starting from the measurements of frequency and voltage at the inverter output, i.e., f_{meas} and V_{meas}, respectively. Based on an error between measurements and respective setpoints, it is possible to carry out the target values of active and reactive powers $P_{ref}^{GSM\text{-}fV}$ and $Q_{ref}^{GSM\text{-}fV}$ through user-defined *P-f* envelope curve and *Q-V* envelope curve, respectively. Considering, for example, the *P-f* envelope curve, it is described by maximum ($P_{ref,max}^{GSM\text{-}PQ}$) and minimum ($P_{ref,min}^{GSM\text{-}PQ}$) active power references and by threshold frequency errors $\pm\Delta f_1$ and $\pm\Delta f_2$. The same characterization is done for the *Q-V* envelope curve. As stated before, the control also gives the possibility to track the reference values coming from the secondary level regulation, i.e., $P_{ref}^{GSM\text{-}PQ}$ and $Q_{ref}^{GSM\text{-}PQ}$, so the two blocks "*P* setpoint aggregation and limits" and "*Q* setpoint aggregation and limits" are implemented. Then, the d-q axis currents set points $i^{GSM}_{d,ref}$ and $i^{GSM}_{q,ref}$ are generated by closed-loop controllers which use the voltage measurement V_{meas}. Then, they are processed by the BESS inverter Current controller as depicted in Figure 1.

2.3. Voltage and Current Inner Control Loops

In this subsection, the inner control loops of the BESS converter are presented in order to provide a detailed description of the entire control system. Voltage control loop is depicted in Figure 4, and as one can see, it is based on PI regulators described by proportional gain K_{PV} and integral gain K_{IV}. Its main control objective is to regulate the output voltage of the inverter by minimizing the errors between the references $v^{BESS}_{d,ref}$, $v^{BESS}_{q,ref}$ and the measurements v^{BESS}_{d}, v^{BESS}_{q}. Moreover, Virtual Impedance strategy is added to the voltage control loop through algebraic manipulation of the α-β voltage reference signals coming from primary controller as follows:

$$v^{\alpha}_{ref,v} = v^{\alpha}_{ref} - \left(R_v i^{BESS}_{\alpha} - X_v i^{BESS}_{\beta} \right) \tag{4}$$

$$v^{\beta}_{ref,v} = v^{\beta}_{ref} - \left(R_v i^{BESS}_{\beta} + X_v i^{BESS}_{\alpha} \right) \tag{5}$$

with R_v and X_v being virtual resistance and reactance, respectively. This modification enables the output impedance to be set (by parameters, or adaptively) and this is usually made predominantly inductive to ensure a strong coupling between active power and frequency, a strong coupling between reactive power and voltage and an inherent decoupling between these two relationships, even in LV microgrids [25]. The outputs of the voltage controller are the inverter output current references $i^{VGM}_{d,ref}$ and $i^{VGM}_{q,ref}$.

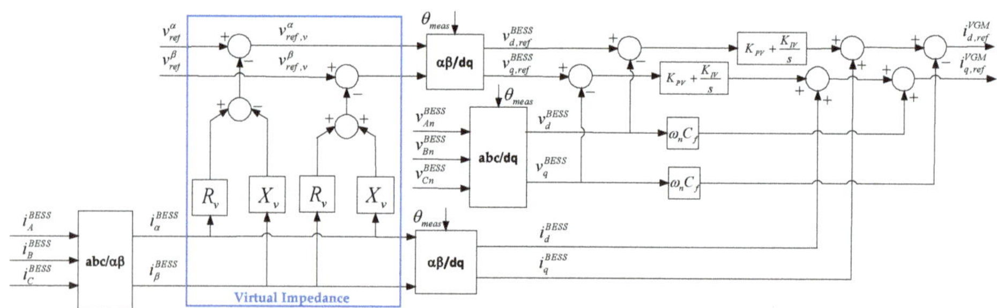

Figure 4. Voltage controller for BESS converter.

Current control loop is instead depicted in Figure 5: due to the fact it is used not only in VGM, but also in GSM operating mode, two selectors are implemented in order to choose the d-q current references coming from the VGM ($i^{VGM}_{d,ref}$ and $i^{VGM}_{q,ref}$) or from the GSM ($i^{GSM}_{d,ref}$ and $i^{GSM}_{q,ref}$). The current controller is based on PI regulators described by proportional gain K_{PC} and integral gain K_{IC} and it acts on the system in order to control the converter output current by minimizing the errors between the references and the measured current described by i^{inv}_{d} and i^{inv}_{q}. The outputs of the current controller are the inverter voltage modulation signals in the d-q reference frame $v^{inv}_{d,ref}$ and $v^{inv}_{q,ref}$.

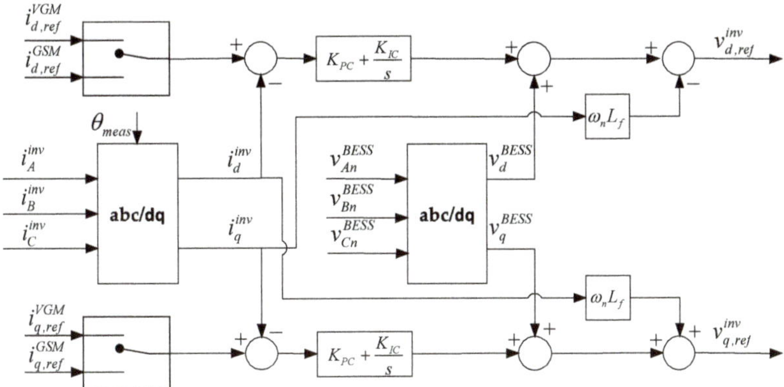

Figure 5. Current controller for BESS converter.

3. DIgSILENT PowerFactory® Model for Testing

3.1. Synchronous Diesel Generator Model for Paralleling Operation

As stated in the introduction section, the proposed BESS control is implemented in the widely used power system simulator DIgSILENT PowerFactory® in order to have reliable and high-fidelity simulation results. Due to the fact that in Grid-Forming operating mode the BESS inverter must be the voltage/frequency master in stand-alone configuration and must allow parallel operations with other DERs, a Synchronous Diesel Generator (SDG) model has been included in the simulated power system; the SDG model is briefly described below starting from the conceptual model depicted in Figure 6.

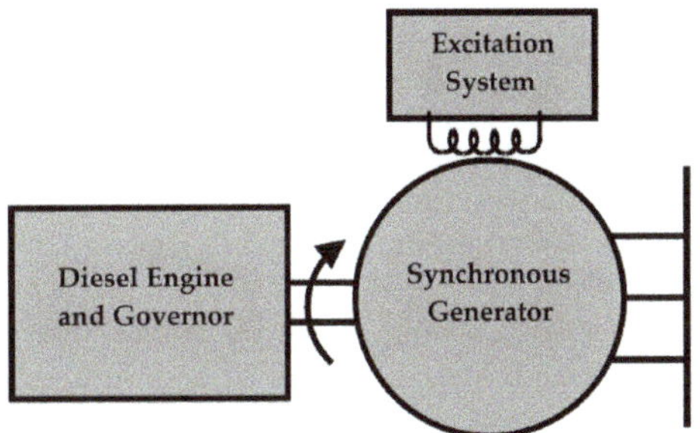

Figure 6. Synchronous Diesel Generator (SDG) conceptual model.

As one can see, it consists of three main elements: Synchronous Generator EMT model is available from PowerFactory library [27], while the Diesel Engine with Governor and Excitation System models are depicted in Figures 7 and 8, respectively.

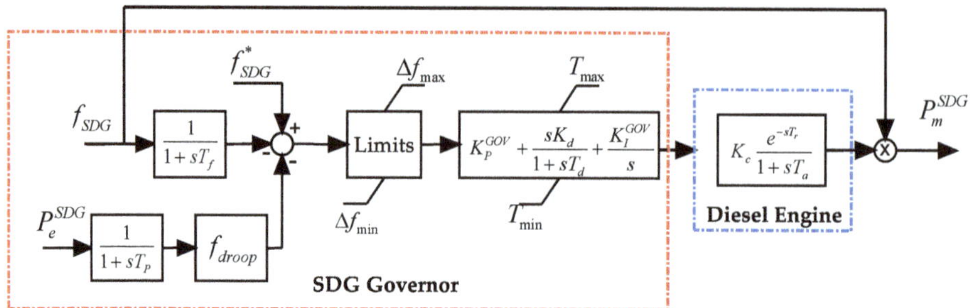

Figure 7. Diesel Engine and Governor model for the SDG.

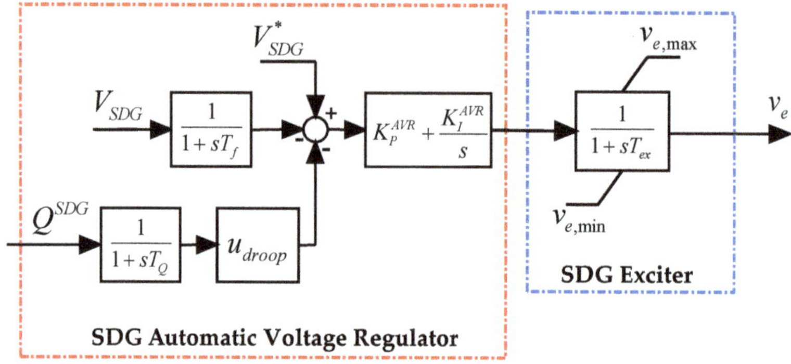

Figure 8. Exciter and Automatic Voltage Regulator model for the SDG.

The frequency-active power regulation of the SDG is realized using a Governor whose inputs are the electrical speed f_{SDG} and the active power measurement P_e^{SDG}. Droop coefficient f_{droop} is implemented to guarantee the correct power sharing in islanded mode and in parallel with the BESS unit. A PID regulator generates the torque input for the diesel generator, which is modeled by the combustion gain K_c, the fuel dynamics time delay T_r, and by a first order dynamic where T_a models the engine fuel system time constant [11]. The voltage-reactive power regulation is performed by an AVR where RMS voltage and reactive power measurements are the inputs and where the droop logic is implemented by the use of the coefficients u_{droop}. The AVR output is then processed by the exciter, which is modeled by a first order dynamic with time constant equal to T_{ex} [28].

3.2. Test MG Layout and Parameters

In order to test the performances of the proposed BESS primary control, the test case MG reported in Figure 9 is used. The main element in the MG is the BESS unit, which is connected to the MG via inverter. The storage component of the BESS is modeled as ideal DC voltage source and so its internal dynamics are neglected. Similarly, the inverter model, provided by DIgSILENT PowerFactory® library, is considered as an ideal controlled AC voltage source. The inverter is interfaced to the MG with a R-L-C filter modeled by the parameters R_f, L_f and C_f, respectively, and with a unitary-ratio transformer T_{BESS}. The SDG exploits the DIgSILENT PowerFactory® synchronous generator model and it is connected to the MG Point of Common Coupling (PCC) using longitudinal impedance $\dot{Z}_{SDG}$. An external grid is connected to the MG through a medium voltage/low voltage transformer T_{MV-LV} in order to test the Grid support and the paralleling functionalities. Load center bus is connected to the PCC with a line $\dot{Z}_{line}$ while $Load_1$ is connected to Load center bus using a cable $\dot{Z}_{cable}$. MG data are reported in Table 1.

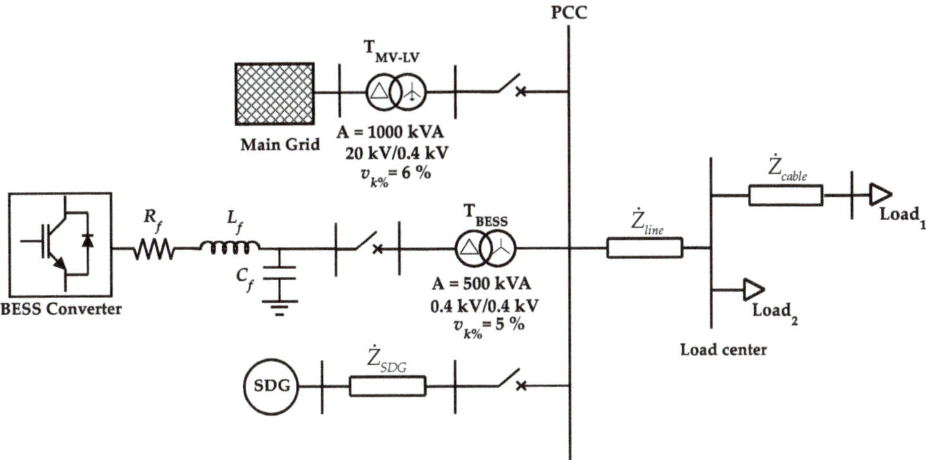

Figure 9. Test case microgrids (MG) one-line diagram.

Table 1. Test case microgrids (MG) parameters.

BESS Converter Data		SDG Data		Load Data		Line and Cables Data at f_n	
A_{BESS}	500 kVA	A_{SDG}	1250 kVA	$P_n(Load_1)$	300 kW	$\dot{Z}_{SDG}$	0.007 + j0.0008 Ω
V_n	400 V (AC-side)	$cos\varphi_n$	0.8	$Q_n(Load_1)$	100 kVAr	$\dot{Z}_{line}$	0.014 + 0.0016 Ω
f_n	50 Hz	V_n	400 V	$P_n(Load_2)$	50 kW	$\dot{Z}_{cable}$	0.0037 + j0.0004 Ω
R_f	0.044 Ω	f_n	50 Hz	$Q_n(Load_2)$	0 kVAr		
L_f	0.088 mH						
C_f	50 µF						

4. Simulations Results

4.1. Black Start and Load Step in a Stand-Alone Configuration

This is the first functionality that the BESS primary control has to satisfy. IEEE 1547.4 Std. gives a clear definition of the black-start, i.e., the ability to start local generation with no external source of power. During this phase, the main grid and the diesel generator are disconnected from the MG and no load is connected. As described in the previous section, the BESS is controlled in Grid-Forming operating mode and the VGM control is bypassed imposing voltage and frequency reference signals. As one can see from Figure 10, the three-phase voltages at the controlled node of the MG perfectly follow the ramp reference signal V_{ref}, and the zoomed box highlights the correct dynamic behavior when the VGM control channel is activated with the switch of V_{ref} from V_{ref}^{Bl-St} to V_{ref}^{VGM} and of θ_{ref} from θ_{ref}^{Bl-St} to θ_{ref}^{VGM} and with the reset of the integrators of the VGM control.

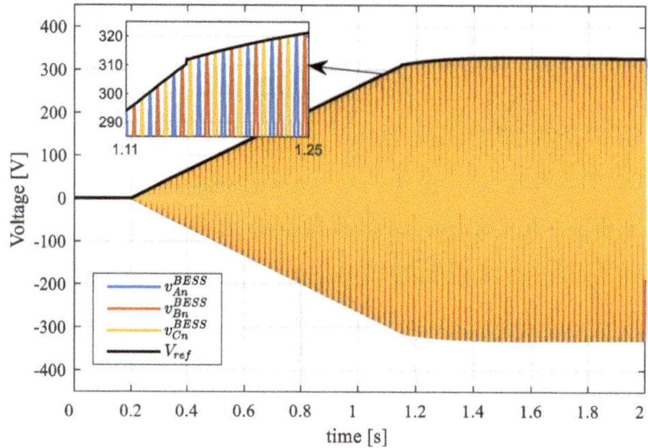

Figure 10. Controlled BESS bus voltage time profiles during black-start procedure.

When the black-start procedure ends (the ramping time is defined through a parameter and in this case it was chosen to last 1 s), a load step event is implemented in the simulation in order to test the dynamic response of the BESS in VGM mode in terms of voltage and frequency in a stand-alone configuration. As one can see from Figure 11, after the load step occurred at $t = 4$ s, frequency and voltage have a time profile in which it is possible to note the effects of the inertia model and of the governor regulating action for the frequency and of rotor flux model and AVR regulating action for the voltage. In this simulation, the droop coefficient m_{droop} and n_{droop} are equal to zero because the power sharing functionality is not required in a stand-alone configuration, and for this reason, after the transient, frequency and voltage return to their rated values (namely 50 Hz for the frequency and 400 V for the voltage). Moreover, it is possible to notice the effect of the virtual impedance implemented in the voltage control loop of the converter; in fact, in the final steady state, the reference value V_{ref}^{VGM} is greater than the controlled inverter output voltage V_{meas}. Finally, Figure 12 reports the active and reactive powers' time profile after load step contingency (P_{load} and Q_{load}), and it is possible to see how the BESS is able to supply the load power request with a fast and smooth dynamic response.

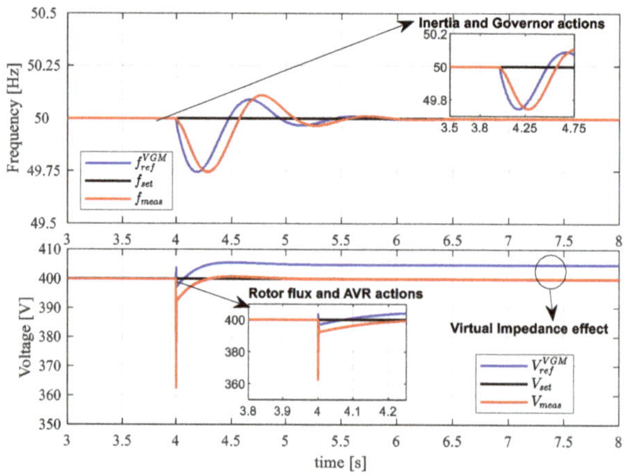

Figure 11. Frequencies (first panel) and RMS (Root Mean Square) voltages (second panel) time profile during load step in stand-alone configuration.

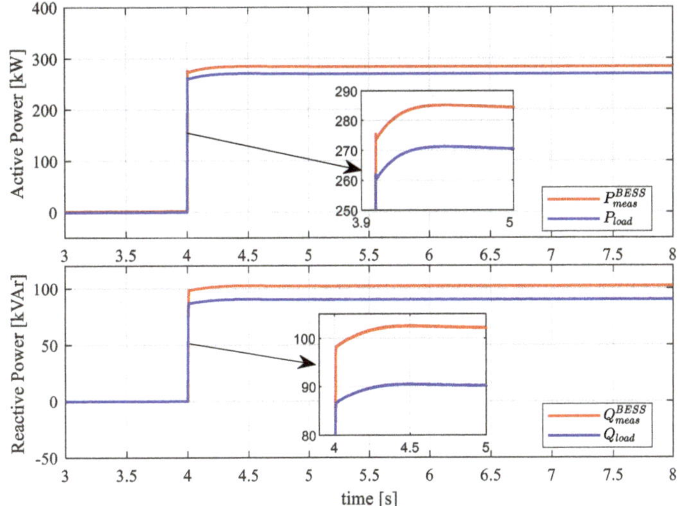

Figure 12. Active (first panel) and reactive (second panel) powers' time profile during load step in stand-alone configuration.

4.2. VGM Operating Mode With Setpoints Variation

As stated in the introduction, when the VGM operating mode is activated, the BESS converter has to provide fast regulation of voltage and frequency. In these simulations, the main grid is connected to the MG and the load request is totally fed by the main grid. The BESS converter is then connected to the MG in Grid-Forming operating mode without active and reactive powers production. In order to highlight the advantages of the proposed VGM control technique, a comparison with the conventional droop control for Grid-Forming operating mode is proposed. When this conventional control technique is adopted, all the macro blocks of the proposed controller (i.e., the AVR, the Rotor Flux Model, the Governor and the Inertia Model) are bypassed avoiding all the dynamics introduced by the virtual generator scheme. So, the control signals V_{ref} and θ_{ref} with the conventional droop control technique become:

$$
\begin{cases}
V_{ref} = V_{set} - n_{droop} Q_{meas}^{BESS} \\
\theta_{ref} = 2\pi \int\limits_0^t \left[f_{set}(s) - m_{droop} P_{meas}^{BESS}(s) \right] ds
\end{cases}
\tag{6}
$$

The first simulation results are depicted in Figure 13: in this scenario, the frequency setpoint f_{set} is fixed at the MG rated value, while the voltage setpoint V_{set} changes from 400 V to 410 V at $t = 2$ s, then from 410 V to 390 V at $t = 6$ s and finally from 390 V to 400 V at $t = 10$ s. As one can see, the set point is correctly followed by the controlled RMS voltage V_{meas} with good time response and minimum overshoots (the droop parameters m_{droop} and n_{droop} are equal to zero in this simulation since there are no other DERs to share the load request). The second panel shows instead reactive powers time profile, and it is possible to see that an increase/decrease of the controlled voltage corresponds to a BESS converter reactive power production/absorption, which is balanced by the main grid, due to the strong coupling between voltage and reactive power. Considering the comparison, the conventional droop control has a faster time response because there is no effect of the virtual generator dynamics, but it has two main drawbacks: (i) voltage steady-state errors due to the absence of an AVR in the droop control to compensate the effect of the virtual impedance implemented in the voltage controller (the greater the output current of the BESS converter, the greater the voltage steady-state error, as apparent in Equations (4) and (5)) and (ii) large initial overshoots in the reactive power time profiles as detailed in Figure 14.

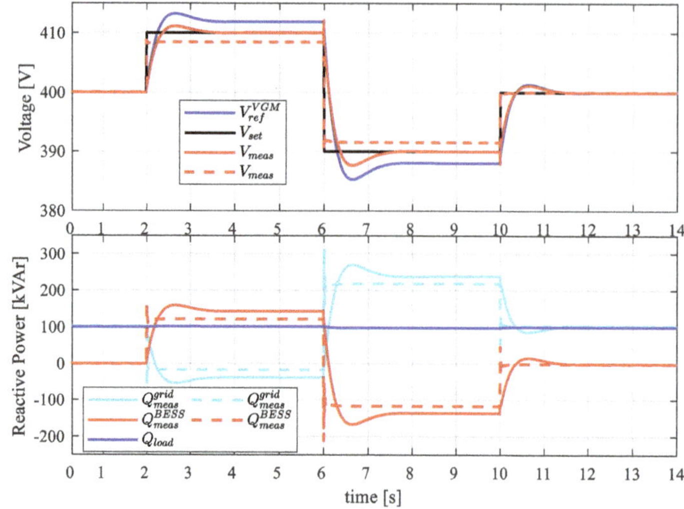

Figure 13. Voltage (first panel) and reactive powers (second panel) time profiles with voltage setpoint variations with Virtual Generator Mode (VGM) (solid lines) and conventional (dotted lines) techniques.

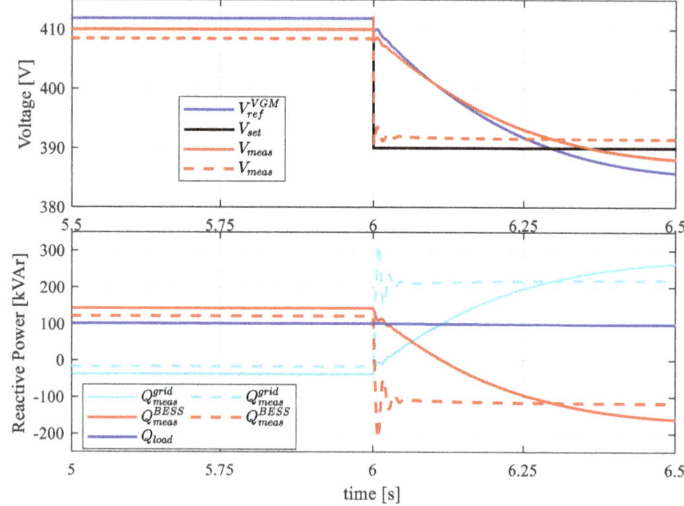

Figure 14. Detail on the comparison between the proposed technique (solid lines) and conventional droop control (dotted lines).

Then, in order to test the possibility of managing the BESS converter active power production by changing the frequency setpoint, in the second scenario the voltage setpoint V_{set} is fixed to the rated value while the frequency setpoint f_{set} changes from 50 Hz to 50.1 Hz V at $t = 2$ s, then from 50.1 Hz to 49.9 Hz V at $t = 6$ s and finally from 49.9 Hz to 50 Hz at $t = 10$ s. Due to synchronism with Main Grid requirement, the frequency droop factor m_{droop} is activated. As one can see from Figure 15, the frequency setpoint is not tracked (first panel), but varying its value it is possible to correctly manage the BESS active power production (second panel), which is correctly balanced by the Main Grid. Also, in this test case, conventional droop control guarantees a faster active power regulation due to the lack of the virtual inertia of the VGM control technique but, as one can see from Figure 16, the droop control creates a greater frequency variation in the very first transient.

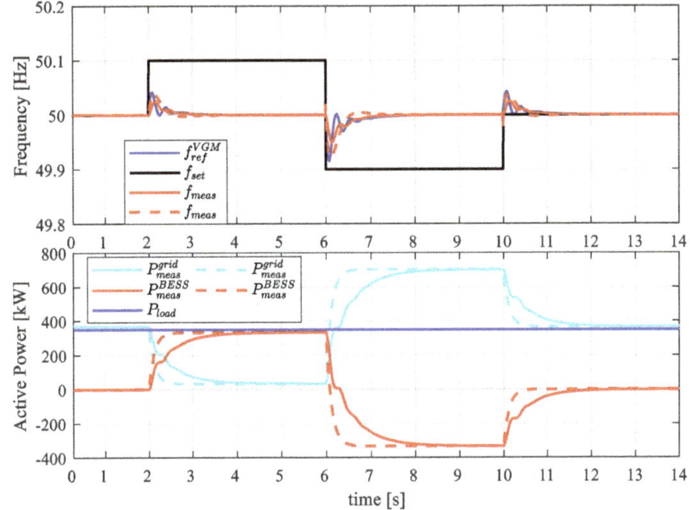

Figure 15. Frequency (first panel) and active powers (second panel) time profiles with frequency setpoint variations with VGM (solid lines) and conventional (dotted lines) techniques.

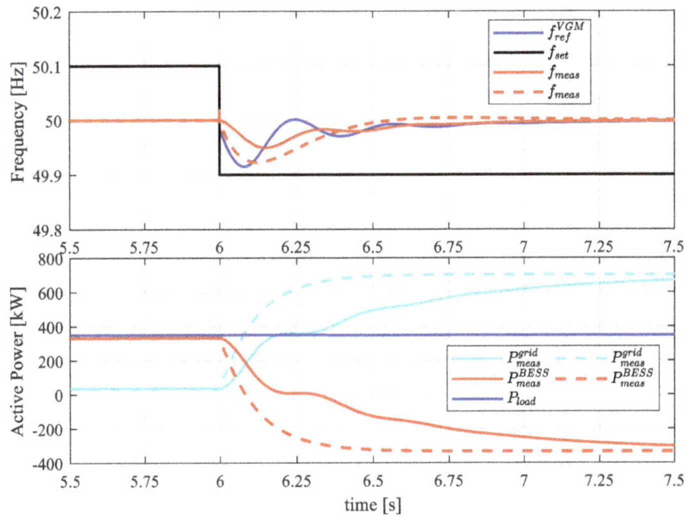

Figure 16. Detail on the comparison between the proposed technique (solid lines) and conventional droop control (dotted lines).

4.3. Paralleling Action and Load Sharing

The aim of this simulation is to show how the BESS converter can be connected to an islanded MG where a diesel generator is already online in Grid-Forming operating mode and to also show the correct power sharing functionality. As one can see from Figure 17, at $t = 0.1$ s a live-start of the BESS converter is activated: as one can see, the controlled voltage (v_{AN}^{BESS}, v_{BN}^{BESS}, v_{CN}^{BESS}) starts to follow the connection bus voltages (v_{AN}^{MG}, v_{BN}^{MG}, v_{CN}^{MG}) and, at t = 0.2 s, the phase reference output of the primary control θ_{ref}^{VGM} is synchronized with the measured phase angle θ_{meas} at the connection bus to the MG. Finally, at t = 0.25 s, the AC-breaker of the BESS converter is closed and in Figure 18 is possible to notice a really limited power transient quantifiable in a transient current less than 1% of the BESS-converter-rated value current.

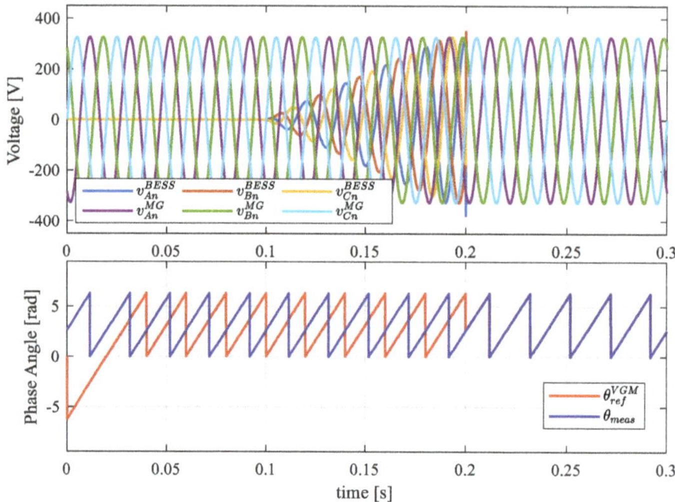

Figure 17. Voltage (first panel) and phase angles (second panel) during synchronization with MG.

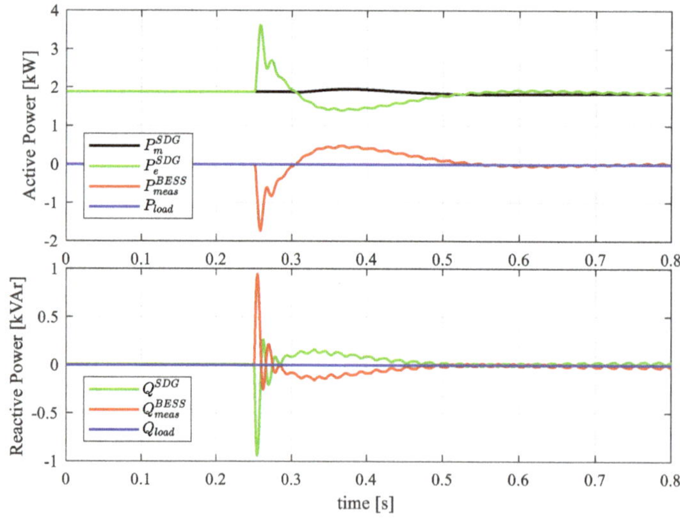

Figure 18. Active (first panel) and reactive (second panel) powers time profile during paralleling procedure.

Then in order to test the proper active and reactive power sharing between the diesel generator and the BESS, a load step contingency is implemented at t = 4 s. Figure 19 shows active and reactive powers time profile and it is possible to highlight the correct active and reactive power sharing according to the rated apparent power of the two units. The proper active power sharing can be guaranteed by the relation $A_{BESS}m_{droop} = A_{SDG}f_{droop}$, while a correct reactive power sharing can be achieved if the droop factors n_{droop} for the BESS and u_{droop} for the diesel generator are correctly set in order to overcome the mismatch in the connection impedances to the MG of the two units. The virtual impedance $R_v + jX_v$ affects the reactive power sharing only during the transient, because in steady state its effect is nulled out by the BESS AVR regulating action on the controlled voltage V_{meas}. Instead, Figure 20 shows frequencies and RMS voltages time profile during the load step contingency and it is possible to highlight the inertial behavior of the BESS converter and voltage and frequency deviations from the respective rated values according to the droop logic in the proposed primary control.

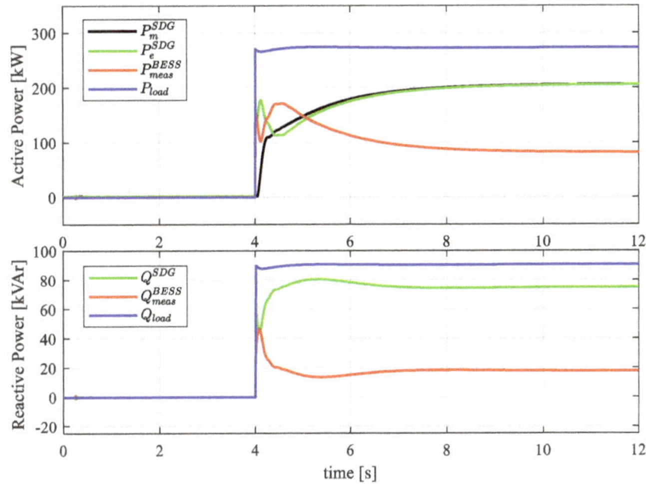

Figure 19. Active (first panel) and reactive (second panel) powers time during load step in paralleling configuration.

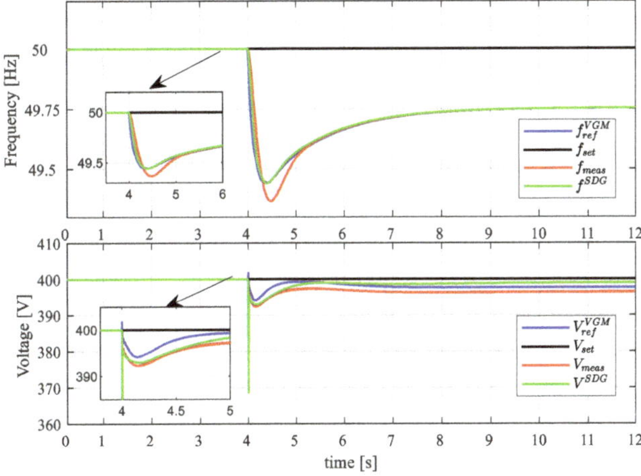

Figure 20. Frequencies (first panel) and RMS voltages (second panel) time profile in paralleling configuration.

4.4. Grid Support Action

In this subsection, the results of the BESS converter controlled in GSM operating mode are presented in detail. The first simulation is carried out in order to test the correct behavior of the proposed primary control in GSM-*fV*, i.e., when the active/reactive power references come from the envelope curves. In order to highlight the positive effects of the BESS converter in support mode, two different MG configurations are considered:

- In the first configuration, the BESS converter is not connected to the MG. Two different loads, namely Load$_1$ and Load$_2$, are connected to the MG with $P_{load,1} = 300$ kW, $Q_{load,1} = 100$ kVAr and $P_{load,2} = 50$ kW (pure resistive load). Loads are supplied by the SDG, while the main grid is disconnected.

- In the second configuration, the BESS is connected to the MG in GSM-*fV* mode ready to provide voltage and frequency support after MG contingencies.

At $t = 2$ s, Load$_1$ is disconnected to the MG and simulation results are depicted in Figures 21 and 22.

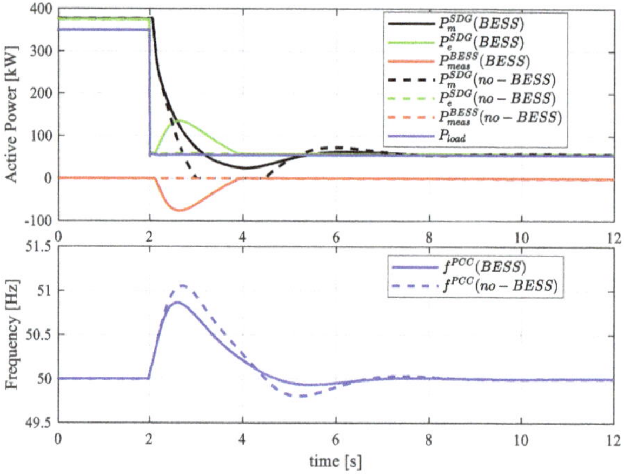

Figure 21. Active powers and Point of Common Coupling (PCC) frequency in Grid Support Mode (GSM)-*fV* mode.

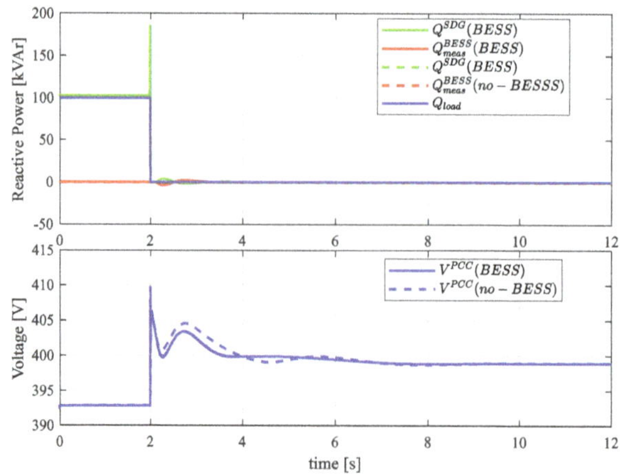

Figure 22. Reactive powers and PCC voltage in GSM-*fV* mode.

As one can see after the load disconnection, MG frequency f^{PCC} and voltage V^{PCC} (measured at PCC) start to increase: if the BESS is connected in GSM mode, when voltage and frequency exceed the threshold values Δf_1 and ΔV_1, respectively, the BESS converter starts to absorb active and reactive power according the implemented envelope curves. It is easy to notice that the dynamic response in terms of frequency and voltage with BESS converter controlled in GSM-*fV* mode (solid lines) is better with respect to the response without BESS connected to the MG (dashed lines). Finally, Figure 23 shows the correct response of the primary controller in GSM-*PQ* mode, i.e., when the active and reactive power references $P_{ref}^{GSM\text{-}PQ}$ and $Q_{ref}^{GSM\text{-}PQ}$ are not provided by envelope curve, but are sent by the system controller. In this configuration, the SDG is not connected, while the MG is connected to the main grid: the primary controller is able to correctly track powers references guaranteeing fast regulating actions.

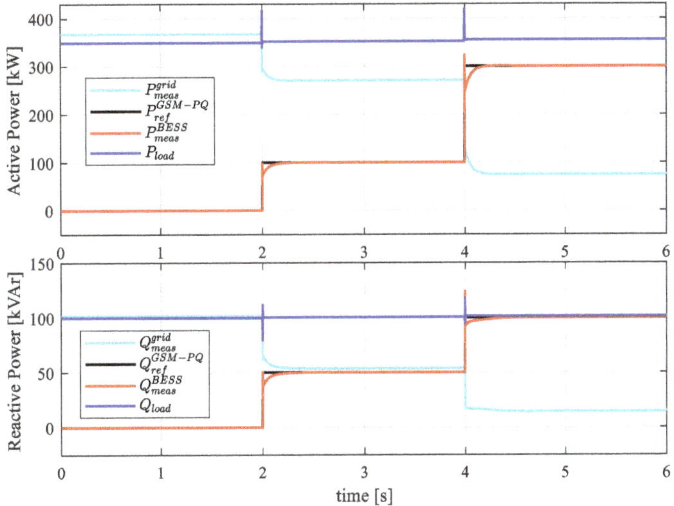

Figure 23. Active and reactive powers time profile in GSM-PQ mode.

5. Conclusions

This paper proposed a comprehensive primary control for BESS converter in order to provide all the regulations required by IEEE Std. 1547 both in grid-connected and islanded configurations. Due to the need to guarantee fast and smooth dynamics of voltage and frequency in islanded configuration, the BESS converter is controlled using a VGM technique in order to have it acting on the system as a real synchronous generator. Simulations are carried out using the dedicated simulation software DIgSILENT PowerFactory® and show that the proposed controller is able to provide the black-start capability, to regulate frequency and voltage independently of the number of paralleled generators, to synchronize and connect BESS converter to the external main grid or to other DERs in with minimum transients and to guarantee a proper active/reactive power sharing among other DERs. Moreover, simulations showed that the proposed primary control has the possibility to provide fast control actions also in grid-connected mode to allow providing frequency and voltage support (GSM-fV) as well as power control following the reference signals from the secondary level control (GSM-PQ). Future works will be focused on the transitions between the two operating modes in order to provide system stability during planned or unplanned islanding or during the reconnection of the main grid.

Author Contributions: M.F., A.T. and A.R. designed the Primary Regulator, A.R. performed the simulations and wrote the paper. P.S., R.P. and A.B. provided supervision to the research activity, provided critical analysis to the results achieved by simulations and revised the manuscript.

Funding: This research received no external funding.

Conflicts of Interest: The authors declare no conflict of interest.

References

1. Mariam, L.; Basu, M.; Conlon, M.F. Microgrid: Architecture, policy and future trends. *Renew. Sustain. Energy Rev.* **2016**, *64*, 477–489. [CrossRef]
2. U.S.D.O. Energy. *DOE Microgrid Workshop Report.* Available online: https://www.energy.gov/about-us (accessed on 27 October 2018).
3. Guerrero, J.M.; Vasquez, J.C.; Matas, J.; de Vicuña, L.G.; Castilla, M. Hierarchical control of droop-controlled AC and DC microgrids—A general approach toward standardization. *IEEE Trans. Ind. Electron.* **2011**, *58*, 158–172. [CrossRef]
4. Luna, A.C.; Meng, L.; Diaz, N.L.; Graells, M.; Vasquez, J.C.; Guerrero, J.M. Online Energy Management Systems for Microgrids: Experimental Validation and Assessment Framework. *IEEE Trans. Power Electron.* **2018**, *33*, 2201–2215. [CrossRef]
5. Lu, X.; Yu, X.; Lai, J.; Wang, Y.; Guerrero, J.M. A Novel Distributed Secondary Coordination Control Approach for Islanded Microgrids. *IEEE Trans. Smart Grid* **2018**, *9*, 2726–2740. [CrossRef]
6. Bidram, A.; Davoudi, A. Hierarchical Structure of Microgrids Control System. *IEEE Trans. Smart Grid* **2012**, *3*, 1963–1976. [CrossRef]
7. IEEE PES Industry Technical Support Task Force. *Impact of IEEE 1547 Standard on Smart Inverters*; Technical Report PES-TR67; IEEE: Piscataway, NJ, USA, 2018.
8. Zeineldin, H.H. A Q–f Droop Curve for Facilitating Islanding Detection of Inverter-Based Distributed Generation. *IEEE Trans. Power Electron.* **2009**, *24*, 665–673. [CrossRef]
9. Zeineldin, H.H.; El-saadany, E.F.; Salama, M.M.A. Distributed Generation Micro-Grid Operation: Control and Protection. In Proceedings of the 2006 Power Systems Conference: Advanced Metering, Protection, Control, Communication, and Distributed Resources, Clemson, SC, USA, 14–17 March 2006; pp. 105–111.
10. Liu, J.; Miura, Y.; Ise, T. Comparison of Dynamic Characteristics Between Virtual Synchronous Generator and Droop Control in Inverter-Based Distributed Generators. *IEEE Trans. Power Electron.* **2016**, *31*, 3600–3611. [CrossRef]
11. Bevrani, H. *Robust Power System Frequency Control*; Springer: Berlin/Heidelberg, Germany, 2009; Volume 85.
12. Rocabert, J.; Luna, A.; Blaabjerg, F.; Rodriguez, P. Control of power converters in AC microgrids. *IEEE Trans. Power Electron.* **2012**, *27*, 4734–4749. [CrossRef]
13. Zhong, Q.; Weiss, G. Synchronverters: Inverters That Mimic Synchronous Generators. *IEEE Trans. Ind. Electron.* **2011**, *58*, 1259–1267. [CrossRef]
14. Zhong, Q.; Nguyen, P.; Ma, Z.; Sheng, W. Self-Synchronized Synchronverters: Inverters Without a Dedicated Synchronization Unit. *IEEE Trans. Power Electron.* **2014**, *29*, 617–630. [CrossRef]
15. Zhong, Q.; Konstantopoulos, G.C.; Ren, B.; Krstic, M. Improved Synchronverters with Bounded Frequency and Voltage for Smart Grid Integration. *IEEE Trans. Smart Grid* **2018**, *9*, 786–796. [CrossRef]
16. Arco, S.D.; Suul, J.A. Virtual synchronous machines—Classification of implementations and analysis of equivalence to droop controllers for microgrids. In Proceedings of the 2013 IEEE Grenoble Conference, Grenoble, France, 16–20 June 2013; pp. 1–7.
17. Fathi, A.; Shafiee, Q.; Bevrani, H. Robust Frequency Control of Microgrids Using an Extended Virtual Synchronous Generator. *IEEE Trans. Power Syst.* **2018**, *33*, 6289–6297. [CrossRef]
18. Hirase, Y.; Noro, O.; Yoshimura, E.; Nakagawa, H.; Sakimoto, K.; Shindo, Y. Virtual synchronous generator control with double decoupled synchronous reference frame for single-phase inverter. *IEEJ J. Ind. Appl.* **2015**, *4*, 143–151. [CrossRef]
19. Chen, Y.; Hesse, R.; Turschner, D.; Beck, H. Improving the grid power quality using virtual synchronous machines. In Proceedings of the 2011 International Conference on Power Engineering, Energy and Electrical Drives, Torremolinos, Spain, 11–13 May 2011; pp. 1–6.
20. Xiang-zhen, Y.; Jian-hui, S.; Ming, D.; Jin-wei, L.; Yan, D. Control strategy for virtual synchronous generator in microgrid. In Proceedings of the 2011 4th International Conference on Electric Utility Deregulation and Restructuring and Power Technologies (DRPT), Weihai, Shandong, China, 6–9 July 2011; pp. 1633–1637.

21. Sakimoto, K.; Miura, Y.; Ise, T. Stabilization of a power system with a distributed generator by a Virtual Synchronous Generator function. In Proceedings of the 8th International Conference on Power Electronics—ECCE Asia, Jeju, Korea, 29 May–2 June 2011; pp. 1498–1505.

22. Shintai, T.; Miura, Y.; Ise, T. Reactive power control for load sharing with virtual synchronous generator control. In Proceedings of the 7th International Power Electronics and Motion Control Conference, Harbin, China, 2–5 Jun 2012; pp. 846–853.

23. Shintai, T.; Miura, Y.; Ise, T. Oscillation Damping of a Distributed Generator Using a Virtual Synchronous Generator. *IEEE Trans. Power Deliv.* **2014**, *29*, 668–676. [CrossRef]

24. Alipoor, J.; Miura, Y.; Ise, T. Power System Stabilization Using Virtual Synchronous Generator With Alternating Moment of Inertia. *IEEE J. Emerg. Sel. Top. Power Electron.* **2015**, *3*, 451–458. [CrossRef]

25. Tuckey, A.; Round, S. Practical application of a complete virtual synchronous generator control method for microgrid and grid-edge applications. In Proceedings of the 2018 IEEE 19th Workshop on Control and Modeling for Power Electronics (COMPEL), Padova, Italy, 25–28 June 2018; pp. 1–6.

26. Lopes, J.P.; Moreira, C.; Resende, F. Control strategies for microgrids black start and islanded operation. *Int. J. Distrib. Energy Resour.* **2005**, *1*, 241–261.

27. D. GmbH. *DIgSILENT PowerFactory 2017 User Manual*; DIgSILENT: Gomaringen, Germany, 2017.

28. Kundur, P.; Balu, N.J.; Lauby, M.G. *Power System Stability and Control*; McGraw-Hill: New York, NY, USA, 1994; Volume 7.

 *energies*

Article

Optimal Control of Multiple Microgrids and Buildings by an Aggregator

Giulio Ferro [1], Riccardo Minciardi [1], Luca Parodi [1,*], Michela Robba [1] and Mansueto Rossi [2]

[1]   DIBRIS, University of Genoa, via Opera Pia 13, 16145 Genova, Italy; giulio.ferro@edu.unige.it (G.F.); riccardo.minciardi@unige.it (R.M.); michela.robba@unige.it (M.R.)
[2]   DITEN, University of Genoa, via A. Magliotto 2, 17100 Savona, Italy; mansueto.rossi@unige.it
*   Correspondence: luca.parodi@edu.unige.it

Received: 6 December 2019; Accepted: 21 February 2020; Published: 27 February 2020

Abstract: The electrical grid has been changing in the last decade due to the presence of renewables, distributed generation, storage systems, microgrids, and electric vehicles. The introduction of new legislation and actors in the smart grid's system opens new challenges for the activities of companies, and for the development of new energy management systems, models, and methods. A new optimization-based bi-level architecture is proposed for an aggregator of consumers in the balancing market, in which incentives for local users (i.e., microgrids, buildings) are considered, as well as flexibility and a fair assignment in reducing the overall load. At the lower level, consumers try to follow the aggregator's reference values and perform demand response programs to contain their costs and satisfy demands. The approach is applied to a real case study.

Keywords: smart grid; optimization; energy management system; interconnected buildings; renewable resources; multi-level; aggregator

---

1. Introduction and State of the Art

International policies for sustainable development have led to an increase in distributed power production based on renewable resources. The electrical grid has been changing in the last decade due to the presence of renewables, distributed generation, storage systems, microgrids, and electric vehicles. These technologies introduce a lot of challenges mainly due to uncertainties and intermittencies associated with their presence [1]. In particular, the massive presence of renewable resources can create some problems to the distribution grid; as a matter of fact, the unpredictability of these sources can cause some power quality issues in the power network (e.g., voltage unbalances and undesired peaks). Usually, traditional controllable generators are called to compensate these fluctuations, operating at different working points from the optimal ones. This function for the traditional generators cause efficiency losses (increasing in operational and maintenance costs) and a decrease of regulation margins for the distribution grid [2]. Moreover, due to the increase of generation from renewables, large fossil-fueled production plants are no longer installed or are reduced in rated power; this affects the capability of the grid to respond to emergency situations.

The characteristics of these new smart grids require an increase of the power reserve to face the sudden request of active/reactive injection/absorption from a distribution system operator (DSO) to compensate for example a sudden drop in the production from a photovoltaic plant. Microgrids and smart buildings can see these issues as an opportunity to provide regulation and reserve services in both directions since they can be considered as prosumers [3–5]. Interconnected buildings can be seen as microgrids or districts that can share thermal and electrical power to satisfy comfort, economic, and technical exigencies. In this framework, demand response (DR) (i.e., the possibility for a traditional load to decrease the active power absorption for a certain period) is an effective and reliable strategy for the successful integration of renewable energy sources, handling the demand pattern using load flexibility whenever the system requires it [6–8]. A significant portion of DR programs can be represented by local users and/or prosumers, which, by themselves, cannot participate in the energy market and offer a reduction of load at a certain price. In fact, the introduction of new legislation and actors in the smart grid's system opens new challenges for the activities of companies, and for the development of new energy management systems, models, and methods. For example, the figure of an aggregator that manages different local consumers and/or producers to reduce power demand or to increase production gives rise to different possible optimization problems [1].

A key point is to integrate the different actors (transmission system operator (TSO), DSO, aggregators, local users, etc.) that can act within collaborative, competitive, or hierarchical rules in a market structure in which the decisions should be taken into account according to successive stages. Each stage is characterized by different players with different objectives, thus representing a portion of a multilayered energy market. An aggregator must take decisions interacting, first of all, with the TSO offering an amount of power to be reduced at a certain price. Then, the aggregator has to collect in its portfolio a large number of customers in order to be paid from the TSO; besides, it must provide incentives to local users, and should avoid customers' dissatisfaction in terms of both economic costs and technical feasibility. At the lower level of the layered market structure, there are users that have two main decisions to be taken: (a) How to provide constraints to the aggregator related to the flexibility that they can technically offer; (b) how to manage their local systems following the requests of the aggregator once they are available.

Distribution network's optimization and control in presence of DR aggregators is particularly interesting and some papers in the recent literature propose new operational management strategies. In particular, it is necessary to define how to schedule load reduction or production increase among the different customers to achieve an overall load reduction and/or shifting. The main effort is devoted to the integration of different actors in a multi-layered market structure. A portion of the recent literature is focused on bi-level programming, in which one or several constraints represent the optimal solution of low-level problems. In other words, the upper-level problem has its own optimization problem to be solved through mathematical programming, in which the analytical solution of the lower level optimization model is inserted as a constraint. In [9], the authors propose a bi-level model for the distribution network and renewable energy expansion planning in a DR framework. The objective function of the upper-level problem minimizes investment, maintenance, energy purchased from substations, and unserved power. The lower level has instead the objective of minimizing the overall payment faced by the consumers. The Karush–Kuhn–Tucker (KKT) complementarity constraints are used to ensure the optimality of the lower-level solution. In [10] the authors present a bi-level architecture based on Model Predictive Control. An upper-level decision maker, by means of its objective function, minimizes its own power losses and costs; moreover, it provides power exchange references related to power-flow constraints. The main differences from our work are that we focus on operational management in which the storage dynamics are represented as well as temperature variation in each building. Moreover, in the present paper, at the upper level, it is considered an objective function that represents an aggregator that receives remuneration from TSO, must provide incentives, and has to manage buildings and microgrids equipped with their own energy management systems (EMSs).

In the same line, in [11] the authors propose a bi-level programming approach, within a game theory approach, for optimizing the electricity tariff offered by an electricity retailer to its customers. The model identifies customers as 'prosumers' (i.e., both producers and consumers) who try to maximize their utility and to minimize their cost of electricity.

The authors in [12] consider a framework in which an aggregator of distributed storage energy systems, electric vehicles, and temperature control loads is present and bids in day-ahead energy and reserve markets. Even in this case, a bi-level optimization model is proposed. In the present paper, a similar framework is considered. However, the main difference, with respect to the previously mentioned contributions, is that microgrids and buildings here are represented with their own EMS. Thus, the dynamic behavior of users is taken into account.

Bi-level programming is used in many other application areas of energy systems in which there are multiple systems and objectives to be considered (see, for example, [13–15]). However, bi-level programming is not the only approach that is used in energy systems for the integration of different systems, actors/agents and decision-makers.

Authors in [16] develop a hierarchical and distributed DR control strategy. They consider a center-free algorithm and information is not collected on all DR devices. A hierarchical EMS for optimal multi-microgrids operation based on multi-agent systems is proposed by the authors of [17]. In [18] the authors proposes a distributed energy allocation mechanism for the DSO market. This mechanism is based on a bi-level iterative auction. The DSO runs an upper-level auction with load aggregators as intermediate agents competing for energy (the locational marginal price is assumed to be known). Aggregators implement lower-level auctions in parallel until market equilibrium is established. In this way, it is possible to converge to a socially optimal solution while maintaining physical grid constraints.

In [19] the authors study the operation of a retailer that acts as an aggregator with price-responsive loads and submits demand bids to the day-ahead and real-time markets. The work presented in [20] uses Benders' decomposition method to reduce the computational burden in a two-stage stochastic model for aggregators in a large-scale case study. In [21], customer aggregators are introduced to supply downstream demand in the most economical way. A two-level optimization problem considers aggregators' flexible energy demand incorporated into the centralized energy dispatch model. The objective function aims to maximize social welfare minimizing the energy purchase cost. In [22] and [23], the authors jointly study energy storage systems (ESSs) and DR. They analyze their impact and possible use in a smart grid.

In [24], the authors focus attention on the integration of residential users because they are in huge numbers and require relatively small power loads. It is shown that residential users can receive financial benefits from reserve provision. A hierarchical structure of users is adopted together with a game-theoretic approach.

In [25], the authors focus attention on plug-in electric vehicles (PEV) in a district and on how they can help in demand-side management. In the experiments, the results show that the proposed system can reduce a significant amount of both electricity cost and peak power.

A crucial point is the estimation of the potential flexibility of different types of consumers for day-ahead and real-time ancillary services provision taking into account customers' reactions to the price signals [26]. Moreover, the use of incentive-based payments as price offer packages is recommended in order to implement DR [27].

In addition to the scheduling of an aggregator's customers, it is fundamental to design the optimal set of customers for an aggregator. In this framework, authors in [28] study the impact of consumer behavior on the portfolio design of a DR aggregator, developing an optimization model.

The present paper considers an aggregator (AGG) that has the knowledge of the power distribution grid because it is also a DSO or works in strict collaboration with the DSO.

The day-ahead balancing market is that session of the market where the TSO requests for some power flexibility in order to deal with the variation of the demand. In this session, the AGG bids the price of energy and the amount of power to be reduced/increased in each time interval during the day and thus it is necessary to coordinate the different customers, trying to satisfy power demand reduction/increase and giving incentives to local users in connection with these requests. For the sake of simplicity, in this paper, only the case of power reduction is considered.

The main contributions provided in this paper can be summarized as follows:

- The development of an optimization-based bi-level architecture for a DSO/aggregator for participation in the balancing market in which the incentives for local users (LUs), together with active and reactive power exchanges, are decision variables.

- The statement of an optimization problem for the optimal management of an aggregator/DSO that considers a simplified prosumers' model for flexibility assessment, providing set points for DR operations of the LUs, ensuring a certain level of fairness.
- The inclusion of detailed dynamical models for the two main classes of LUs (microgrids and smart buildings) including electrical and thermal elements.

The structure of the paper is organized as follows: In Section 2 the proposed control architecture is presented. Section 3 describes the AGG's optimization problem while Section 4 states the LUs' optimization problem. Section 5 presents the obtained results in a case study. Finally, Section 6 reports some concluding remarks.

2. The Proposed Decision Architecture

In this work, we propose a two-level hierarchical architecture for the problem, as depicted in Figure 1.

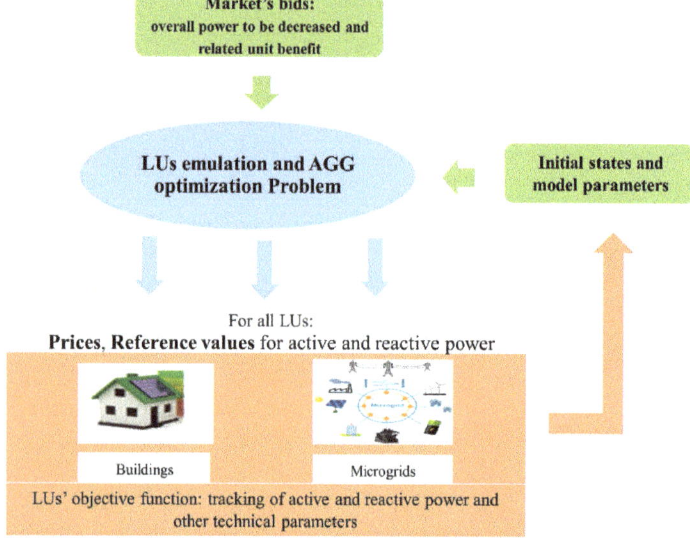

Figure 1. The proposed decision architecture of the Aggregator (AGG)–Local Users (LUs) problem.

The higher level represents the AGG, aiming at minimizing different terms of cost, that include: (a) Incentives to be paid to LUs, (b) fees/benefits (to/from the main grid) for not reaching the agreed reduction, (c) a term aiming at a fair distribution of the power reduction among the LUs. The AGG has to define active and reactive power exchanges with the LUs, as well as the incentives (that are expressed as a function of active power reduction).

At the lower level, LUs try to track the power reference values given by the solution of the AGG problem. Unlike the AGG, each LU uses a detailed model of its components.

Before the process starts, the AGG has perfect knowledge about the balancing market results (power reductions requests, incentives, and fees), and receives the maximum flexibility affordable by LUs calculated in the day-ahead on the basis of technical and economic considerations. Then, the AGG solves its optimization problem by using a simplified (emulated) model for each LU and assuming to have access to an estimate of the storage element associated with each LU. Instead, each LU does not have any information about the electrical grid and the other users, thus avoiding competitive behaviors. In fact, no competition between different LUs is allowed.

It is important to note that the above two-level optimization procedure is intended to operate within a receding horizon control scheme. In particular, as reported in Figure 2, an optimization horizon (T) is defined over which it is assumed that reliable forecasts about future uncertain inputs and parameters (in this case, power demand, generation from renewables, prices) are available. Such forecasts are provided by prediction models which are considered to be external to the proposed control scheme. Of course, the optimization horizon is the length over which the optimization problem is run. In a receding horizon scheme, in the first run, the optimal control actions for the entire length of the optimization horizon are found, but only the solution corresponding to the first time interval is actually applied. Then, at the beginning of the second time interval, new information from sensors in field (concerning the system state) and updated forecasts of the external inputs are acquired. On this basis, a new optimization run is carried out, and so on until the horizon over which the system is actually operated or simulated (simulation horizon) is completely covered (see Figure 2).

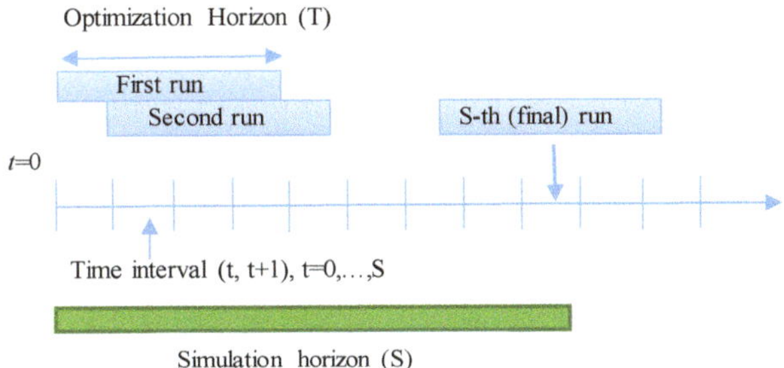

Figure 2. A pictorial representation of the receding horizon control scheme.

The optimization problem considered in this paper refers to a single iteration of the receding horizon control procedure and is focused on the possibility of efficiently solving the problem via a two-level approach. Clearly, in a receding horizon scheme, at each new time interval, new information is acquired and the two-level optimization procedure is again carried out.

3. The AGG System Model and Optimization Problem

3.1. Representation of the Network Constraints

The considered physical system is composed by buildings and microgrids, both including renewables and storage systems, and connected to the main grid.

In order to formally represent the system model, the following sets are considered:

- $N = \{1, \ldots, n\}$ set of indexes associated to the distribution grid nodes;
- $B = \{1, \ldots, b\}$ set of indexes associated to LUs of type building;
- $M = \{b+1, \ldots, b+m\}$ set of indexes associated to LUs of type microgrid;
- $A = B \cup M$ set of the indexes associated to all users;
- $H_j = \{1, \ldots, h_j\}$ set of indexes associated to the cogeneration plants of microgrid j, $j \in M$;
- $RR_j = \{1, \ldots, l_j\}$ set of indexes associated to the renewable sources of LU j, $j \in A$;
- $R_j = \{1, \ldots, r_j\}$ set of indexes associated to rooms in building j, $j \in B$.

It is necessary to consider the power flow equations for active and reactive power, which affect the power exchanges among LUs connected to the grid. As in [29,30], in the linearized form and in discrete time, these equations can be expressed as follows

$$p_{u,z,t} = G_{u,z}(v_{u,t} - v_{z,t}) + B_{u,z}(\delta_{u,t} - \delta_{z,t}) \quad u,z \in N, u \neq z, t = 0, \ldots, T-1 \tag{1}$$

$$q_{u,z,t} = B_{u,z}(v_{u,t} - v_{z,t}) - G_{u,z}(\delta_{u,t} - \delta_{z,t}) \quad u, z \in N, u \neq z, t = 0, \ldots, T - 1 \tag{2}$$

where:

- $G_{u,z}$ and $B_{u,z}$ are conductance and susceptance parameters for line (u, z);
- $v_{u,t}$ and $\delta_{u,t}$ are the voltage magnitude and phase at node u, respectively;
- $p_{u,z,t}$ and $q_{u,z,t}$ are the active and the reactive power flows (from u to z), respectively.

T is the number of time intervals considered, that is, the optimization horizon.

The quantities $v_{u,t}$, $p_{u,z,t}$ and $q_{u,z,t}$, are a-dimensional and are expressed with reference to a common voltage/power base value (i.e., as p. u. values).

The balance equation for active power at each node p is given by

$$\sum_{j \in A} CONN_{u,j} D_{j,t} + \sum_{\substack{z \in N \\ z \neq u}} P_{u,z,t} = 0 \quad u \in N, u \neq 1, \ t = 0, \ldots, T - 1 \tag{3}$$

where $CONN_{u,j}$ is equal to 1 if the LU j is directly connected to u and 0 otherwise. $D_{j,t}$ is the power flow [kW] to the j-th LU from the unique node to which is connected (assumed positive if it is sent to the user j). $P_{u,z,t}$ is the power flow [kW], unrestricted in sign, from node u to node z (i.e., the p.u. value ($p_{u,z,t}$) multiplied by the base value).

Note that (3) holds for $u \neq 1$. In fact, conventionally node 1 is defined as the connection node to the main grid, to which no LU is directly connected. For this node, instead of (3), the following equation holds:

$$\sum_{\substack{z \in N \\ z \neq 1}} P_{1,z,t} - P_{grid,t} = 0 \quad t = 0, \ldots, T - 1 \tag{4}$$

where $P_{grid,t}$ is the power flow [kW] between node 1 and the main grid (considered as positive when the power is entering the node).

Similarly, the balance equation for reactive power is given by:

$$\sum_{j \in A} CONN_{u,j} Q_{j,t} + \sum_{\substack{z \in N \\ z \neq u}} Q_{u,z,t} = 0 \quad u \in N, u \neq 1, \ t = 0, \ldots, T - 1 \tag{5}$$

where $Q_{j,t}$ is the reactive power exchange [kvar] between node u and the LUs and $Q_{p,q,t}$ represents reactive power flows from node u to node z. In this case too, node 1 has a different equation, namely:

$$\sum_{\substack{z \in N \\ z \neq 1}} Q_{1,z,t} - Q_{grid,t} = 0 \quad t = 0, \ldots, T - 1 \tag{6}$$

where $Q_{grid,t}$ is the reactive power flow [kvar] between node 1 and the main grid.

In this model, the following constraints must be fulfilled:

$$v_u^{MIN} \leq v_{u,t} \leq v_u^{MAX} \quad u \in N, \ t = 0, \ldots, T - 1 \tag{7}$$

$$Q_{grid}^{MIN} \leq Q_{grid,t} \leq Q_{grid}^{MAX} \quad t = 0, \ldots, T - 1 \tag{8}$$

$$P_{grid}^{MIN} \leq P_{grid,t} \leq P_{grid}^{MAX} \quad t = 0, \ldots, T - 1 \tag{9}$$

$$D_j^{MIN} \leq D_{j,t} \leq D_j^{MAX} \quad j \in A, \ t = 0, \ldots, T - 1 \tag{10}$$

$$Q_j^{MIN} \leq Q_{j,t} \leq Q_j^{MAX} \quad j \in A, \ t = 0, \ldots, T - 1 \tag{11}$$

where the upper and lower bounds have a straightforward meaning.

It is supposed that the AGG only requests for a variation (decrease) of active power. To this end, the AGG, in order to determine incentives and to divide among its customers the (active) load reduction, has to quantify the reductions' possibilities of each LU using limited information about the local systems' states. For this reason, an emulation is necessary (i.e., an approximate simulation) of the dynamic behavior of each LU, as described in the following sub-sections.

3.2. Emulation of the LUs Corresponding to Buildings

This subsection reports the model used for the emulation of a building, embedded in the statement of the higher-level problem. This model includes a storage system, renewables, and heat pumps. The state of charge (SOC) dynamics of storage systems (one for each building) can be represented by the following equations

$$SOC_{j,t+1} = loss_j SOC_{j,t} - \frac{\eta_j(P_{S,j,t})P_{S,j,t}\Delta}{CAP_j} \quad j \in B, \ t = 0,\dots,T-1 \tag{12}$$

$$\eta_j(P_{S,j,t}) = \begin{cases} \eta_{c,j} \ if \ P_{S,j,t} < 0 \\ 1/\eta_{d,j} \ otherwise \end{cases} \quad j \in B \tag{13}$$

where CAP_j [kWh] is the capacity of the storage in building j, $\eta_{c,j}$, $\eta_{d,j}$ are efficiency parameters [adim] (≤ 1) in charging and discharging modes, $loss_j$ [adim] is a loss coefficient due to the internal losses and Δ is the time discretization interval [h]. Instead, $P_{S,j,t}$ is the power flow (unrestricted in sign and assumed as positive when energy is drawn from the storage) from/to the storage [kW], and $SOC_{j,t}$ [adim] is the storage state of charge (expressed as a fraction with respect to the capacity).

The electrical balance equation includes the sum of power feeding the building (i.e., power from storage, renewables, power from the main grid) that is equal to the power demand $L_{j,t}$ minus a possible decrease $D_{AGG,j,t}$ [kW] requested by the AGG (being of course $D_{AGG,j,t} \leq L_{j,t}$). That is,

$$P_{RES,j,t} + P_{S,j,t} + D_{j,t} = L_{j,t} - D_{AGG,j,t} \quad j \in B, \ t = 0,\dots,T-1 \tag{14}$$

with

$$L_{j,t} = q_{HP,j,t} + q_{E,j,t} \quad j \in B, \ t = 0,\dots,T-1 \tag{15}$$

where $P_{RES,j,t}$ is the power from renewables for building j, $q_{HP,j,t}$ is the electrical power feeding the heat pumps, and $q_{E,j,t}$ the remaining forecasted electrical load.

As regards reactive power, it is possible to write:

$$Q_{RES,j,t} + Q_{S,j,t} + Q_{j,t} = Q_{F,j,t} \quad j \in B, \ t = 0,\dots,T-1 \tag{16}$$

where $Q_{RES,j,t}$, $Q_{S,j,t}$, and $Q_{F,j,t}$ are the reactive power [kvar] associated with renewables and storage systems and the forecasted demand, respectively, and $Q_{j,t}$ is the reactive power [kvar] exchange with the main grid.

It is supposed that the AGG has no detailed information about the temperature in the various rooms in each building. Thus, a single variable $T_{j,t}$ [K] is used to represent the temperature of the overall building j. The dynamics of such temperatures is represented as in [31], namely:

$$T_{j,t+1} = T_{j,t} + \frac{\Delta}{C_{TH,j}}[\eta_s q_{HP,j,t} - \frac{1}{R_{TH,ext,j}}(T_{j,t} - T_{ext,t})] \quad j \in B, \ t = 0,\dots,T-1 \tag{17}$$

where $C_{TH,j}$ is the thermal capacitance in building j [kWh/K], $T_{ext,t}$ is the external temperature [K], $\eta_s q_{HP,j,t}$ is the thermal power [kW] provided by heat pumps, $R_{TH,ext,j}$ is the thermal resistance [K/kW] between building j and the external environment, and η_s [adim] is given by:

$$\eta_s = \sigma \eta_{s,heat} - (1-\sigma)\eta_{s,cool} \quad j \in B \tag{18}$$

where σ is a given constant specifying whether the heat pump operates in heating mode ($\sigma = 1$) or in the cooling mode ($\sigma = 0$), and $\eta_{s,heat}$ and $\eta_{s,cool}$ are efficiency parameters [adim].

The following upper and lower bound constraints must be taken into account:

$$T_j^{MIN} \leq T_{j,t} \leq T_j^{MAX} \quad j \in B, \ t = 0, \ldots, T-1 \tag{19}$$

$$0 \leq q_{HP,j,t} \leq q_{HP}^{MAX} \quad j \in B, \ t = 0, \ldots, T-1 \tag{20}$$

$$SOC_j^{MIN} \leq SOC_{j,t} \leq SOC_j^{MAX} \quad j \in B, \ t = 0, \ldots, T-1 \tag{21}$$

$$P_{S,j,t}^{MIN} \leq P_{S,j,t} \leq P_{S,j,t}^{MAX} \quad j \in B, \ t = 0, \ldots, T-1 \tag{22}$$

$$Q_{S,j,t}^{MIN} \leq Q_{S,j,t} \leq Q_{S,j,t}^{MAX} \quad j \in B, \ t = 0, \ldots, T-1 \tag{23}$$

$$Q_{RES,j}^{MIN} \leq Q_{RES,j,t} \leq Q_{RES,j}^{MAX} \quad j \in B, \ t = 0, \ldots, T-1 \tag{24}$$

where all bounds for control and state variables are known parameters, being $P_{S,j,t}^{MIN}$ and $Q_{S,j,t}^{MIN}$ negative values. Note that, whereas $P_{RES,j,t}$ is a given quantity (known at least as a forecast), $Q_{RES,j,t}$ may be considered as a decision variable since its value may be adjusted, for any time interval, by means of an electronic system based on the use of inverters.

Finally, the following constraints must be satisfied:

$$D_{AGG,j,t} \leq D_{AGG,j,t}^{MAX} \quad j \in B, \ t = 0, \ldots, T-1 \tag{25}$$

$$\sum_{t=0}^{T-1} D_{AGG,j,t} \leq D_{AGG,TOT,j} \quad j \in B, \tag{26}$$

where the upper bounds $D_{AGG,j,t}^{MAX}$ and $D_{AGG,TOT,j}$ are given for each building.

3.3. Emulation of the LUs Corresponding to Microgrids

This subsection presents the model used to emulate microgrids within the statement of the higher-level problem. Microgrids can be modeled in a similar way as for buildings. However, the system model is in this case a bit more complicated because the thermal balance must be considered, given the presence of cogeneration plants in microgrids.

Specifically, the electrical balance is given by:

$$P_{RES,j,t} + P_{S,j,t} + P_{EL,j,t} + D_{j,t} = q_{HP,j,t} + q_{E,j,t} - D_{AGG,j,t} \quad j \in M, \ t = 0, \ldots, T-1 \tag{27}$$

$$Q_{RES,j,t} + Q_{S,j,t} + Q_{j,t} = Q_{F,j,t} \quad j \in M, \ t = 0, \ldots, T-1 \tag{28}$$

where $P_{EL,j,t}$ is the power produced from fossil fuel plants.

The thermal balance in the heating operation mode is given by:

$$P_{TH,BOIL,j,t} + P_{TH,RES,j,t} + P_{TH,j,t} \geq D_{H,j,t} \quad j \in M, \ t = 0, \ldots, T-1 \tag{29}$$

where $P_{TH,BOIL,j,t}$, $P_{TH,RES,j,t}$, $P_{TH,j,t}$ represent thermal power from the gas-fed boiler, renewables, and cogeneration plant, respectively, and $D_{H,j,t}$ is the thermal demand for heat.

Instead, in the cooling operation mode, the balance is:

$$P_{TH,RES,j,t} + P_{TH,j,t} \geq D_{H,j,t} + P_{TH,CHI,j,t} \quad j \in M, \ t = 0, \ldots, T-1 \tag{30}$$

where $P_{TH,CHI,j,t}$ is the thermal power needed to feed chillers. Moreover, in the cooling mode, also the following constraints need to be satisfied:

$$\eta_{s,cool}q_{HP,j,t} + \chi P_{TH,CHI,j,t} \geq D_{C,j,t} \quad j \in M, \ t = 0, \ldots, T-1 \tag{31}$$

where $D_{C,j,t}$ is the cooling power demand, and χ is the efficiency of the chiller [adim].

Besides, also in the case of microgrids, constraints (12),(13) (written of course for $j \in M$) and (20)÷(26) must be fulfilled.

Finally, owing to the finite capacity of the plants, the following upper bounds must be considered:

$$P_{EL,j,t} \leq P_{EL}^{MAX} \quad j \in M, \ t = 0, \ldots, T-1 \tag{32}$$

$$P_{TH,CHI,j,t} \leq P_{TH,CHI,j}^{MAX} \quad j \in M, \ t = 0, \ldots, T-1 \tag{33}$$

3.4. The Higher-Level Optimization Problem

In this subsection, the statement of the higher-level optimization problem solved by the AGG will be provided. This problem will embed, as constraints, all equations and inequalities that have been introduced in the previous subsections in order to emulate the LUs' behavior.

The cost function of the higher-level decision problem consists of several terms.

(A) The first of them is C_{inc} [€], that is the cost relevant to incentives that must be provided to the LUs in order to diminish their own load. Let $C_{AGG,j,t}$ [€/h], the incentive that the AGG gives to the generic j-*th* LU per unit time. The following structure of $C_{AGG,j,t}$ as a function of $D_{AGG,j,t}$ is assumed:

$$C_{AGG,j,t} = a_j(D_{AGG,j,t})^2 + b_j D_{AGG,j,t} + c_j \quad j \in A, \ t = 0, \ldots, T-1 \tag{34}$$

where a_j, b_j and c_j are given parameters of suitable physical dimension. Then, C_{inc} can be expressed as:

$$C_{inc} = \sum_{t=0}^{T-1} \sum_{j \in A} C_{AGG,j,t} \Delta \tag{35}$$

(B) The second term (actually a benefit) is B_{DR} [€], that is the benefit from the external grid due to the load reduction. To express this term, note that the benefit recieved from the market for the generic time interval is given by a known unit cost $C_{Market,t}$ [€/kWh] multiplied by the reduction of power DR_t [kW] at the point of connection with the main grid. It is supposed that an overall (maximal) reduction MR_t $t = 0, \ldots, T-1$ [kW] has been agreed in advance. The reduction of power DR_t can be formalized as the difference between the forecasted request $P_{grid,da,t}$ (in the day-ahead) and the new actual request $P_{grid,t}$ that has to be determined. That is:

$$DR_t = P_{grid,da,t} - P_{grid,t} \quad t = 0, \ldots, T-1 \tag{36}$$

It is important to note that $P_{grid,da,t}$ is a given value that cannot be updated, whereas $P_{grid,t}$ is a decision variable appearing in (4).

Moreover, if the power reduction is lower than MR_t, there is a fee that must be paid. Thus the overall benefit obtained through a set of reductions DR_t $t = 0, \ldots, T-1$ is given by:

$$B_{DR} = \sum_{t=0}^{T-1} \left[C_{Market,t} DR_t - C_{fee,t}(MR_t - DR_t) \right] \Delta \tag{37}$$

where $C_{fee,t}$ is a unit cost coefficient [€/kWh]. Namely, this coefficient represents the cost associated to a unit of unsatisfied energy demand. The following constraints hold as regards DR_t, namely:

$$0 \leq DR_t \leq MR_t \quad t = 0,\ldots,T-1 \tag{38}$$

On this basis, B_{DR} can be written simply as:

$$B_{DR} = \sum_{t=0}^{T-1} \left[\left(C_{Market,t} + C_{fee,t} \right) DR_t - C_{fee,t} MR_t \right] \tag{39}$$

(C) Finally, there is a term called "Assignment Fairness" (C_{AF} [kW^2]) that is introduced to avoid too unbalanced power reductions among the population of LUs. This term can be written as:

$$C_{AF} = \sum_{t=0}^{T-1} \sum_{j \in A} \left[D_{AGG,j,t} - DR_t \frac{L_{j,t}}{\sum_{l \in A} L_{l,t}} \right]^2 \tag{40}$$

At this point it is possible to formalize the overall AGG decision problem as follows. The higher-level (AGG) decision problem:

$$\min J = C_{inc} - B_{DR} + \omega C_{AF} \tag{41}$$

s.t. constraints (1)÷(40) and

$$\alpha_{min} C_{Market,t} D_{AGG,j,t} \leq C_{AGG,j,t} \leq \alpha_{max} C_{Market,t} D_{AGG,j,t} \quad j \in A, \ t = 0,\ldots,T-1 \tag{42}$$

In the expression of the cost (41), ω [€/kW^2] is a tradeoff coefficient whose value should be tuned on the basis of the real case study considered.

Constraint (42) must be inserted in order to avoid excessive discrepancies between the benefit provided by the market to the AGG (with reference to a specific LU j) and the overall incentive provided by the AGG to that LU. In (42) α_{min} and α_{max} are fixed parameters (<1) and represent minimum and maximum fractions of the benefit to the LUs.

Having so represented the optimization problem to be solved by the AGG, we are now in a position to introduce problems that have to be solved by each LU.

4. The LUs' Optimization Problem

The LUs track the references coming from the solution of the AGG problem, but have detailed local information and may decide how to manage production plants and storage systems in order to accomplish the AGG's requests.

4.1. The Model Used for the Optimization Problem of a Building

In the detailed model considered for the optimization of a single building, each room is individually considered. The dynamics of the temperature of the i-th room of building j ($\widetilde{T}_{i,j,t}$ [K]), is represented as:

$$\widetilde{T}_{i,j,t+1} = \widetilde{T}_{i,j,t} + \frac{\Delta}{C_{i,j}} \left[\eta_s q_{i,j,t} - \frac{1}{\widetilde{R}_{TH,ext,i,j}} \widetilde{A}_{ext,i,j} \left(\widetilde{T}_{i,j,t} - T_{ext,t} \right) - \sum_{\substack{r \in R_j \\ r \neq i}} \frac{1}{\widetilde{R}_{TH,i,r,j}} \widetilde{A}_{i,r,j} \left(\widetilde{T}_{i,j,t} - \widetilde{T}_{r,j,t} \right) \right]$$

$$j \in B, \ t = 0,\ldots,T-1, \ i \in R_j$$

$$\tag{43}$$

where:

- $C_{i,j}$ is the thermal capacitance of room i in building j [kWh/K];

- $T_{ext,t}$ is external temperature [K];
- $\widetilde{R}_{TH,ext,i,j}$ is the thermal resistance between room i in building j and the external environment [k/kW];
- $\widetilde{R}_{TH,i,r,j}$ is the thermal resistance between room i and room r in building j [K/kW];
- $\widetilde{A}_{i,r,j}$ is an entry of the adjacency matrix: It is equal to 1 if room i is adjacent to room r and 0 otherwise [adim];
- $\widetilde{A}_{ext,i,j}$ is a coefficient equal to 1 if room i is adjacent to external environment and 0 otherwise [adim];
- $\eta_s\, q_{i,j,t}$ is the thermal power (unrestricted in sign) provided by heat pumps in room i in building j [kW].

Moreover, let

$$\widetilde{q}_{HP,j,t} = \sum_{i \in R_j} q_{i,j,t} \quad j \in B,\ t = 0,\ldots,T-1 \tag{44}$$

where $\widetilde{q}_{HP,j,t}$ has the same physical meaning as $q_{HP,j,t}$ but now it corresponds to the power provided to the heat pumps in building j that is actually determined solving the problem for the detailed model of that building.

Even for the detailed model of the single building the presence of a single storage element is considered, whose dynamics can still be represented via (12)–(13).

The power balance equation is given by:

$$P_{RES,j,t} + \widetilde{P}_{S,j,t} + D_{LU,j,t} = \widetilde{L}_{j,t} - D_{AGG,j,t} \quad j \in B,\ t = 0,\ldots,T-1 \tag{45}$$

which has the same meaning as (14) but where:

- $\widetilde{P}_{S,j,t}$ is the power flow form the storage element corresponding to the solution of the local problem;
- $D_{LU,j,t}$ is the actual power flow from the main grid to the LU j, determined by the solution of the local problem;
- $\widetilde{L}_{j,t}$ is the building load, expressed in a more detailed way with respect to (14), that is:

$$\widetilde{L}_{j,t} = \widetilde{q}_{HP,j,t} + P_{veh,j,t} + P_{wash,j,t} + \widetilde{q}_{E,j,t} \quad j \in B,\ t = 0,\ldots,T-1 \tag{46}$$

Including:

- fixed consumption $\widetilde{q}_{E,j,t}$;
- power necessary to feed electrical vehicles $P_{veh,j,t}$;
- power necessary to feed washing machines $P_{wash,j,t}$.

However, there is a fundamental difference between (45)–(46) with respect to (14)–(15). In fact, whereas $L_{j,t}$, $q_{HP,j,t}$, and $q_{E,j,t}$ are considered as fixed in (14)–(15), quantities $\widetilde{L}_{j,t}$, $P_{veh,j,t}$, and $P_{wash,j,t}$ are decision variables within the problem of the LU corresponding to building j. The washing machines (partly) and the electrical vehicle loads are considered as deferrable, provided that the following constraints are fulfilled:

$$\sum_{t=0}^{T-1} P_{veh,j,t}\Delta = E_{TOT,veh,j} \quad j \in B \tag{47}$$

$$P_{wash,j,t} = P_{wash,j,t}^{fix} + P_{wash,j,t}^{def} \quad j \in B,\ t = 0,\ldots,T-1 \tag{48}$$

$$\sum_{t=0}^{T-1} P_{wash,j,t}^{def}\Delta = E_{TOT,wash,j}^{def} \quad j \in B \tag{49}$$

where $E_{TOT,veh,j}$, is the total electrical energy demand for electrical vehicles [kWh]. $P_{wash,j,t}$ is the power demand for the washing machines that is split in two terms: one fixed $P_{wash,j,t}^{fix}$ (not adjustable) and one deferrable $P_{wash,j,t}^{def}$. $E_{TOT,wash,j}^{def}$ is the total deferrable energy demand [kWh] for washing machines.

4.2. The Optimization Problem for a Building

Each LU j receives form the AGG, as a part of the solution of its problem, the load reductions $D_{AGG,j,t}t = 0,\ldots,T-1$ attributed to the specific LU, and the (reference) values $D_{j,t}t = 0,\ldots,T-1$, corresponding to the power flow between the grid and that LU. Then, the LU problem can be formalized as the minimization of the quadratic deviation between the sequence $D_{LU,j,t}$ of the actual power flow from the main grid and the LU, and the reference sequence $D_{j,t}$.

Optimization Problem For The Single Building j

$$min J_{B,j} = \sum_{t=0}^{T-1}\left(D_{LU,j,t} - D_{j,t}\right)^2 \tag{50}$$

s.t. constraints (12)–(13) and (43)÷(49), written for the index j of interest.

4.3. The Model Used for the Optimization Problem of a Microgrid

We consider a model with the following entities: Microturbines (representing the cogeneration plants), photovoltaics, electric vehicles charging stations, heat pumps and chillers. Even for the detailed model of the single microgrid, the presence of a single storage element is considered, whose dynamics can still be represented via (12)–(13).

In this model, the electric power balance is given by:

$$\sum_{h \in H_j} P_{EL,h,j,t} + \sum_{l \in RR_j} P_{RES,l,j,t} + \widetilde{P}_{S,j,t} + D_{LU,j,t} = \widetilde{L}_{j,t} - D_{AGG,j,t} \quad j \in M, \ t = 0,\ldots,T-1 \tag{51}$$

$$\widetilde{L}_{j,t} = \widetilde{q}_{HP,j,t} + \widetilde{q}_{E,j,t} + P_{veh,j,t} \quad j \in M \tag{52}$$

$$\sum_{l \in RR_j} Q_{RES,l,j,t} + \widetilde{Q}_{S,j,t} + Q_{LU,j,t} = Q_{D,j,t} \quad j \in M, \ t = 0,\ldots,T-1 \tag{53}$$

where:

- $P_{EL,h,j,t}$ is the active power coming from the controllable microturbine h in microgrid j;
- $P_{RES,l,j,t}$ is the active power that is injected from local renewable energy sources of kind l;
- $\widetilde{P}_{S,j,t}$ is the active power (unrestricted in sign) drawn from the storage;
- $D_{LU,j,t}$ is the power coming from the grid;
- $\widetilde{q}_{HP,j,t}$ is the power requested by heat pumps;
- $\widetilde{q}_{E,j,t}$ is the fixed electrical demand;
- $P_{veh,j,t}$ is the power consumed by electric vehicles.

Powers $P_{veh,j,t}$ is considered as a deferrable load and is subject to constraints similar to those in Equation (47).

Furthermore, $Q_{RES,l,j,t}$ and $\widetilde{Q}_{S,j,t}$, are the corresponding reactive power flows. Note that it is assumed that it is possible to regulate the reactive power from the controllable microturbines, imposing them as equal to zero. Thus, there is no decision variable corresponding to these powers. Instead, it is assumed that the reactive power coming from the renewable energy sources, as well as the reactive power coming from the storage, can be regulated. Thus, such powers are considered as decision variables. Besides, $Q_{D,j,t}$ is a fixed amount of reactive power corresponding to the heat pumps and the electric load (the reactive power for electric vehicles is assumed to be zero).

Besides, the following constraints must be taken into account:

$$P_{EL,h,j,t} \leq P_{EL,h,j}^{MAX} \quad j \in M, \ t = 0,\ldots,T-1 \tag{54}$$

The thermal balance constraints, for the heating mode, correspond to:

$$\widetilde{P}_{TH,BOIL,j,t} + \sum_{l\in RR_j} P_{TH,RES,l,j,t} + \sum_{h\in H_j} P_{TH,h,j,t} \geq D_{H,j,t} \quad j \in M, \; t = 0,\ldots,T-1 \tag{55}$$

whereas, for the cooling mode, the constraints are:

$$\sum_{l\in RR_j} P_{TH,RES,l,j,t} + \sum_{h\in H_j} P_{TH,h,j,t} \geq D_{H,j,t} + \widetilde{P}_{TH,CHI,j,t} \quad j \in M, \; t = 0,\ldots,T-1 \tag{56}$$

$$\eta_{s,cool}\widetilde{q}_{HP,j,t} + \chi\widetilde{P}_{TH,CHI,j,t} \geq D_{C,j,t} \quad j \in M, \; t = 0,\ldots,T-1 \tag{57}$$

$$\widetilde{P}_{TH,CHI,j,t} \leq P^{MAX}_{TH,CHI,j} \quad j \in M, \; t = 0,\ldots,T-1 \tag{58}$$

where:

- $\widetilde{P}_{TH,BOIL,j,t}$ is the thermal power produced by the (unique) gas-fed boiler;
- $P_{TH,RES,l,j,t}$ is the thermal power from the *l-th* renewable energy source;
- $P_{TH,h,j,t}$ is the thermal power produced by the *h-th* co-generative microturbine;
- $\widetilde{P}_{TH,CHI,j,t}$ is the thermal power consumed by the (unique) chiller;

It is important to note that the thermal balance and the electrical power balance are coupled owing to the presence of co-generative microturbines. Namely, for a given microturbine h, it is assumed that the following relationship holds:

$$P_{TH,h,j,t} = \mu_{h,j} P_{EL,h,j,t} \quad j \in M, \; t = 0,\ldots,T-1 \tag{59}$$

where $\mu_{h,j}$ is a given parameter characteristic of the considered microturbine. More complex relationships could be used instead of (59) but, for the sake of simplicity, in this paper a purely linear model is adopted.

4.4. The Optimization Problem for a Microgrid

Even in this case, the optimization problem can be stated as the minimization of the quadratic deviation of the actual behavior with respect to the reference value given by the upper problem, as regards both active ($D_{j,t}t = 0,\ldots,T-1$) and reactive ($Q_{j,t}t = 0,\ldots,T-1$) powers.

Optimization problem for the single microgrid j:

$$minJ_{M,j} = \sum_{t=0}^{T}\left(D_{LU,j,t} - D_{j,t}\right)^2 + \left(Q_{LU,j,t} - Q_{j,t}\right)^2 \quad j \in M \tag{60}$$

s.t. constraints (12)–(13), (47) and (51)÷(59)

5. Application to a Case Study

The developed bi-level architecture has been applied to a case study in the Liguria Region, Italy. In particular, the considered case study is related to Savona Municipality data, which is located in the north western part of Italy. In particular, on the territory, it is already present a microgrid at the campus level [29], which includes heat pumps, buildings and renewables, a laboratory microgrid, and a sustainable building that, with renewables and storage systems, can be operated in islanded mode and considered as a microgrid. Moreover, the municipality is planning to replicate the microgrid in different areas and to constitute an energy community in the context of a so-called "smart city project". For these areas data have been estimated on the basis of territorial characteristics and the Savona Campus Microgrid.

Specifically, six different local users have been considered. Three of them are building type LUs, while others are microgrid type LUs. Figure 3 shows a scheme of the considered case study. The network is composed by a slack node (i.e., the connection to the main grid, green), plus other 5 (blue) nodes: two with building type LUs connected, and three with microgrid type LUs.

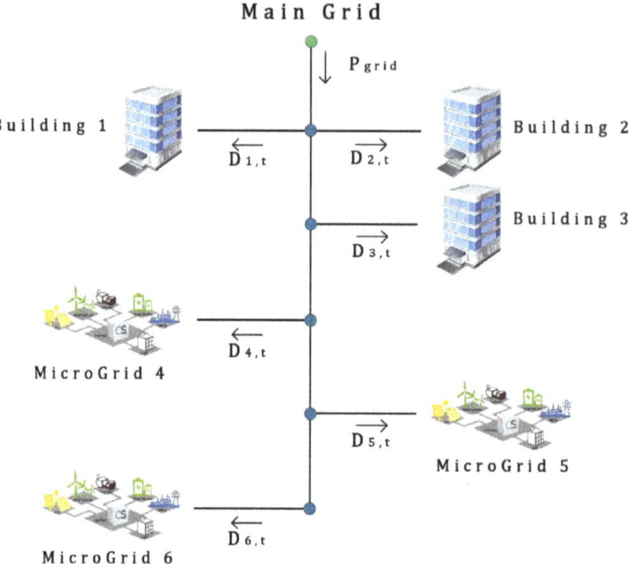

Figure 3. Network architecture of the case study.

The model has been implemented with the optimization modeling software LINGO and the results are presented in the following sub-sections. Due to the high-computational burden, each problem runs in about 4 to 6 min.

5.1. The Available Data and Demands

The upper level represents the AGG and emulates the LUs' behavior in order to determine and distribute the demand reduction. The considered power production from renewables for the different local users (only from PV plants) is reported in Figure 4, while electrical ($L_{j,t}$) and thermal ($D_{H,j,t}$) demands (the latter is considered only for microgrids) of two LUs (one building and one microgrid) are reported in Figures 5 and 6. Electrical and thermal demands for the other LUs have profiles similar to those reported in Figures 5 and 6, respectively. Note that a 24 h optimization horizon has been considered, whereas the time discretization interval is 1 h.

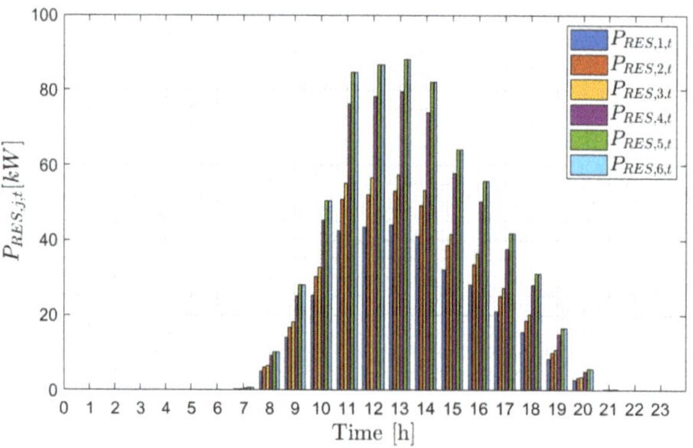

Figure 4. Renewable power production of the LUs.

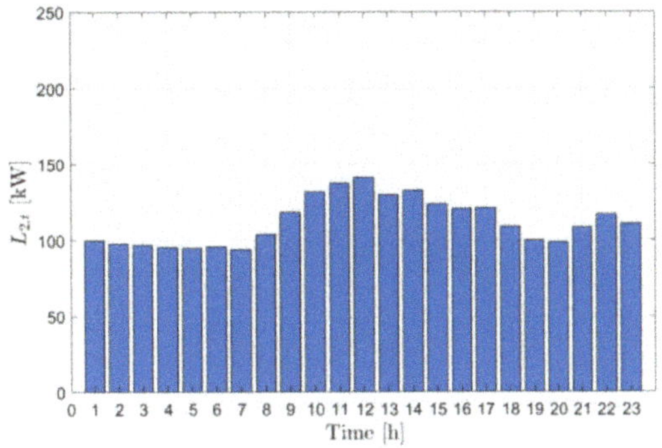

Figure 5. Electrical demand of the LU number 2 (Building).

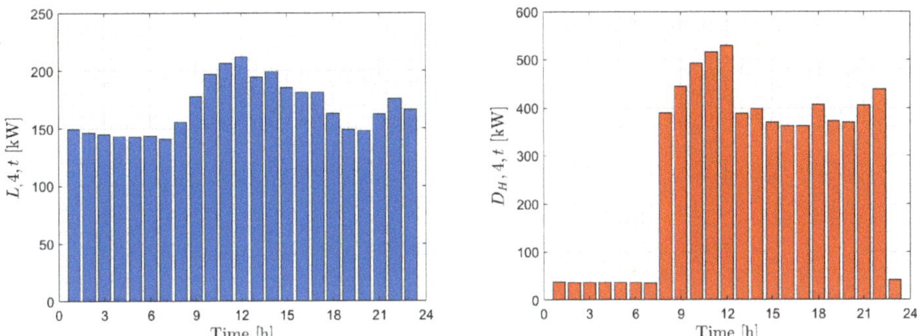

Figure 6. Electrical and thermal demand of the LU number 4 (microgrid).

5.2. Results for the AGG Optimization Problem

The optimal results for the AGG's optimization problem are presented in Figures 7–9.

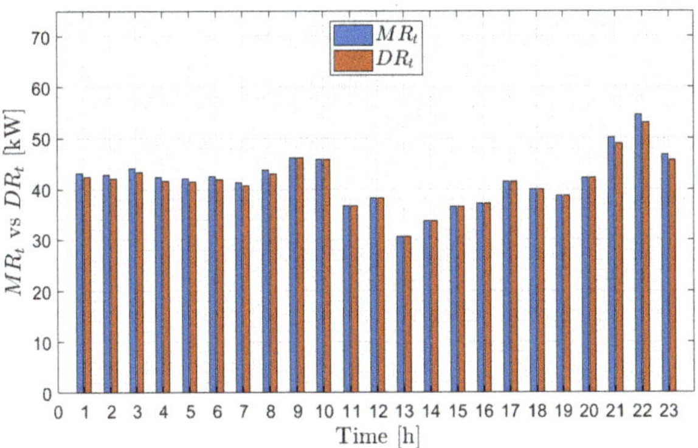

Figure 7. Power reduction at the grid node.

The optimal solution has a power reduction DR_t as presented in Figure 7. It highlights that the power reduction requested at the grid node (MR_t) is almost satisfied along the entire optimization interval, with a slight difference during the first and the last hours due to the absence of renewable power.

Besides, Figure 8 reports the values for the power exchanged between the main grid and each LU ($D_{j,t}$), while Figure 9 shows the optimal results related to the power reduction $D_{AGG,j,t}$. It can be noted that the overall power reduction is split in rates which are maintained almost equal along the whole day. This is mainly due to the presence of the "Assignment Fairness" term in the cost function minimized in the higher level problem.

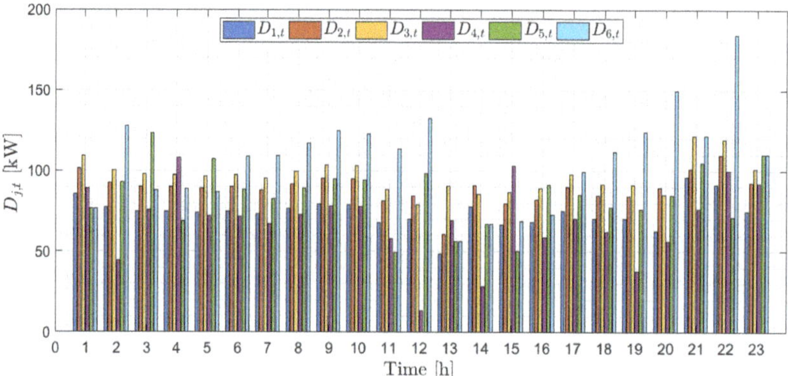

Figure 8. Values of the power exchanged by the LUs and the main grid.

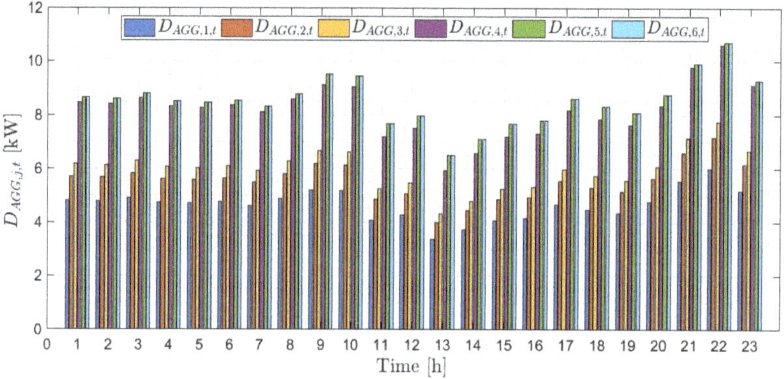

Figure 9. Optimal power reduction resulted from upper-level calculation.

Once the upper-level optimization problem is solved, its solution is used as a reference pattern for the lower-level decision problems.

5.3. Lower-Level Results: Building-Type LUs

The results of the lower-level optimization problems are presented in Figures 10 and 11. It can be noted that the power reference pattern set by the AGG is tracked almost exactly in the solution of the lower-level problems (Figure 10).

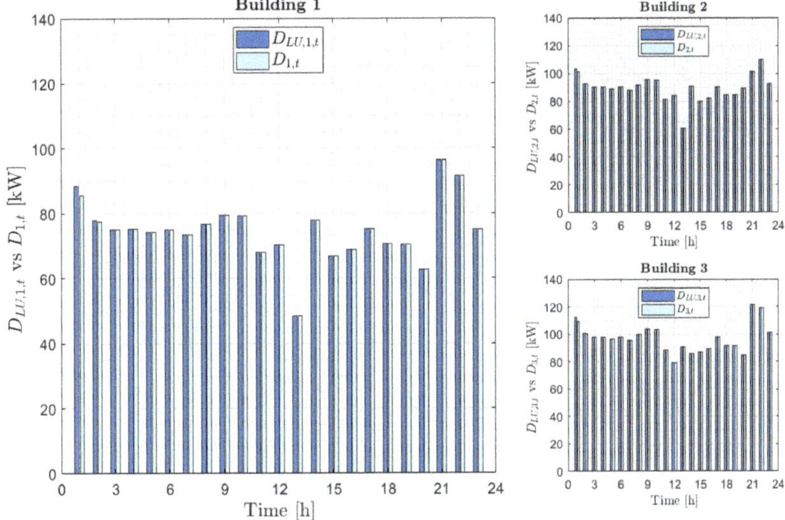

Figure 10. Power exchange between the nodes and the LUs (buildings) compared with their reference values.

Figure 11 presents the electrical balance (51) at each LU where the largest portion of the load is satisfied by purchasing from the grid since there is not any other power sources but the PV plants and the storage. Note that, for the sake of simplicity in the representation of the power pattern, the power exchanged with the storage $\widetilde{P}_{S,j,t}$ is represented as:

$$\widetilde{P}_{S,j,t} = \widetilde{P}_{S,inj,j,t} - \widetilde{P}_{S,abs,j,t} \quad j \in B, \ t = 0,\dots,T-1 \tag{61}$$

where $\widetilde{P}_{S,inj,j,t}$ the first term in the r.h.s. represents the power drawn from the storage and injected into the building, whereas $\widetilde{P}_{S,abs,j,t}$ represents the power absorbed by the storage.

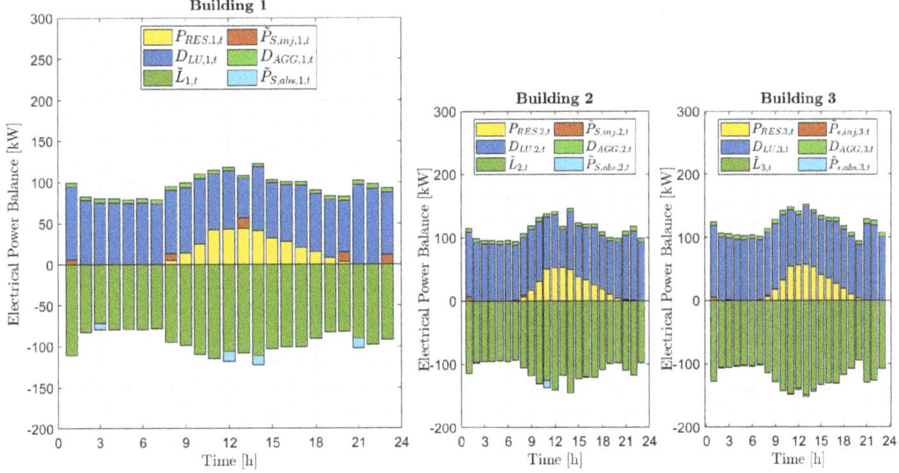

Figure 11. Electrical balance in the LUs (buildings).

5.4. Lower-Level Results: Microgrid Type LU

The second type of LUs are microgrids. They have higher flexibility due to the presence of a larger pool of energy production technologies. This leads to a solution (Figure 12) which perfectly fits the reference data.

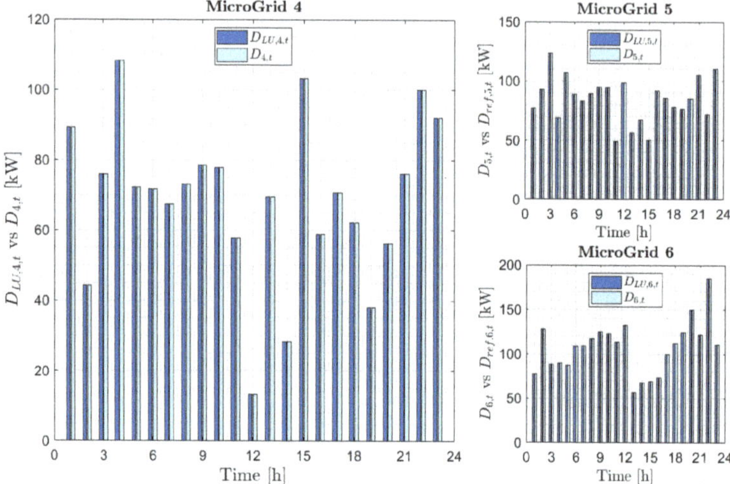

Figure 12. Power exchange between the nodes and the LUs (microgrids) compared with their reference values.

In Figures 13 and 14 both the electrical (51) and thermal balances (56)are reported, highlighting the variety of the energy production pool. Specifically, the presence of two microturbines for each of the three microgrids, one with electrical nominal power of 35 and 65 kW (C30 and C65, respectively). The fraction of the load satisfied through the energy bought from the grid is definitely lower than in the case of buildings because the microturbines are co-generative plants, which are able to satisfy also thermal demands.

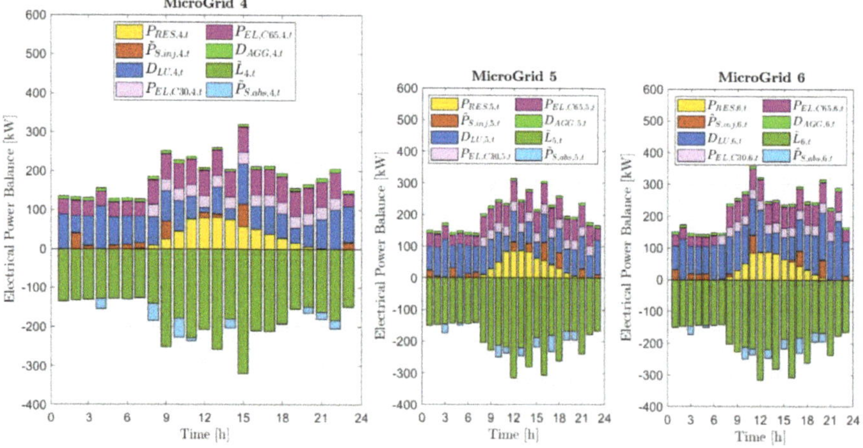

Figure 13. Electrical balance in the LUs (microgrids).

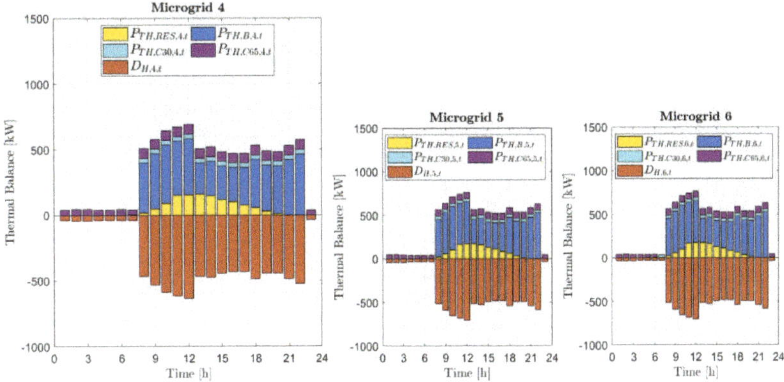

Figure 14. Thermal balance in the LUs (microgrids).

As expected, the higher flexibility of the microgrids provides a better solution in regards to the objective function, since the electrical profile totally fits the reference one (see Figure 12).

6. Conclusions

Demand response is a subject of utmost importance and it can provide a noteworthy improvement to the performance of the grid. Of course, this requires the development and the application of viable and effective approaches for energy management. In this framework, an approach was proposed in this paper based on a bi-level architecture in which an AGG of consumers plays the role of a higher-level decision-maker. The problem solved at the AGG level is relevant to the balancing market, and incentives for LUs (i.e., buildings and microgrids) are considered as well as their flexibility. The higher-level, representing the AGG, determines the reference values, to be considered by the LUs, minimizing the overall cost function considered. At the lower level, the local users try to follow the reference values and, through demand response programs, try to contain their costs and to satisfy their demands. The microgrids are characterized by higher flexibility with respect to the other LUs (buildings), owing to the presence of several production plants. The proposed architecture has been developed within the following assumptions in regards to the (bidirectional) interaction between the two levels. First of all, the higher level provides the reference patterns for the lower-level problems. Besides, there is information flow from the lower level to the upper level in regards to the storage initial state and the maximum affordable reduction for each LU. The application of the presented decision scheme could be viewed within a receding-horizon framework, in which, at each time step, only the solution referring to the next time discretization interval is applied. Then, at the next time step, a new solution to the same problem, concerning an optimization interval of the same length as above, is determined after having received the information about the current system state. Further developments will be the implementation of more detailed models for single plants and the use of different renewable energy sources. Another possible extension could the application of the model to grids with different architectures to determine if the load flexibility can be improved.

Author Contributions: Conceptualization, M.R. (Michela Robba), G.F., L.P., R.M. and M.R. (Mansueto Rossi); Methodology, M.R. (Michela Robba), G.F., L.P., R.M. and M.R. (Mansueto Rossi); Software, M.R. (Michela Robba), G.F., L.P., and M.R. (Mansueto Rossi); Validation, M.R. (Michela Robba), G.F., L.P., R.M. and M.R. (Mansueto Rossi); Formal Analysis, M.R. (Michela Robba), G.F., L.P., R.M. and M.R. (Mansueto Rossi); Investigation, M.R. (Michela Robba), G.F. and L.P.; Resources, M.R. (Michela Robba), L.P. and M.R. (Mansueto Rossi); Data Curation, M.R. (Michela Robba), L.P. and M.R. (Mansueto Rossi); Writing-Original Draft Preparation, M.R. (Michela Robba), G.F., L.P., R.M. and M.R. (Mansueto Rossi); Writing-Review & Editing, M.R. (Michela Robba), G.F., L.P., R.M. and M.R. (Mansueto Rossi); Visualization, M.R. (Michela Robba), G.F., L.P., R.M. and M.R. (Mansueto Rossi); Supervision, M.R. (Michela Robba), G.F., L.P., R.M. and M.R. (Mansueto Rossi); Project Administration, M.R. (Michela Robba), R.M.; All authors have read and agreed to the published version of the manuscript.

Funding: This research received no external funding. The outputs of the optimization results have been used as input data for the development of an Energy Management System for sustainable districts that will be developed within the PICK-UP innovation project funded by Liguria Region within the Innovation Pole on Energy Technologies and Sustainable Development (EASS).

Conflicts of Interest: The authors declare no conflict of interest.

Nomenclature

Sets:

$T = \{1, \ldots, t\}$ set of time instants;

$N = \{1, \ldots, n\}$ set of indexes associated to the distribution grid nodes;

$B = \{1, \ldots, b\}$ set of indexes associated to LUs of type building;

$M = \{b+1, \ldots, b+m\}$ set of indexes associated to LUs of type microgrid;

$A = B \cup M$ set of the indexes associated to all users;

$H_j = \left\{1, \ldots, h_j\right\}$ set of indexes associated to the cogeneration plants of microgrid j, $j \in M$;

$RR_j = \left\{1, \ldots, l_j\right\}$ set of indexes associated to the renewable sources of LU j, $j \in A$;

$R_j = \left\{1, \ldots, r_j\right\}$ set of indexes associated to rooms in building j, $j \in B$.

Decision Variables:

$p_{u,z,t}$ = active power flow for line (u,z) at time t [p.u.];

$P_{u,z,t}$ = active power flow for line (u,z) at time t [kW];

$q_{u,z,t}$ = reactive power flow for line (u,z) at time t [p.u.];

$Q_{u,z,t}$ = reactive power flow for line (u,z) at time t [kvar];

$v_{u,t}$ = voltage magnitude at node u at time t [p.u.];

$\delta_{u,t}$ = phase magnitude at node u at time t [p.u.];

$D_{j,t}$ = active power exchange from the node u to the LU at time t [kW];

$Q_{j,t}$ = reactive power exchange from the node u to the LU at time t [kvar];

$P_{grid,t}$ = active power exchange form the slack node to the main grid at time t [kW];

$Q_{grid,t}$ = reactive power exchange from the slack node to the main grid at time t [kvar];

$SOC_{j,t}$ = State of Charge of the storage element j at time t;

$Q_{RES,j,t}$ = reactive power from renewables of the j-th LU at time t [kvar];

$P_{S,j,t}$ = active power exchange with the storage of the j-th LU at time t [kW];

$Q_{S,j,t}$ = reactive power exchange with the storage of the j-th LU at time t [kvar];

$D_{AGG,j,t}$ = active power decrease requested by the AGG to the j-th LU at time t [kW];

$q_{HP,j,t}$ = electrical power provided by heat pumps in the j-th LU at time t [kW];

$L_{j,t}$ = active power demand of the j-th LU at time t [kW];

$T_{j,t}$ = temperature of the building j at time t [K];

$P_{EL,j,t}$ = power produced from cogeneration plants in microgrid j at time t [kW];

$P_{TH,BOIL,j,t}$ = thermal power produced by the boiler in microgrid j at time t [kW];

$P_{TH,j,t}$ = thermal power produced by the cogeneration plants in microgrid j at time t [kW];

DR_t = actual power reduction at the slack node at time t [kW];

$C_{AGG,j,t}$ = incentive that the AGG gives to the generic j-th LU per unit time [€/h];

$D_{LU,j,t}$ = actual power flow (calculated at the lower level) from the grid to the LU j at time t [kW];

$\widetilde{T}_{i,j,t}$ = temperature of the i-th room of building j at time t [K];

$\widetilde{q}_{HP,j,t}$ = same physical meaning as $q_{HP,j,t}$ in the lower-level problem [kW];

$\widetilde{P}_{S,j,t}$ = same physical meaning as $P_{S,j,t}$ in the lower-level problem [kW];

$\widetilde{L}_{j,t}$ = same physical meaning as $L_{j,t}$ in the lower-level problem [kW];

$P_{veh,j,t}$ = power necessary to feed electrical vehicles in the j-th LU at time t [kW];

$P_{wash,j,t}$ = power necessary to feed washing machines in building j at time t [kW];

$P^{def}_{wash,j,t}$ = deferrable fraction of $P_{wash,j,t}$ [kW];

$Q_{LU,j,t}$ = actual reactive power flow (calculated at the lower level) from the grid to the microgrid j at time t [kvar];

$P_{EL,h,j,t}$ = active power coming from the cogenerative microturbine h in microgrid j at time t [kW];

$P_{TH,h,j,t}$ = thermal power produced by the *h-th* cogenerative microturbine in microgrid j at time t [kW];

$P_{TH,B,j,t}$ = thermal power produced by the (unique) gas-fed boiler in microgrid j at time t [kW];

$Q_{RES,l,j,t}$ = reactive power coming from the renewable plant l in microgrid j at time t [kvar];

Parameters

$G_{u,z}$ = conductance for line (u,z);

$B_{u,z}$ = susceptance for line (u,z);

$CONN_{u,j}$ = coefficient expressing the connection of LU j and node u;

x^{MIN} = minimum value of a physical quantity x;

x^{MAX} = maximum value of a physical quantity x;

Δ = time interval length [h]

$P_{RES,j,t}$ = active power from renewables of the *j-th* LU at time t; [kW]

$loss_j$ = loss coefficient in the storage of the *j-th* LU;

CAP_j = capacity of the storage of the *j-th* LU [kWh];

$\eta_j\left(P_{S,j,t}\right)$ = efficiency parameter of the storage of the *j-th* LU;

$\eta_{c,j}$ = charging efficiency parameter of the storage of the *j-th* LU;

$\eta_{d,j}$ = discharging efficiency parameter of the storage of the *j-th* LU;

$Q_{F,j,t}$ = reactive power demand of the *j-th* LU at time t [kvar];

$D_{AGG,TOT,j}$ = maximum active power decrease over the entire day in the j-th LU [kW];

$q_{E,j,t}$ = active power load without the heat pumps in the j-th LU at time t [kW];

$T_{ext,t}$ = external temperature at time t [K];

$C_{TH,j}$ = thermal capacitance in building j [kWh/K];

$R_{TH,ext,j}$ = resistance between building j and the external environment [K/kW];

η_s = heat pump efficiency coefficient;

$\eta_{s,heat}$ = heat pump efficiency coefficient in heating mode;

$\eta_{s,cool}$ = heat pump efficiency coefficient in heating mode;

σ = constant specifying whether the heat pump operates in heating or cooling mode;

χ = efficiency of the chiller;

$P_{TH,RES,j,t}$ = thermal power produced by renewables in microgrid j at time t [kW];

$D_{H,j,t}$ = thermal demand (heating) in microgrid j at time t [kW];

$D_{C,j,t}$ = thermal demand (cooling) in microgrid j at time t [kW];

$P_{TH,CHI,j,t}$ = thermal power needed to feed the (unique) chiller in microgrid j at time t [kW];

ω = tradeoff coefficient [€/kW^2];

a_j = parameter of the cost function of the j-th LU [€/h·kW^2];

b_j = parameter of the cost function of the j-th LU [€/kWh];

c_j = parameter of the cost function of the j-th LU [€/h];

MR_t = overall (maximal) power reduction at the slack node at time t [kW];

$P_{grid,da,t}$ = day-ahead power forecast regarding the slack node at time t [kW];

$C_{Market,t}$ = benefit from the market for the generic time interval at time t [€/kWh];

$C_{fee,t}$ = unit cost coefficient at time t [€/kWh];

$C_{i,j}$ = thermal capacitance of room i in building j [kWh/K];

$q_{i,j,t}$ = electrical power provided by heat pumps in room i in building j at time t [kW];

$\widetilde{R}_{TH,ext,i,j}$ = thermal resistance between room i in building j and the external environment [K/kW];

$\widetilde{A}_{ext,i,j}$ = entry of the adjacency matrix;

$\widetilde{R}_{TH,i,r,j}$ = is the thermal resistance between room i and room r in building j [K/kW];

$\widetilde{A}_{i,r,j}$ = entry of the adjacency matrix regarding rooms i and r in building j;

$\widehat{q}_{E,j,t}$ = fixed electrical demand in the lower-level problem in the j-th LU at time t [kW];

$E_{TOT,veh,j}$ = total electrical energy demand for electrical vehicles in the *j-th* LU [kWh];

$P^{fix}_{wash,j,t}$ = fixed fraction of $P_{wash,j,t}$ [kW];

$E^{def}_{TOT,wash,j}$ = total deferrable energy demand for washing machines in building j [kWh];

$\mu_{h,j}$ = parameter characteristic of the *h-th* cogenerative microturbine;

$P_{RES,l,j,t}$ = active power coming from the renewable plant l in microgrid j at time t [kW];

$Q_{D,j,t}$ = fixed amount of reactive power corresponding to the heat pumps and the electric load [kvar];
$P_{TH,RES,l,j,t}$ = thermal power from the *l-th* renewable energy source in microgrid *j* at time *t* [kW];
Objective function:
J = objective function of the AGG optimization problem;
$\hat{J}_{B,j}$ = objective function of building j;
$\hat{J}_{M,j}$ = objective function of microgrid j;
C_{inc} = cost of providing incentives to the LUs [€];
B_{DR} = benefit from the external grid due to the load reduction [€];
C_{AF} = "Assignment Fairness" [kW^2];

References

1. Carreiro, A.M.; Jorge, H.M.; Antunes, C.H. Energy management systems aggregators: A literature survey. *Renew. Sustain. Energy Rev.* **2017**, *73*, 1160–1172. [CrossRef]
2. Hill, C.A.; Such, M.C.; Chen, D.; Gonzalez, J.; Grady, W.M.K. Battery energy storage for enabling integration of distributed solar power generation. *IEEE Trans. Smart Grid* **2012**, *3*, 850–857. [CrossRef]
3. Goncalves Da Silva, P.; Ilic, D.; Karnouskos, S. The Impact of Smart Grid Prosumer Grouping on Forecasting Accuracy and Its Benefits for Local Electricity Market Trading. *IEEE Trans. Smart Grid* **2014**, *5*, 402–410. [CrossRef]
4. Wang, B.; Sechilariu, M.; Locment, F. Intelligent DC microgrid with smart grid communications: Control strategy consideration and design. *IEEE Trans. Smart Grid* **2012**, *3*, 2148–2156. [CrossRef]
5. Grijalva, S.; Tariq, M.U. Prosumer-based smart grid architecture enables a flat, sustainable electricity industry. In Proceedings of the ISGT 2011, Anaheim, CA, USA, 17–19 January 2011; pp. 1–6.
6. Fontenot, H.; Dong, B. Modeling and control of building-integrated microgrids for optimal energy management—A review. *Appl. Energy* **2019**, *254*, 113689. [CrossRef]
7. Li, D.; Chiu, W.Y.; Sun, H.; Poor, H.V. Multiobjective optimization for demand side management program in smart grid. *IEEE Trans. Ind. Informatics* **2018**, *14*, 1482–1490. [CrossRef]
8. Srinivasan, S.; Palmie, G.; Srinivasan, S.; Palmieri, G.; Del Vecchio, C.; Glielmo, L. Demand Side Management for heating controls in Microgrids. *IFAC-PapersOnLine* **2016**, *49*, 611–616.
9. Asensio, M.; Muñoz-Delgado, G.; Contreras, J. Bi-level approach to distribution network and renewable energy expansion planning considering demand response. *IEEE Trans. Power Syst.* **2017**, *32*, 4298–4309. [CrossRef]
10. Ferro, G.; Laureri, F.; Minciardi, R.; Robba, M. Optimal Integration of Interconnected Buildings in a Smart Grid: A Bi-level Approach. In Proceedings of the 2017 UKSim-AMSS 19th International Conference on Computer Modelling & Simulation (UKSim), Cambridge, UK, 5–7 April 2017; pp. 155–160.
11. Kovács, A. Bilevel programming approach to demand response management with day-ahead tariff. *J. Mod. Power Syst. Clean Energy* **2019**, *7*, 1632–1643. [CrossRef]
12. Jia, Y.; Mi, Z.; Yu, Y.; Song, Z.; Sun, C. A bilevel model for optimal bidding and offering of flexible load aggregator in day-ahead energy and reserve markets. *IEEE Access* **2018**, *6*, 67799–67808. [CrossRef]
13. Fernandez-Blanco, R.; Arroyo, J.M.; Alguacil, N.; Guan, X. Incorporating Price-Responsive Demand in Energy Scheduling Based on Consumer Payment Minimization. *IEEE Trans. Smart Grid* **2015**, *7*, 817–826. [CrossRef]
14. Najafi, A.; Falaghi, H.; Contreras, J.; Ramezani, M. A Stochastic Bilevel Model for the Energy Hub Manager Problem. *IEEE Trans. Smart Grid* **2016**, *8*, 2394–2404. [CrossRef]
15. Feijoo, F.; Das, T.K. Emissions control via carbon policies and microgrid generation: A bilevel model and Pareto analysis. *Energy* **2015**, *90*, 1545–1555. [CrossRef]
16. Wei, W.; Wang, D.; Jia, H.; Wang, C.; Zhang, Y.; Fan, M. Hierarchical and distributed demand response control strategy for thermostatically controlled appliances in smart grid. *J. Mod. Power Syst. Clean Energy* **2017**, *5*, 30–42. [CrossRef]
17. Bui, V.H.; Hussain, A.; Kim, H.M. A multiagent-based hierarchical energy management strategy for multi-microgrids considering adjustable power and demand response. *IEEE Trans. Smart Grid* **2018**, *9*, 1323–1333. [CrossRef]

18. Nazif Faqiry, M.; Das, S. Distributed bilevel energy allocation mechanism with grid constraints and hidden user information. *IEEE Trans. Smart Grid* **2017**, *10*, 1869–1879. [CrossRef]
19. Saez-Gallego, J.; Kohansal, M.; Sadeghi-Mobarakeh, A.; Morales, J.M. Optimal Price-Energy demand bids for aggregate price-responsive loads. *IEEE Trans. Smart Grid* **2018**, *9*, 5005–5013. [CrossRef]
20. Soares, J.; Canizes, B.; Ghazvini, M.A.F.; Vale, Z.; Venayagamoorthy, G.K. Two-Stage Stochastic Model Using Benders' Decomposition for Large-Scale Energy Resource Management in Smart Grids. *IEEE Trans. Ind. Appl.* **2017**, *53*, 5905–5914. [CrossRef]
21. Shao, C.; Ding, Y.; Wang, J.; Song, Y. Modeling and integration of flexible demand in heat and electricity integrated energy system. *IEEE Trans. Sustain. Energy* **2018**, *9*, 361–370. [CrossRef]
22. Asensio, M.; De Quevedo, P.M.; Muñoz-Delgado, G.; Contreras, J. Joint distribution network and renewable energy expansion planning considering demand response and energy storage-part I: Stochastic programming model. *IEEE Trans. Smart Grid* **2018**, *9*, 655–666. [CrossRef]
23. Asensio, M.; De Quevedo, P.M.; Muñoz-Delgado, G.; Contreras, J. Joint distribution network and renewable energy expansion planning considering demand response and energy storage-part II: Numerical results. *IEEE Trans. Smart Grid* **2018**, *9*, 667–675. [CrossRef]
24. Chen, S.; Cheng, R.S. Operating reserves provision from residential users through load aggregators in smart grid: A game theoretic approach. *IEEE Trans. Smart Grid* **2017**, *10*, 1588–1598. [CrossRef]
25. Cheng, P.-H.; Huang, T.-H.; Chien, Y.-W.; Wu, C.-L.; Tai, C.-S.; Fu, L.-C. Demand-side management in residential community realizing sharing economy with bidirectional PEV while additionally considering commercial area. *Int. J. Electr. Power Energy Syst.* **2020**, *116*, 105512. [CrossRef]
26. De Zotti, G.; Pourmousavi, S.A.; Morales, J.M.; Madsen, H.; Poulsen, N.K. Consumers' Flexibility Estimation at the TSO Level for Balancing Services. *IEEE Trans. Power Syst.* **2019**, *34*, 1918–1930. [CrossRef]
27. Aghajani, G.R.; Shayanfar, H.A.; Shayeghi, H. Demand side management in a smart micro-grid in the presence of renewable generation and demand response. *Energy* **2017**, *126*, 622–637. [CrossRef]
28. Salah, F.; Henriquez, R.; Wenzel, G.; Olivares, D.E.; Negrete-Pincetic, M.; Weinhardt, C. Portfolio design of a demand response aggregator with satisficing consumers. *IEEE Trans. Smart Grid* **2019**, *10*, 2475–2484. [CrossRef]
29. Delfino, F.; Ferro, G.; Robba, M.; Rossi, M. An Energy Management Platform for the Optimal Control of Active and Reactive Powers in Sustainable Microgrids. *IEEE Trans. Ind. Appl.* **2019**, *55*, 7146–7156. [CrossRef]
30. Bracco, S.; Brignone, M.; Delfino, F.; Pampararo, F.; Rossi, M.; Ferro, G.; Robba, M. An Optimization Model for Polygeneration Microgrids with Renewables, Electrical and Thermal Storage: Application to the Savona Campus. In Proceedings of the 2018 IEEE International Conference on Environment and Electrical Engineering and 2018 IEEE Industrial and Commercial Power Systems Europe (EEEIC/I&CPS Europe), Palermo, Italy, 12–15 June 2018; pp. 1–6.
31. Yao, J.; Costanzo, G.T.; Zhu, G.; Wen, B. Power Admission Control with Predictive Thermal Management in Smart Buildings. *IEEE Trans. Ind. Electron.* **2015**, *62*, 2642–2650. [CrossRef]

Article

A Building Energy Management System Based on an Equivalent Electric Circuit Model

Giovanni Bianco [1,2,*], Stefano Bracco [1], Federico Delfino [1], Lorenzo Gambelli [1], Michela Robba [2] and Mansueto Rossi [1]

[1] Department of Naval, Electrical, Electronic and Telecommunication Engineering, University of Genoa, Savona Campus, Via Magliotto 2, 17100 Savona, Italy; stefano.bracco@unige.it (S.B.); federico.delfino@unige.it (F.D.); loregambelli@gmail.com (L.G.); mansueto.rossi@unige.it (M.R.)

[2] Department of Informatics, Bioengineering, Robotics and Systems Engineering, University of Genoa, Savona Campus, Via Magliotto 2, 17100 Savona, Italy; michela.robba@unige.it

* Correspondence: giovanni.bianco@edu.unige.it

Received: 31 January 2020; Accepted: 31 March 2020; Published: 3 April 2020

Abstract: In recent decades, many EU and national regulations have been issued in order to increase the energy efficiency in different sectors and, consequently, to reduce environmental pollution. In the building sector, energy efficiency interventions are usually based on the use of innovative insulated materials and on the installation of cogeneration and tri-generation units, as well as solar technologies. New and retrofitted buildings are more and more commonly being called "smart buildings", since they are characterized by the installation of electric and thermal power generation units, energy storage systems, and flexible loads; the presence of such technologies determines the necessity of installing Building Energy Management Systems (BEMSs), which are used to optimally manage their operation. The present paper proposes a BEMS for a smart building, equipped with plants based on renewables (photovoltaics, solar thermal panels, and geothermal heat pump), where the heating and cooling demand are satisfied by a Heating, Ventilation and Air Conditioning System (HVAC) fed by a geothermal heat pump. The developed BEMS is composed of two different modules: an optimization tool used to optimally manage the HVAC plant, in order to guarantee a desired level of comfort inside rooms, and a simulation tool, based on an equivalent electric circuit model and used to evaluate the thermal dynamic behavior of the building. The paper describes the two modules and shows the main results of the validation phase that has been conducted on a real test-case represented by the Smart Energy Building (SEB) located at the Savona Campus of the University of Genoa, Italy.

Keywords: building energy management system; simulation; optimization; equivalent electric circuit

---

1. Introduction

All over the world, several countries are introducing low carbon and energy-saving policies to reduce the impact of the greenhouse gas effect and to face the growing energy demand of end-users. This allows for the development of more efficient energy systems and the spread of Distributed Generation (DG) technologies. Renewables sources (RES), co(tri)generation plants, and energy storage units, both thermal and electrical, characterize DG installations [1]. Moreover, smart grid and microgrid projects arise to integrate sustainable energy generation units, flexible loads, smart mobility systems and complex ICT infrastructures in the industrial, tertiary, and residential sectors. However, the dissemination of non-programmable and small-size dynamic generators is leading to balance and operational issues for distribution grids [2]. Consequently, it becomes necessary to develop software tools, called "Energy Management Systems" (EMSs), which can be used to optimally operate the aforementioned infrastructures with the aim of satisfying the energy needs of end-users without compromising the safe operation of electric and thermal networks. In such a context, flexibility is a key factor that both generation and consumption units have to guarantee: first of all, this implies that dispatchable power plants have to be able to follow the demand profile of end-users and the aleatory production of non-programmable renewable sources; secondly, demand response strategies have to be investigated to manage flexible loads [3].

The implementation of energy efficiency interventions, also based on the installation of the aforesaid technologies, becomes a challenging solution in commercial and residential buildings, which are one of the most energy-consuming systems: for instance, in the US, Heating, Ventilation and Air Conditioning Systems (HVAC) represent 27% of the overall consumed energy [4], and contribute to 30% of the total global carbon dioxide emissions [5]. Consequently, buildings can play a key role, both as flexible loads and in the reduction of CO_2 production, if sustainable and cost-effective technological solutions are taken into account during the design and construction phase, as well as in the case of retrofitting. Smart new-built or retrofitted buildings are operated daily by software tools called "Building Energy Management Systems" (BEMSs), which are based on optimization models characterized by different objectives: the minimization of operating costs and/or of carbon dioxide emissions, the maximization of self-consumption to take full advantage of local energy production from renewable sources, etc. [6]. One of the main goal of BEMSs is that of guaranteeing an acceptable comfort level to building occupants through a cost-effective management of HVAC plants and other equipment (lighting systems, waste disposal, rainwater harvesting, etc.) [7–9]. Moreover, it is important to note that nowadays a progressively higher number of buildings are assembled in clusters: in new residential and industrial districts, it is a good practice to install centralized heating and cooling systems based on high efficiency cogeneration and trigeneration technologies. By using this method, thermal and electrical networks, fed by the aforesaid systems, provide energy to the different buildings and, in many cases, renewable energy power plants (solar and wind) and storage systems are installed too. In these cases, the district becomes a "sustainable energy district" where each building is equipped with a BEMS and the whole energy infrastructure is managed by a centralized EMS; obviously, a close correlation among BEMSs and the EMS exists.

In literature, two main approaches are presented for the formalization of a BEMS: day-ahead and real-time management [10]. In the first case, the optimization tool determines the scheduling of dispatchable generation units and flexible loads for each hour of the following day on the basis of the forecast of loads (thermal and electrical) and renewable sources. In the second case, the optimization algorithms are often based on the model predictive control (MPC) technique [11–14].

Another important aspect is related to the energy analysis and thermal load simulation of a building. In fact, optimization models generally include the system model as a constraint; however, this should be accurate in order to represent the real system but also simple to be embedded in a decision problem. In literature, different approaches and tools are proposed to simulate a building in steady-state and dynamic conditions. In most cases, researchers use commercial software tools, which are complex environments able to model each space of a building in a very detailed way; the most used pieces of software are EnergyPlus, DesignBuilder, and IES. In [15], Ogunsola et al. use EnergyPlus to simulate a data-set in order to validate a reduced order thermal load model; in [16], the author estimate the overall thermal load of a Near to Zero Emissions Building (NZEB) in Lebanon for residential application using DesignBuilder; while the detailed thermal load of an educational building in the UK has been simulated by the authors with IES in [17]. Generally, the 3D model of the building is analyzed and each structural element (walls, ceilings, etc.) is modeled. A very detailed thermal analysis can be conducted but computational times become very long; consequently, these tools cannot be easily integrated within BEMSs, above all those which are real-time executed.

For the reasons explained above, many researchers prefer to develop tools which are usually based on the lumped-parameter theory, which permits to model a complex system, such as a building, as an "elementary system", i.e., a spatial entity, contained within a control volume delimited by a control surface [18]. The elementary system can exchange mass and/or energy (heat and/or work) with the environment and is described by state variables that depend on both the time and only a spatial coordinate, if a one-dimensional flow modelling is chosen. In this way, a building can be modelled using different levels of detail, as a one equivalent room or as a set of equivalent rooms with each one representing similar spaces as a whole, [19,20], not considering each single room and its interaction with the near ones. These approaches avoid computational complexity issues but are less accurate for the comfort evaluation of every single ambient [21]; they are more suitable to compute the building flexibility in demand response applications [22], in which the evaluation of the state of comfort in each room is not so necessary with respect to other aspects, such as the management of flexible loads and power plants.

Other literature studies propose more complex and accurate simulation models that evaluate the thermal load profile of the building as a function of weather and occupancy level forecast. For instance, in [23], Park et al. use EnergyPlus to model two different residential buildings with a high accuracy level but a high computational cost; [24] provides an analysis of the different capabilities of the building energy performance of simulation programs and models. These models simulate all the rooms of the considered building and are validated through experimental data measured on field. In other words, the building is modeled as a group of interconnected elementary systems, each one representing a room. Most of the models use equivalent electric circuits to study the thermal dynamic behavior of buildings; for instance, walls are modeled as a combination of resistors and capacitors characterized by resistance and capacitance values depending on the geometry (thickness of different layers) and materials (masonry and insulation layers, etc.). Due to the complexity of these models, the estimation of resistance and capacitance values is not so simple, especially in the case of buildings characterized by a huge number of rooms with a complex space partition. Moreover, the inclusion of these very detailed simulators within BEMSs is a hard task and in some cases the simulation model needs to be simplified [25].

The present study aims at defining a BEMS able to optimally schedule the HVAC system of a smart building, of which both the thermal and cooling energy needs are satisfied by exploiting the on-site geothermal heat pump. The proposed optimization model can manage both the heat pump and the fan coils at the room level. Moreover, differently from other approaches proposed in literature (e.g. [26]), which assume that the building is a single zone to optimize, in this work, all the building rooms are taken into account.

As is well known, a ground-source heat pump is a system used to heat and/or cool a building transferring heat from or to the ground (characterized by a nearly constant temperature during the year), respectively during winter or summer months [18]. Consequently, ground-source heat pumps are more efficient than air-source heat pumps because they take advantage of the relatively constant ground temperature. Heat exchangers for specific applications with ground water or geothermal closed loops can be adopted in the aforesaid systems. In open loop systems, well or surface body water is used as the heat exchange fluid that circulates directly through the heat pump system; on the other hand, closed systems consist of underground continuous piping loops filled with a water and anti-freeze solution, which constitutes the working fluid that is used to transfer heat from/to the ground to/from the geothermal heat pump, and are typically installed inside the building. Closed loop systems can be horizontal, vertical, or pond/lake. Horizontal loops are a cost-effective solution for residential installations, particularly for new buildings where sufficient land is available; vertical loops are preferable for commercial buildings and schools, whereas pond/lake systems are installed when an adequate body of water is available.

The developed BEMS acts on the HVAC plant, sending operating commands to the heat pump and to the fan coils arranged in each room of the building. The BEMS is based on a multi-objective MILP (Mixed Integer Linear Programming) mathematical model. The objective function to be minimized takes into account both operating costs of the HVAC system and the discomfort cost. On the other hand, the constraints of the problem include the mathematical model developed to simulate the dynamic thermal behavior of the building, based on an equivalent electric circuit approach. The model has been validated on a smart building of the Savona Campus of the University of Genoa [27].

In summary, the novelty of the proposed BEMS is characterized by the following aspects with respect to the literature:

- A detailed system model for the building, characterized by a control variable for each room associated to the different fan coils;
- A detailed model for the overall heating cooling system characterized by a geothermal heat pump and a storage system;
- A simulation model fitted with real data coming from a smart building connected to a smart Microgrid;
- A model for the optimal scheduling of fan coil circuit operation considering the perceived comfort in each room of the building;
- The possibility of being used in optimization applications due to its low computational cost, without losing the room level accuracy.

The remainder of the paper is organized as follows: in Section 2, the system's architecture is presented and the main equations of the BEMS are described. In Section 3, the description of the case study is reported and the main validation results of the simulation model are highlighted. Section 4 is focused on the analysis of optimal results obtained by running the BEMS for a typical day, whereas Section 5 reports the main conclusions of the study.

2. The BEMS's Architecture

The proposed BEMS is characterized by different modules (see Figure 1):

- A forecasting module used to predict the energy production from renewable sources and ambient conditions (external and ground temperature and solar radiation);
- An optimization module able to manage the Heating, Ventilation and Air Conditioning System (HVAC);
- A simulation module able to evaluate the thermal dynamic behavior of the building through an equivalent electric circuit model.

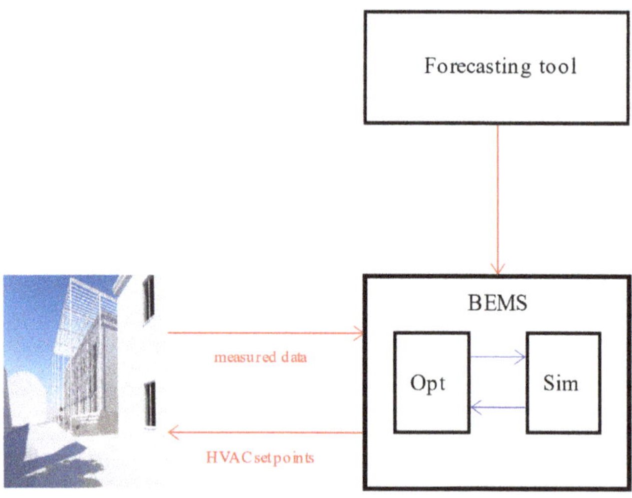

Figure 1. The proposed Building Energy Management System (BEMS) architecture.

The BEMS receives as input both the on-field measurements collected by the supervisory control and data acquisition (SCADA) and the output of the forecasting tool. Moreover, the simulation and the optimization tools are strictly related since the equations used to simulate the building thermal dynamics represent some of the constraints of the optimization model. In the following, the structure of the simulator and the optimization tool are described.

2.1. The Simulation Module

In this study, a grey-box modeling approach has been used to analyze the building thermal behavior. This kind of model is a reasonable tradeoff between purely physically based models, called white-box, which require many physical parameters, and black-box models, which need a large amount of data for the training phase to accurately represent the real physical behavior of the system [28]. In particular, an equivalent electric model is proposed here, using a well-known electric/thermal analogy where current represents heat flux and voltage indicates temperature. In literature, similar approaches are used to evaluate the thermal behavior of a building, but with different levels of complexity [21]. The majority of the models presented in the literature assume only one or a few temperature nodes to define the thermal behavior of the whole building and do not take into account temperature differences among the internal rooms; in other terms, they do not consider the presence of several internal walls and external walls characterized by different boundary conditions. On the other hand, the present model is more detailed since it can be used to evaluate the temperature profile, as a function of time, for each room of the building. The equivalent electric model is based on an Resistor Capacitor (RC) circuit, where resistors are used to model conductive and convective thermal resistances while capacitors are introduced to consider the accumulation of thermal energy inside the building.

2.1.1. Room Temperature Model

To determine the temperature inside each room, it is necessary to model all the thermal flows exchanged between the room and each adjacent space, which can be represented by another room, the ground or the outside. Dynamic energy balances have been written to model the thermal energy accumulation inside the rooms of the building, using first-order differential equations that have been discretized in order to insert them into the optimization model. In Figure 2, the equivalent electric circuit for the generic k^{th} room is shown.

For the kth room of the building, the internal temperature $T^i_{k,t+1}$ at time $t + 1$ can be evaluated through the following energy balance:

$$T^i_{k,t+1} = T^i_{k,t} + \frac{1}{C^i_k} \cdot \left(\dot{Q}^{ew,i}_{k,t} + \dot{Q}^{c,i}_{k,t} + \dot{Q}^{f,i}_{k,t} + \dot{Q}^{iw,i}_{k,j,t} + \dot{Q}^{w,i}_{k,t} + \dot{Q}^{co}_{k,t} + \dot{Q}^{ac}_{k,t} \right) \cdot \Delta t \quad k \in K, t = 0, \ldots, N-1 \tag{1}$$

where the thermal fluxes are computed by applying the Ohm's law, which correlates heat fluxes (currents) and temperatures (voltages) through appropriate resistances. In particular, the considered thermal resistances are convective resistances ($R^{ew,i}_k$, $R^{c,i}_k$, $R^{f,i}_k$, $R^{iw}_{k,j}$) between the room temperature ($T^i_{k,t}$) and the temperature of the internal faces of the external walls ($R^{ew,i}_k$, $T^{ew,i}_{k,t}$), ceilings ($R^{c,i}_k$, $T^{c,i}_{k,t}$), floors ($R^{f,i}_k$, $T^{f,i}_{k,t}$), and separation walls with adjacent rooms ($R^{iw}_{k,j}$, $T^{iw}_{k,j,t}$). C^i_k indicates that the equivalent thermal capacity of room kth and Δt is the time interval. In particular, the thermal fluxes reported in (1) indicate the heat exchange between: external wall/room ($\dot{Q}^{ew,i}_{k,t}$), ceiling/room ($\dot{Q}^{c,i}_{k,t}$), floor/room ($\dot{Q}^{f,i}_{k,t}$), adjacent rooms/room ($\dot{Q}^{iw}_{k,j,t}$), and windows/room ($\dot{Q}^{w,i}_{k,t}$). The windows are modeled through just one thermal resistance (R^w_k) since it is assumed that their thermal capacity can be considered negligible. So, the following thermal flux expressions derive:

$$\dot{Q}^{ew,i}_{k,t} = \frac{T^{ew,i}_{k,t} - T^i_{k,t}}{R^{ew,i}_k} \quad k \in K, t = 0, \ldots, N-1 \tag{2}$$

$$\dot{Q}_{k,t}^{c,i} = \frac{T_{k,t}^{c,i} - T_{k,t}^{i}}{R_{k}^{c,i}} \quad k \in K, t = 0, \ldots, N-1 \tag{3}$$

$$\dot{Q}_{k,t}^{f,i} = \frac{T_{k,t}^{f,i} - T_{k,t}^{i}}{R_{k}^{f,i}} \quad k \in K, t = 0, \ldots, N-1 \tag{4}$$

$$\dot{Q}_{k,j,t}^{iw} = \sum_{j=1}^{J^k} \frac{T_{k,j,t}^{iw} - T_{k,t}^{i}}{R_{k,j}^{iw}} \quad k \in K, j \in J^k, t = 0, \ldots, N-1 \tag{5}$$

$$\dot{Q}_{k,t}^{w,i} = \frac{T_{t}^{a} - T_{k,t}^{i}}{R_{k}^{w}} \quad k \in K, t = 0, \ldots, N-1 \tag{6}$$

Finally, $\dot{Q}_{k,t}^{co}$ is related to the convective fraction of the thermal flux provided by involuntary heat generators within the room (i.e., occupants or personal computers) [29], while $\dot{Q}_{k,t}^{fc}$ is the thermal power (19) supplied by fan coils of the HVAC system.

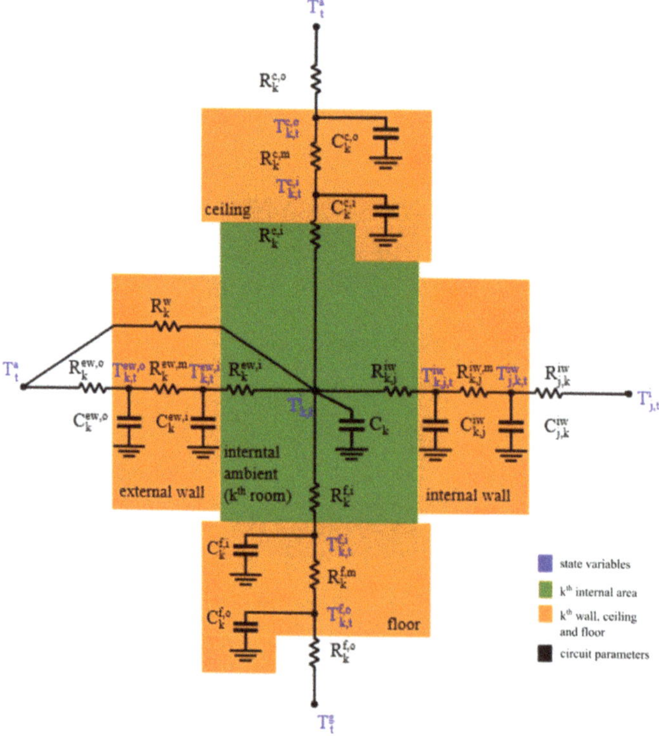

Figure 2. The equivalent circuit model for the generic kth room.

2.1.2. Wall, Ceiling, and Floor Temperature Model

In the model, each wall (internal and external) and each horizontal masonry element (slab, ceiling, floor) is modeled with a 3R2C thermal network, as shown in Figure 2. Each node of the network is characterized by a temperature, which represents a state variable of the whole system [30]. Due to the chosen electrical model, two nodes describe each structural part (wall or horizontal masonry) of the kth

room: the first related to its external side (external walls $T_{k,t}^{ew,o}$, ceilings $T_{k,t}^{c,o}$, floors $T_{k,t}^{f,o}$, internal walls $T_{j,k,t}^{iw}$) and the latter to its internal side (external walls $T_{k,t}^{ew,i}$, ceilings $T_{k,t}^{c,i}$, floors $T_{k,t}^{f,i}$, internal walls $T_{k,j,t}^{iw}$).

The temperature $T_{k,t+1}^{ew,o}$ and $T_{k,t+1}^{ew,i}$ of the two nodes related to the external wall of the kth room can be evaluated as:

$$T_{k,t+1}^{ew,o} = T_{k,t}^{ew,o} + \frac{1}{C_k^{ew,o}} \cdot \left(\frac{T_t^a - T_{k,t}^{ew,o}}{R_k^{ew,o}} + \frac{T_{k,t}^{ew,i} - T_{k,t}^{ew,o}}{R_k^{ew,m}} + \dot{Q}_{k,t}^{ew,sol} \right) \cdot \Delta t \tag{7}$$
$$k \in K, t = 0, \ldots, N-1$$

$$T_{k,t+1}^{ew,i} = T_{k,t}^{ew,i} + \frac{1}{C_k^{ew,i}} \cdot \left(\frac{T_{k,t}^i - T_{k,t}^{ew,i}}{R_k^{ew,i}} + \frac{T_{k,t}^{ew,o} - T_{k,t}^{ew,i}}{R_k^{ew,m}} + \dot{Q}_{k,t}^{ew,rad} \right) \cdot \Delta t \tag{8}$$
$$k \in K, t = 0, \ldots, N-1$$

where $\dot{Q}_{k,t}^{ew,sol}$ represents the solar gain on the external wall, whereas $\dot{Q}_{k,t}^{ew,rad}$ identifies the radiative gain on the internal face of the external wall due to involuntary heat generators [29]. The parameter $R_k^{ew,m}$ is defined as the thermal conductive resistance of the external wall situated between the temperature of the internal side of the external wall ($T_{k,t}^{ew,i}$) and the temperature of the external side of the external wall ($T_{k,t}^{ew,o}$), whereas $R_k^{ew,o}$ is the convective resistance between the outside temperature (T_t^a) and the external side of the external wall ($T_{k,t}^{ew,o}$); finally, $C_k^{ew,o}$ and $C_k^{ew,i}$ are defined as the thermal capacity values of the external wall, respectively located on the external side of the external wall ($C_k^{ew,o}$) and on the internal side of the external wall ($C_k^{ew,i}$). Similar equations are used to calculate the temperature values of the nodes that characterize the ceilings (9), (10), the floors (11), (12) and the internal walls (13), (14); note that for the evaluation of $T_{k,t}^{f,o}$ no solar gains are considered.

$$T_{k,t+1}^{c,o} = T_{k,t}^{c,o} + \frac{1}{C_k^{c,o}} \cdot \left(\frac{T_t^a - T_{k,t}^{c,o}}{R_k^{c,o}} + \frac{T_{k,t}^{c,i} - T_{k,t}^{c,o}}{R_k^{c,m}} + \dot{Q}_{k,t}^{c,sol} \right) \cdot \Delta t \qquad k \in K, t = 0, \ldots, N-1 \tag{9}$$

$$T_{k,t+1}^{c,i} = T_{k,t}^{c,i} + \frac{1}{C_k^{c,i}} \cdot \left(\frac{T_{k,t}^i - T_{k,t}^{c,i}}{R_k^{c,i}} + \frac{T_{k,t}^{c,o} - T_{k,t}^{c,i}}{R_k^{c,m}} + \dot{Q}_{k,t}^{c,rad} \right) \cdot \Delta t \qquad k \in K, t = 0, \ldots, N-1 \tag{10}$$

$$T_{k,t+1}^{f,o} = T_{k,t}^{f,o} + \frac{1}{C_k^{f,o}} \cdot \left(\frac{T_t^g - T_{k,t}^{f,o}}{R_k^{f,o}} + \frac{T_{k,t}^{f,i} - T_{k,t}^{f,o}}{R_k^{f,m}} \right) \cdot \Delta t \qquad k \in K, t = 0, \ldots, N-1 \tag{11}$$

$$T_{k,t+1}^{f,i} = T_{k,t}^{f,i} + \frac{1}{C_k^{f,i}} \cdot \left(\frac{T_{k,t}^i - T_{k,t}^{f,i}}{R_k^{f,i}} + \frac{T_{k,t}^{f,o} - T_{k,t}^{f,i}}{R_k^{f,m}} + \dot{Q}_{k,t}^{f,rad} \right) \cdot \Delta t \qquad k \in K, t = 0, \ldots, N-1 \tag{12}$$

$$T_{j,k,t+1}^{iw} = T_{j,k,t}^{iw} + \frac{1}{C_{j,k}^{iw}} \cdot \left(\frac{T_{j,t}^i - T_{j,k,t}^{iw}}{R_{j,k}^{iw}} + \frac{T_{k,j,t}^{iw} - T_{j,k,t}^{iw}}{R_{k,j}^{iw,m}} + \dot{Q}_{j,k,t}^{iw,rad} \right) \cdot \Delta t \; k \in K, j = J^k, t = 0, \ldots, N-1 \tag{13}$$

$$T_{k,j,t+1}^{iw} = T_{k,j,t}^{iw} + \frac{1}{C_{k,j}^{iw}} \cdot \left(\frac{T_{k,t}^i - T_{k,j,t}^{iw}}{R_{k,j}^{iw}} + \frac{T_{j,k,t}^{iw} - T_{k,j,t}^{iw}}{R_{j,k}^{iw,m}} + \dot{Q}_{k,j,t}^{iw,rad} \right) \cdot \Delta t \; k \in K, j = J^k, t = 0, \ldots, N-1 \tag{14}$$

2.1.3. HVAC System Model

As shown in Figure 3, in the analyzed case, the HVAC system is composed of a geothermal heat pump (*ghp*) feeding a thermal storage (*st*) connected to a fan coil (*fc*) circuit. The geothermal heat pump has as primary source the heat extracted by the geothermal probes; the heat pump absorbs a constant active power ($P^{el,ghp}$) from the grid when it is in operation (binary variable δ_t^{ghp} equal to 1) and produces thermal power ($P_t^{th,ghp}$). The coefficient of performance (COP) correlates the thermal power to the electrical one, as follows:

$$P_t^{th,ghp} = \delta_t^{ghp} \cdot COP \cdot P^{el,ghp} \qquad t = 0,\ldots,N-1 \tag{15}$$

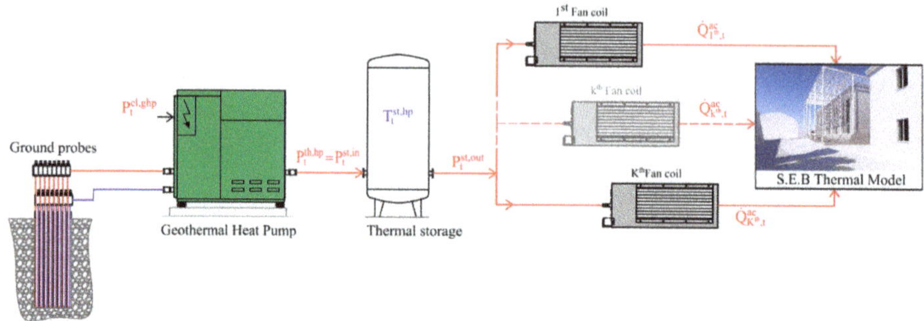

Figure 3. Scheme of the Heating, Ventilation and Air Conditioning System (HVAC) plant.

The thermal storage is a water tank, used to store the thermal energy produced by the heat pump; as a consequence, the thermal power provided to the storage ($P_t^{st,in}$) is assumed to be equal to the thermal power produced by the heat pump:

$$P_t^{st,in} = P_t^{th,ghp} \qquad t = 0,\ldots,N-1 \tag{16}$$

The storage is a dynamic component that can be used to increase the thermal inertia of the overall system, playing a fundamental role in terms of load flexibility. It is described by its thermal capacity (C^{st}) and its temperature (T_t^{st}), which can be calculated by means of the following energy balance equation:

$$T_{t+1}^{st} = T_t^{st} + \frac{1}{C^{st}} \cdot \left(P_t^{st,in} - P_t^{st,out} \right) \cdot \Delta t \qquad t = 0,\ldots,N-1 \tag{17}$$

The thermal power output of the storage ($P_t^{st,out}$) is equal to the sum of the thermal fluxes provided by fan coils installed inside the building:

$$P_t^{st,out} = \sum_{k=1}^{K} \dot{Q}_{k,t}^{fc} \qquad k \in K, t = 0,\ldots,N-1 \tag{18}$$

It is important to highlight that all the fan coils installed in the same room are considered as a single equivalent device, since they receive the same control signal from the BEMS. The thermal flux transferred to the room by the fan coils ($\dot{Q}_{k,t}^{fc}$) is defined as proposed in [31]:

$$\dot{Q}_{k,t}^{fc} = \delta_{k,t}^{fc} \cdot \varepsilon^{fc} \cdot \dot{m}_k^{air} \cdot c_p^{air} \cdot \left(T_t^{st} - T_{k,t}^i \right) \quad k \in K, t = 0,\ldots,N-1 \tag{19}$$

where ε^{fc} is the fan coil efficiency and $\dot{m}_k^{air}$ is the air mass flow rate of the equivalent fan coil in the kth room. The binary variable $\delta_{k,t}^{fc}$ assumes a unit value when the fan coils are activated.

2.2. The Optimization Module

The main goal of the developed BEMS is that of optimizing the operation of the HVAC plant in order to guarantee the thermal comfort in each different room of the building. The objective function to be minimized (*ObjF*) is given by the sum of two terms: the total operating cost of the heat pump (due to the electricity absorbed from the grid) and the sum of square errors between the actual room temperature $T^i_{k,t}$ and the optimal comfort temperature $T^{i\ *}_{k,t}$. The latter of these two terms is calculated by applying the Fanger's comfort model to each room [32]:

$$ObjF = \sum_{t=0}^{N-1} \left[\alpha \cdot \Delta t \cdot P^{el,ghp}_t \cdot p^{el}_t + \beta \cdot \mu^{oc}_t \cdot \sum_{k=1}^{K} \left(T^i_{k,t} - T^{i\ *}_{k,t} \right)^2 \right] \qquad k \in K, t = 0,\dots,N-1 \tag{20}$$

where Δt indicates the time step of the optimization problem, whereas p^{el}_t is the electricity price. The (20) is a multi-objective cost function. Since the optimal comfort temperature tracking cost and the electricity cost are not numerically commensurable, it introduced a numeric coefficient β to make the two different costs comparable. It might be also seen as a weight to the various objectives (i.e., the decision maker can give more importance to a specific objective or another one). The α parameter is always equal to 1 except for when one wants to evaluate only the comfort temperature tracking ($\alpha = 0$), while μ^{oc}_t is the indicator of occupancy (equal to 1 from 8.00 a.m. to 6.00 p.m. and equal to 0 in the remaining hours).

The constraints of the optimization problem are described by the equations of the simulation model (1-19) together with the following ones, which fix lower and upper bounds for the storage temperature as a function of the adopted technology:

$$T^{st,min} \leq T^{st}_t \leq T^{st,max} \qquad t = 0,\dots,N-1 \tag{21}$$

The problem is Mixed-Integer Non-Linear Programming (MINLP), since continuous and binary variables are present, and all the constraints are linear except for (19).

The constraints (19) have been modified in order to obtain a MILP problem. In particular, the aforementioned non-linearity has been solved with a big-M approach [33] rewriting the constraints (19) as follows:

$$\dot{Q}^{fc}_{k,t} = \varepsilon^{fc} \cdot \dot{m}^{air}_k \cdot c^{air}_p \cdot \left(z^2_t - z^1_{k,t} \right) \qquad k \in K, t = 0,\dots,N-1 \tag{22}$$

$$z^1_{k,t} \leq \delta^{fc}_{k,t} \cdot T^{i,max} \qquad k \in K, t = 0,\dots,N-1 \tag{23}$$

$$z^1_{k,t} \leq T^i_{k,t} \qquad k \in K, t = 0,\dots,N-1 \tag{24}$$

$$z^1_{k,t} \geq T^i_{k,t} - \left(1 - \delta^{fc}_{k,t} \right) \cdot T^{i,max} \qquad k \in K, t = 0,\dots,N-1 \tag{25}$$

$$z^1_{k,t} \geq 0 \qquad k \in K, t = 0,\dots,N-1 \tag{26}$$

$$z^2_{k,t} \leq \delta^{fc}_{k,t} \cdot T^{st,max} \qquad k \in K, t = 0,\dots,N-1 \tag{27}$$

$$z^2_{k,t} \leq T^{st}_t \qquad k \in K, t = 0,\dots,N-1 \tag{28}$$

$$z^2_{k,t} \geq T^{st}_t - \left(1 - \delta^{fc}_{k,t} \right) \cdot T^{st,max} \qquad k \in K, t = 0,\dots,N-1 \tag{29}$$

$$z^2_{k,t} \geq 0 \qquad k \in K, t = 0,\dots,N-1 \tag{30}$$

where $T^{i,max}$ and $T^{st,max}$ are the big-M values, representing the maximum values that $T^i_{k,t}$ and T^{st}_t can reach in real operation.

The decision variables of the optimization problem are:

- Room, wall, ceiling, floor, storage temperatures ($T^i_{k,t}$, $T^{ew,i}_{k,t}$, $T^{c,i}_{k,t}$, $T^{f,i}_{k,t}$, $T^{iw}_{k,j,t}$, $T^{ew,o}_{k,t}$, $T^{c,o}_{k,t}$, $T^{f,o}_{k,t}$, T^{st}_t);
- Storage thermal power input/output ($P^{st,in}_t$, $P^{st,out}_t$);
- Exchanged thermal fluxes ($\dot{Q}^{ew,i}_{k,t}$, $\dot{Q}^{c,i}_{k,t}$, $\dot{Q}^{f,i}_{k,t}$, $\dot{Q}^{iw,i}_{k,j,t}$, $\dot{Q}^{w,i}_{k,t}$);
- Heat pump thermal power ($P^{th,ghp}_t$);

- Thermal power provided by fan coils ($\dot{Q}_{k,t}^{fc}$);
- The binary variables (δ_t^{ghp}, $\delta_{k,t}^{fc}$), which constitute the control variables related to the on/off operation of the heat pump and fan coils;
- The auxiliary variables referred to the room temperature ($z_{k,t}^1$) and to the storage temperature ($z_{k,t}^2$) used to linearize the fan coil constraint (19).

The state variables are:

- Rooms, walls, ceilings, floors and storage temperatures ($T_{k,t}^i$, $T_{k,t}^{ew,i}$, $T_{k,t}^{c,i}$, $T_{k,t}^{f,i}$, $T_{k,j,t}^{iw}$, $T_{k,t}^{ew,o}$, $T_{k,t}^{c,o}$, $T_{k,t}^{f,o}$, T_t^{st}).

The main input data of the optimization model are the time profiles of:

- The convective fraction of the thermal fluxes provided by involuntary heat generators within the room, radiative gains due to involuntary heat generators and the solar gains on the external walls ($\dot{Q}_{k,t}^{co}$, $\dot{Q}_{k,t}^{ew,sol}$, $\dot{Q}_{k,t}^{ew,rad}$, $\dot{Q}_{k,t}^{c,sol}$, $\dot{Q}_{k,t}^{c,rad}$, $\dot{Q}_{k,t}^{f,rad}$, $\dot{Q}_{j,k,t}^{iw,rad}$);
- The external ambient temperature (T_t^a) and the ground temperature (T_t^g);
- The optimal comfort temperature ($T_{k,t}^{i\,*}$).

3. The Case Study

The developed BEMS is applied to a real case study represented by the Smart Energy Building (SEB) at the Savona Campus of the University of Genoa (Liguria Region, north-west Italy) [27]. This section reports the description of the building and the experimental validation of the simulation model.

3.1. The Smart Energy Building

The SEB, Figure 4, is an environmentally sustainable building connected to a microgrid (called Smart Polygeneration Microgrid [27]) as a "prosumer" and is equipped with renewable power plants and characterized by energy efficiency measures.

Figure 4. The Smart Energy Building.

The building has two floors and 22 rooms, it is characterized by a floor area of 510 m^2 (for each floor) and it is 10 m high. It is made of reinforced concrete with Predalles prestressed slabs and the ground floor elevation has been adopted to prevent flood water entering the building; high performance thermal insulation materials for building applications (ventilated facades and claddings) and acoustic insulation systems are applied. The building is certified as A4, which is the most efficient "energy class" in accordance with the Italian Classification of Building Energy Efficiency. Moreover, the SEB is a ZEB (Zero Emission Building) since its electrical loads are satisfied by a Photovoltaic (PV) field and the storage systems of the microgrid, while its thermal and cooling loads are covered by a geothermal heat pump. The SEB hosts: three laboratories, a gym, two classrooms, and some offices. The following technologies are installed in the SEB: a geothermal heat pump, a thermal storage system, an air handling unit, a PV field (21.25 kW peak power), extremely low consumption LED lamps, a rainwater harvesting system, and a hydroponic system.

The geothermal plant installed in the SEB is characterized by a close-loop vertical configuration: eight borehole heat exchangers (Figure 5 left and center) are buried about 120 m deep in the soil. The geothermal heat pump (Figure 5 right) is a Clivet WSHN-XEE2 MF 14.2; it is hydraulically connected to a storage water tank having a volume of 500 liters.

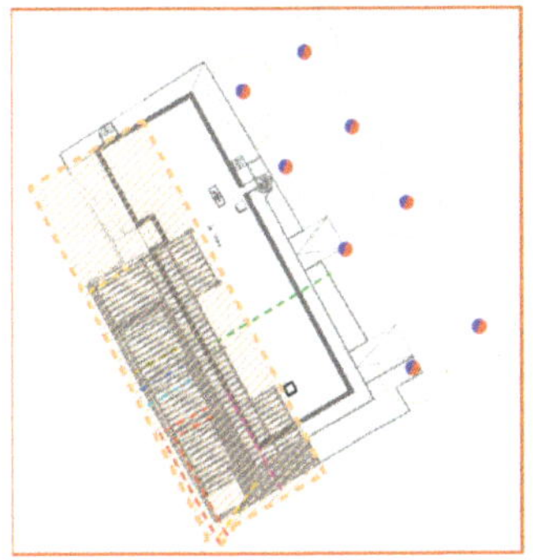

Figure 5. The geothermal plant.

The SEB is equipped with field sensors for measuring humidity and temperature in each room. The building has an embedded communication system and can be managed remotely by the control room of the Smart Polygeneration Microgrid.

3.2. The Validation of the Simulation Model

The accuracy of the simulation model has been evaluated by comparing the values obtained by the simulator and the actual data measured on the field at the Savona Campus. The availability of the measured values was guaranteed by the supervisory control and data acquisition (SCADA) system of the SEB. The variables measured by the SCADA system are as follows: the electric power absorbed by the heat pump, the air temperatures in all the rooms, and the on/off operation of fan coils. The sampling time of the electric power values has been 1 minute, whereas room temperature values have been collected every 15 minutes to avoid the overload of the SCADA system; nevertheless, a sample time of 15 minutes is acceptable to appreciate the thermal inertia of the building. Finally, the values of solar radiation and external ambient temperature derive from the measurements of a weather station located in the city of Savona and operated by the Regional Agency for the Protection of the Ligurian Environment (ARPAL).

The validation of the simulator has been performed by using data collected both during nighttime and daytime periods. Since during the night the HVAC plant is off, the experimental data referring to this period reflect only the thermal behaviour of the building without considering fan coil power injection, and so permit to better characterize the dynamics of the walls; on the other hand, daytime data have been used to evaluate how the simulation model is able to represent the HVAC operation. During the daytime, the HVAC system operates to maintain the desired indoor temperature within a slot of temperature-wide one degree. The mid temperature value of the slot is the optimal comfort temperature, and this value is set in each room by a central controller and can be modified by occupants within a range of two degrees depending on their personal feelings. The controller is governed by means an heuristic rule, this control logic operates turning on the HVAC plant until the internal rooms temperatures reach the high value of the slot and then it turns off the plant until the temperature values becomes smaller than the low limit of the slot, as an hysteresis cycle.

The validation of the simulator has been successfully conducted for each room of the building, but the results described in the following refer to three typical rooms, featured by different intended uses, which well characterize the whole building. The aforementioned rooms, depicted in Figure 6a,b, are as follows: the U-Gym (located on the ground floor and purple-colored in Figure 6a), one laboratory, and one office (located on the first floor and respectively pink and cyan-colored in Figure 6b).

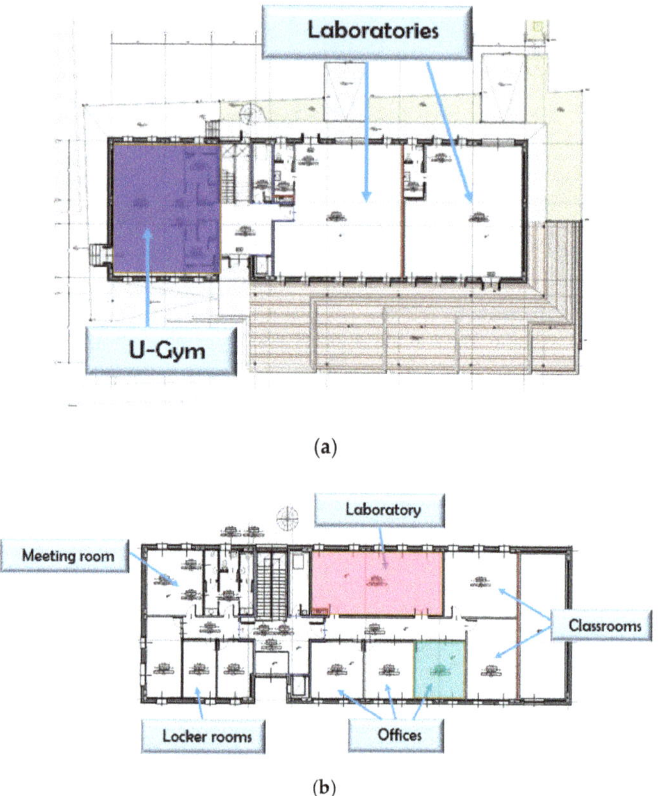

Figure 6. (**a**) Smart Energy Building (SEB) ground floor; (**b**) SEB first floor.

The nighttime temperature values have been measured between 9 p.m. and 6 a.m., when the HVAC system is off, during the month of November 2019. Figure 7 shows the comparison between the predicted and the measured temperature values for the examined three rooms, referring to one day. It is important to highlight that the stepped profile of measured temperatures is due to the sensitivity of the measurement sensors, which can only detect a temperature change of 0.2 °C. In order to quantify the accuracy of the simulation model, some statistical indicators have been calculated for each day of the analyzed month. In particular, in Figure 8, the values of the mean percentage error are reported, whereas Figures 9 and 10 show, respectively, the maximum percentage error and the root mean square error. The results show that in 80% of cases the mean percentage error remains below 2.5% while the maximum error is below 4%.

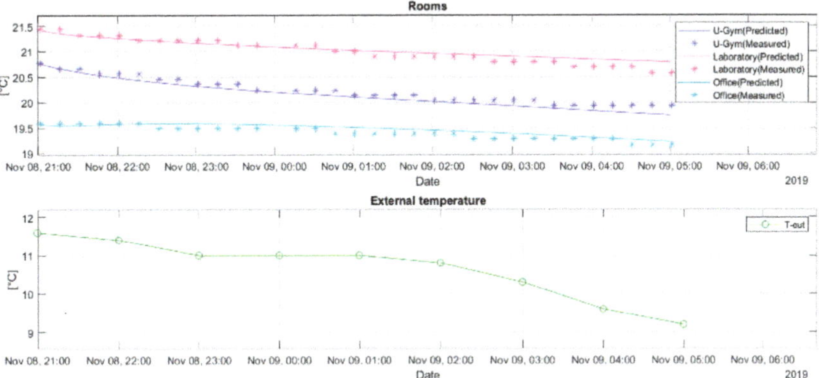

Figure 7. Nighttime room and external temperatures.

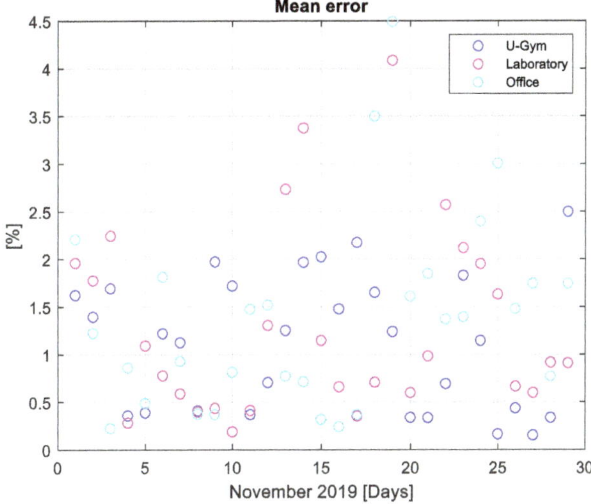

Figure 8. Nighttime temperature mean percentage error.

The second validation has been conducted considering data collected during daytime periods to evaluate the accuracy of the model during the operation of the fan coils system. The operation scheduling of the HVAC plant has been acquired through the SCADA system by using the operation setpoints as input for the simulator. Referring to a generic day, the resulting temperatures compared to the actual measured temperatures are plotted in Figure 11a, while Figure 11b depicts the external temperature referred to that day, which is possible to observe as the curve related to the typical on/off operation behavior of the fan coils.

In Table 1, some statistical indicators are calculated for each considered room referring to the analyzed day. In particular, the mean percentage error, the maximum percentage error, and the root mean square error.

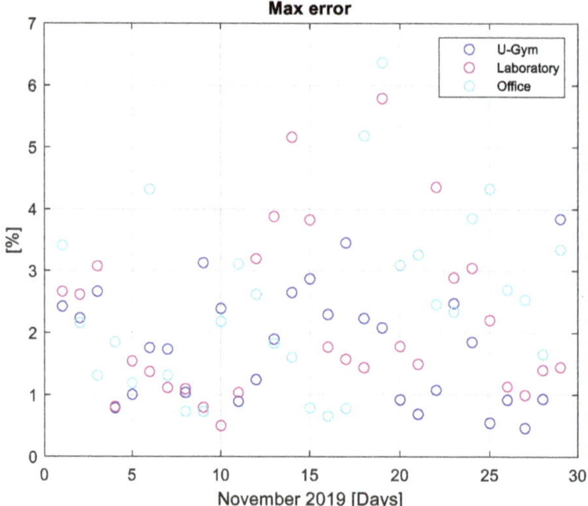

Figure 9. Nighttime temperature maximum percentage error.

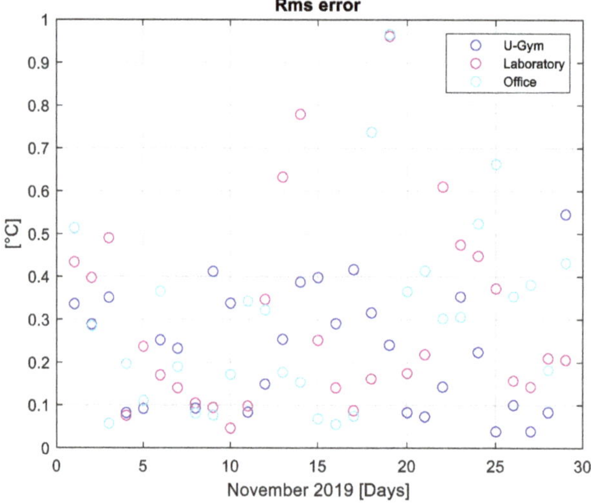

Figure 10. Nighttime temperature root mean square error.

Table 1. Discomfort indicator values.

Room Name	Mean Error [%]	Max Error [%]	Rms Error [°C]
U-Gym	1.2	3.5	0.3
Laboratory	1.2	6.1	0.4
Office	1.0	4.7	0.3

The mean percentage error and the root mean square error have also been calculated for the whole month and the results are depicted in Figures 12 and 13.

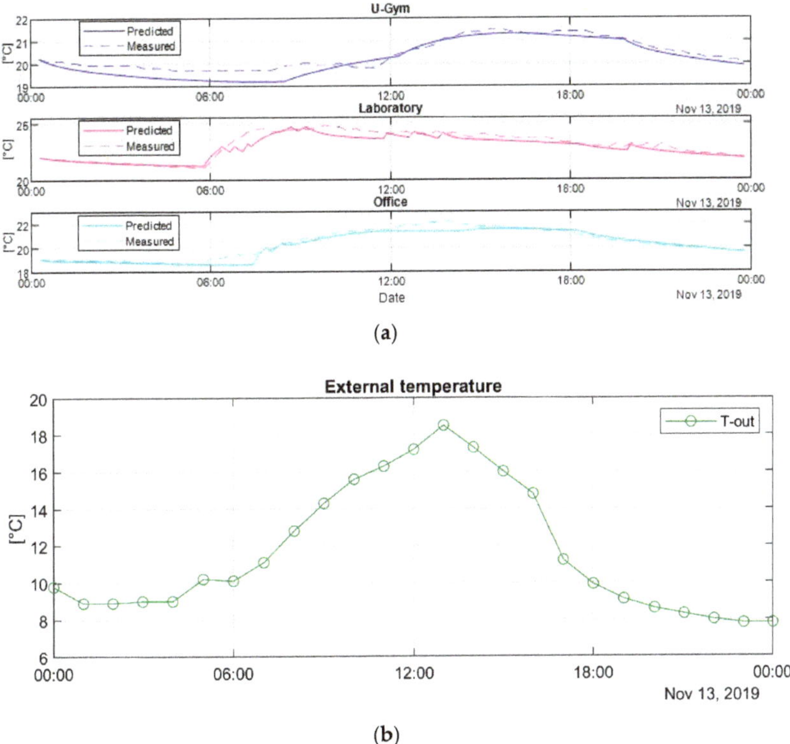

(**a**)

(**b**)

Figure 11. (**a**) Daytime room temperatures. (**b**) Daytime external temperatures.

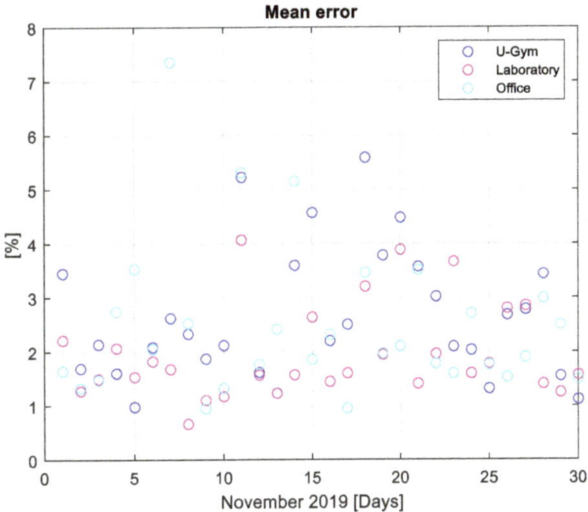

Figure 12. Daytime temperatures mean percentage error.

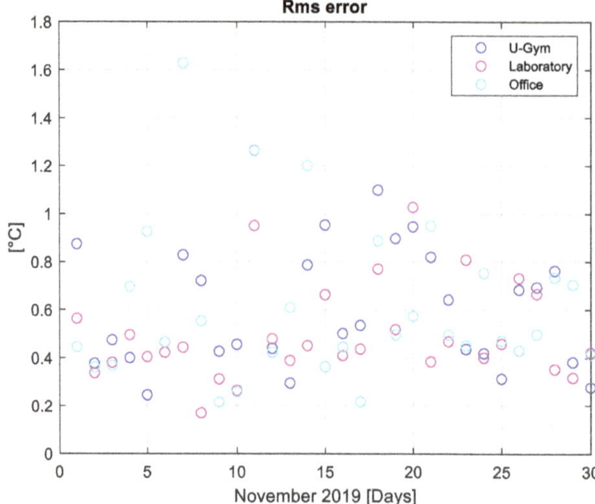

Figure 13. Daytime temperature root mean square error.

4. Optimization Results

The BEMS described in Section 2 has been applied to the Smart Energy Building according to a day-ahead optimization logic. The aim is to optimally schedule the operation of the HVAC system to guarantee the thermal comfort of the occupants of each different room and at the same time to reduce the electrical energy consumption of the heat pump. The time step (Δt) of the optimization problem has been chosen equal to 15 minutes and the horizon of the optimization problem (N) is one day ($N = 96$); consequently, a whole day is analyzed starting from 12.00 a.m. until 12.00 a.m. of the following day. The optimal comfort temperature values $\left(T_{k,t}^{i\,*}\right)$ are tracked only between 8.00 a.m. and 6.00 p.m., particularly during the period of occupancy of the building. In these hours, the value of the optimal comfort temperature is assumed to be equal to 21 [°C], whereas the external ambient temperature is assumed to be constant and equal to 10 [°C] for all the following simulations.

The optimization problem has been formalized in the YALMIP environment [34], using CPLEX 12.7 as solver. The optimization has been computed on a PC Intel i7, 16 GB RAM. The average computation time was 434 seconds.

4.1. Optimization without Minimizing Heat Pump Operating Costs (Case 1)

The results here reported refer to the case of the objective function having $\alpha = 0$ and $\mu_t^{oc} = 1$. In this case, the BEMS controls the HVAC system with the only goal of tracking the comfort temperature inside the different rooms. In Figure 14, the temperature profiles of the three chosen rooms are shown. It is possible to highlight how the temperature profiles are tracking the setpoint of the temperature.

The level of comfort is evaluated, in accordance with Fanger's model, by means of the predicted percent dissatisfied (PPD) indicator, which evaluates the percentage of thermally dissatisfied people who feel too cool or too warm [35]. The PPD values are depicted in Figure 15, whereas the mean PPD values are reported in Table 2.

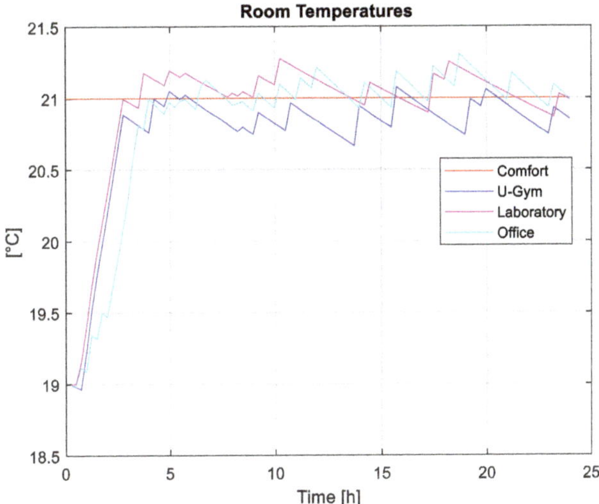

Figure 14. Room temperature profiles in the case of only comfort temperature tracking.

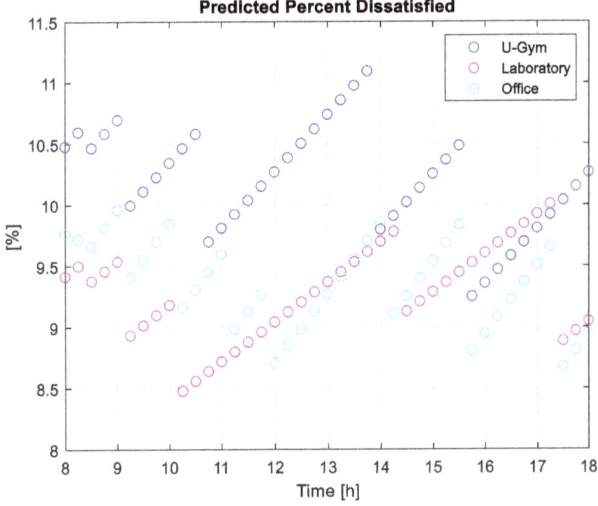

Figure 15. Predicted percent dissatisfied values.

Table 2. Discomfort indicator values.

Room Name	Average Predicted Percent Dissatisfied [%]
U-Gym	10.1
Laboratory	9.3
Office	9.4

4.2. Overall Optimization (Case 2)

The results described in the following refer to the case of the objective function having $\alpha = 1$. In this case, the problem has a multi-objective function that presents a conflict between two opposite objectives:

the tracking of the optimal comfort temperature and the reduction in the energy consumptions of the HVAC plant. Figure 16 depicts the temperature profiles of the three rooms.

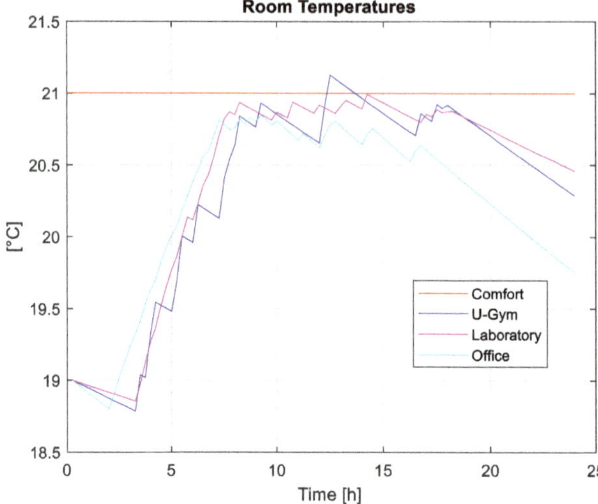

Figure 16. Room temperature profiles in the case of comfort temperature tracking and energy consumption minimization.

Figure 16 highlights how the room temperatures attempt to track the optimal comfort temperature during the period of occupancy of the building (8 a.m. − 6 p.m.); the temperature profiles follow the comfort temperature less effectively than in the previous case, because, in this case, the objective function has two conflictual goals, the comfort tracking and the minimization of HVAC operating costs. The average daily PPD values for the three rooms are reported in Table 3, whereas, in Figure 17, the profile of PPD is plotted during the occupancy period.

Table 3. Discomfort indicator values.

Room Name	Average Predicted Percent Dissatisfied [%]
U-Gym	10.2
Laboratory	10.1
Office	10.9

The mean values of PPD for the three rooms are within the acceptability limits imposed by the law, that is 20% [36]. Figure 17 shows how all the values of PPD never overcome 12%.

To compare the previously analyzed cases with the standard case in which the HVAC operation is not optimized by means of the BEMS tool, a case in which the proposed BEMS is not implemented within the control system of the thermal plant is simulated (case 0). In Figure 18, it is possible to see the temperature profiles of the three considered rooms.

Figure 18 shows that the internal temperatures perform a hysteresis cycle within a band that has the optimal comfort temperature as its average value; this is the result of the heuristic rule that is imposed in standard operating conditions by the control of the HVAC system. The PPD values are reported in Figure 19 during the occupancy period, whereas, in Table 4, the daily average values for the three rooms are shown.

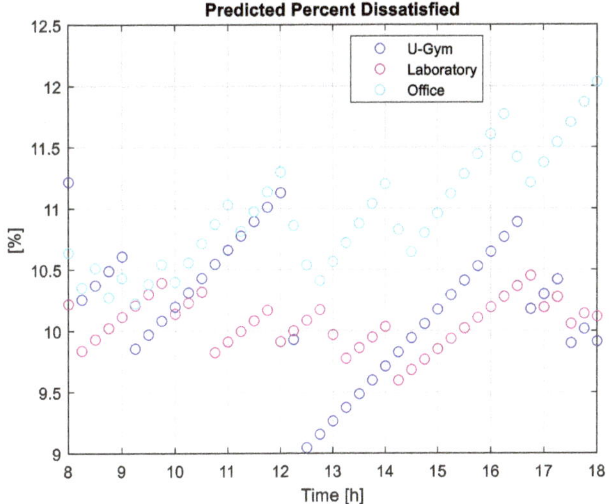

Figure 17. Predicted percent dissatisfied values.

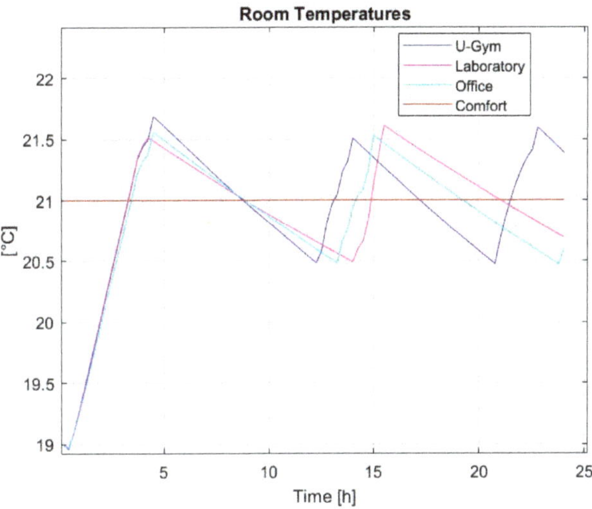

Figure 18. Room temperature profiles in the case of standard HVAC plant control logic.

Table 4. Discomfort indicator values.

Room Name	Average Predicted Percent Dissatisfied [%]
U-Gym	11.05
Laboratory	11.11
Office	11.06

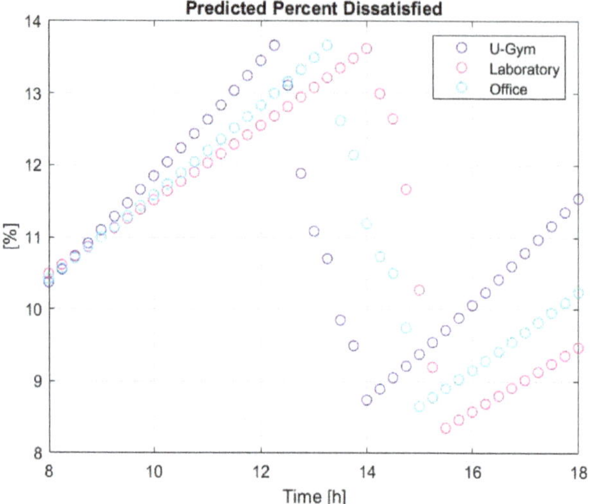

Figure 19. Predicted percent dissatisfied values.

The PPD values for the three rooms are in the acceptability range. As depicted in Figure 19, the maximum PPD value never overcomes 14%.

Figure 20 plots the geothermal heat pump scheduling profile for the three examined scenarios; each bar in Figure 20 indicates that the device is on for a time interval of 15 minutes.

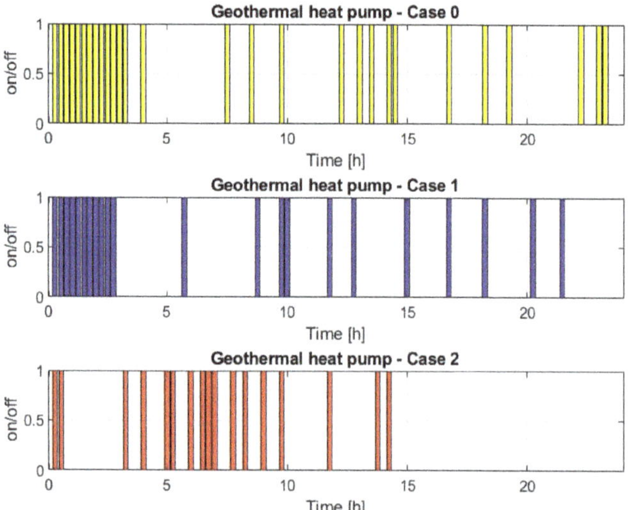

Figure 20. Geothermal heat pump operation scheduling, in the three different cases.

The operating costs of the HVAC plant are reported in Table 5 for the three different cases analyzed. In case 0, Figure 18, it is considered that the HVAC system operates under control of the heuristic rule presented in Section 3.2; instead, in case 1 and 2, referred respectively to Figures 14 and 16, the HVAC operates integrated with the BEMS optimization tool. In Table 5, the overall mean PPD values for each case are also collected.

Table 5. HVAC plant daily operating costs and predicted percent dissatisfied (PPD) mean values.

Case	Cost [€]	PPD [%]
Case 0	36.75	11.07
Case 1	28.87	9.61
Case 2	22.31	10.4

As depicted in Figure 20, the HVAC behavior in the three cases is clearly different. More precisely, in case 0 and 1, the HVAC system starts to operate at the begging of the simulation in order to fill the gap between the actual temperature and the optimal comfort setpoint. During the remainder of the simulation, the HVAC operates to keep the temperature within the optimal temperature slot. The difference between case 0 and case 1 is that, in case 0, the control logic is the heuristic rule, presented in Section 3.2, since the proposed BEMS is not implemented, while, in case 1, the proposed BEMS is implemented and it optimizes the management of the HVAC plant to reach the optimal comfort temperature in the most efficient way; indeed, between the two cases, there is a significant difference in terms of electricity costs and also the PPD value is lower in case 1, as reported in Table 5, whereas, in case 2, the proposed BEMS operates also to reduce the energy consumption. The results highlight a lower daily electricity cost considering case 0 and case 1, while a higher value of PPD is detected considering case 1 in which the main goal of the BEMS is the comfort of the occupants, the PPD value of case 2 is still lower than the one of case 0.

5. Conclusions and Future Developments

A simulator for the evaluation of the thermal dynamic behavior of a building at room level has been developed basing on an equivalent electric model and by focusing on the operation of the HVAC system. The simulator has been successfully validated through real data collected on field at the Savona Campus of the Genoa University, with mean error values of around 1%. Moreover, a BEMS for the day-ahead optimal control of the HVAC system has been proposed. Future developments will investigate the possibility to adapt the BEMS to a receding horizon control scheme with the aim of using it in a demand response application.

Author Contributions: Conceptualization, G.B., S.B., F.D., L.G., M.R. (Michela Robba) and M.R. (Mansueto Rossi); Methodology, G.B., M.R. (Michela Robba) and M.R. (Mansueto Rossi); software, G.B. and L.G.; validation, G.B. and L.G.; data curation, M.R. (Mansueto Rossi); writing—original draft preparation, G.B. and S.B.; writing—review and editing, G.B. and S.B.; project administration, F.D., M.R. (Michela Robba) and M.R. (Mansueto Rossi); funding acquisition, F.D., M.R. (Michela Robba) and M.R. (Mansueto Rossi). All authors have read and agreed to the published version of the manuscript.

Funding: This research received external funding by Liguria Region and the six involved companies (ABB, SOFTECO, GRUPPO SIGLA, MAPS, FLAIRBIT, RULEX), the University of Genoa and the National Council of Research.

Acknowledgments: The work is part of the activities developed within the PICK UP project funded by Liguria Region and that involves six companies (ABB, SOFTECO, GRUPPO SIGLA, MAPS, FLAIRBIT, RULEX), the University of Genoa and the National Council of Research. The authors would like to thank all the partners for their assistance with the collection and elaboration of data.

Conflicts of Interest: The authors declare no conflict of interest.

References

1. Delfino, F.; Ferro, G.; Minciardi, R.; Robba, M.; Rossi, M.; Rossi, M. Identification and optimal control of an electrical storage system for microgrids with renewables. *Sustain. Energy Grids Netw.* **2019**, *17*, 100183. [CrossRef]
2. Delfino, F.; Ferro, G.; Robba, M.; Rossi, M. An Energy Management Platform for the Optimal Control of Active and Reactive Power in Sustainable Microgrids. *IEEE Trans. Ind. Appl.* **2019**, *55*, 7146–7156. [CrossRef]
3. Rocha, P.; Siddiqui, A.; Stadler, M. Improving energy efficiency via smart building energy management systems: A comparison with policy measures. *Energy Build.* **2015**, *88*, 203–213. [CrossRef]

4. Simma, K.C.J.; Mammoli, A.; Bogus, S.M. Real-time occupancy estimation using WiFi network to optimize HVAC operation. *Proc. Procedia Comput. Sci.* **2019**, *155*, 495–502. [CrossRef]
5. Ahmad, M.W.; Mourshed, M.; Rezgui, Y. Trees vs Neurons: Comparison between random forest and ANN for high-resolution prediction of building energy consumption. *Energy Build.* **2017**, *147*, 77–89. [CrossRef]
6. Kim, N.K.; Shim, M.H.; Won, D. Building Energy Management Strategy Using an HVAC System and Energy Storage System. *Energies* **2018**, *11*, 2690. [CrossRef]
7. Boodi, A.; Beddiar, K.; Benamour, M.; Amirat, Y.; Benbouzid, M. Intelligent systems for building energy and occupant comfort optimization: A state of the art review and recommendations. *Energies* **2018**, *11*, 2604. [CrossRef]
8. Hurtado, L.A.; Nguyen, P.H.; Kling, W.L.; Zeiler, W. Building energy management systems-Optimization of comfort and energy use. In Proceedings of the 2013 48th International Universities' Power Engineering Conference (UPEC), Dublin, Ireland, 2–5 September 2013; pp. 1–6.
9. Zhang, D.; Shah, N.; Papageorgiou, L.G. Efficient energy consumption and operation management in a smart building with microgrid. *Energy Convers. Manag.* **2013**, *74*, 209–222. [CrossRef]
10. Shakeri, M.; Shayestegan, M.; Abunima, H.; Reza, S.M.S.; Akhtaruzzaman, M.; Alamoud, A.R.M.; Sopian, K.; Amin, N. An intelligent system architecture in home energy management systems (HEMS) for efficient demand response in smart grid. *Energy Build.* **2017**, *138*, 154–164. [CrossRef]
11. Li, X.; Malkawi, A. Multi-objective optimization for thermal mass model predictive control in small and medium size commercial buildings under summer weather conditions. *Energy* **2016**, *112*, 1194–1206. [CrossRef]
12. Tang, R.; Wang, S. Model predictive control for thermal energy storage and thermal comfort optimization of building demand response in smart grids. *Appl. Energy* **2019**, *242*, 873–882. [CrossRef]
13. Bahramnia, P.; Rostami, S.M.H.; Wang, J.; Kim, G. Modeling and controlling of temperature and humidity in building heating, ventilating, and air conditioning system using model predictive control. *Energies* **2019**, *12*, 4805. [CrossRef]
14. Serale, G.; Fiorentini, M.; Capozzoli, A.; Bernardini, D.; Bemporad, A. Model Predictive Control (MPC) for enhancing building and HVAC system energy efficiency: Problem formulation, applications and opportunities. *Energies* **2018**, *11*, 631. [CrossRef]
15. Ogunsola, O.T.; Song, L.; Wang, G. Development and validation of a time-series model for real-time thermal load estimation. *Energy Build.* **2014**, *76*, 440–449. [CrossRef]
16. Samarji, T.; Jouni, A.; Karaki, A. Net zero energy buildings: Application in Lebanon on a typical residential building. In Proceedings of the 2012 International Conference on Renewable Energies for Developing Countries (REDEC), Beirut, Lebanon, 28–29 November 2012; pp. 1–7.
17. Barbhuiya, S.; Barbhuiya, S. Thermal comfort and energy consumption in a UK educational building. *Build. Environ.* **2013**, *68*, 1–11. [CrossRef]
18. Delfino, F.; Procopio, R.; Rossi, M.; Bracco, S.; Brignone, M.; Robba, M. *Microgrid Design and Operation: Toward Smart Energy in Cities*; Artech House: Norwood, MA, USA, 2018.
19. Tang, R.; Wang, S.; Xu, L. An MPC-based optimal control strategy of active thermal storage in commercial buildings during fast demand response events in smart grids. *Energy Procedia* **2019**, *158*, 2506–2511. [CrossRef]
20. Li, X.; Wen, J. Review of building energy modeling for control and operation. *Renew. Sustain. Energy Rev.* **2014**, *37*, 517–537. [CrossRef]
21. Ji, Y.; Xu, P.; Duan, P.; Lu, X. Estimating hourly cooling load in commercial buildings using a thermal network model and electricity submetering data. *Appl. Energy* **2016**, *169*, 309–323. [CrossRef]
22. De Coninck, R.; Helsen, L. Quantification of flexibility in buildings by cost curves - Methodology and application. *Appl. Energy* **2016**, *162*, 653–665. [CrossRef]
23. Park, S.H.; Chung, W.J.; Yeo, M.S.; Kim, K.W. Evaluation of the thermal performance of a Thermally Activated Building System (TABS) according to the thermal load in a residential building. *Energy Build.* **2014**, *73*, 69–82. [CrossRef]
24. Crawley, D.B.; Hand, J.W.; Kummert, M.; Griffith, B.T. Contrasting the capabilities of building energy performance simulation programs. *Build. Environ.* **2008**, *43*, 661–673. [CrossRef]
25. Missaoui, R.; Joumaa, H.; Ploix, S.; Bacha, S. Managing energy Smart Homes according to energy prices: Analysis of a Building Energy Management System. *Energy Build.* **2014**, *71*, 155–167. [CrossRef]

26. Vogler-Finck, P.J.C.; Wisniewski, R.; Popovski, P. Reducing the carbon footprint of house heating through model predictive control—A simulation study in Danish conditions. *Sustain. Cities Soc.* **2018**, *42*, 558–573. [CrossRef]

27. Bracco, S.; Delfino, F.; Laiolo, P.; Morini, A. Planning & open-air demonstrating smart city sustainable districts. *Sustainability* **2018**, *10*, 1–14.

28. Braun, J.E.; Chaturvedi, N. An inverse gray-box model for transient building load prediction. *HVAC R Res.* **2002**, *8*, 73–99. [CrossRef]

29. ASHRAE 2017. *ASHRAE Fundamental Handbook SI*; ASHRAE: Atlanta, GA, USA, 2017; ISBN 6785392187.

30. Shamsi, M.H.; O'Grady, W.; Ali, U.; O'Donnell, J. A generalization approach for reduced order modelling of commercial buildings. *Energy Procedia* **2017**, *122*, 901–906. [CrossRef]

31. Mantovani, G.; Ferrarini, L.; Member, S. With Model Predictive Control Techniques. *IEEE Trans. Ind. Electron.* **2015**, *62*, 2651–2660. [CrossRef]

32. Fanger, P.O. Assessment of man's thermal comfort in practice. *Br. J. Ind. Med.* **1973**, *30*, 313–324. [CrossRef]

33. Savola, T.; Fogelholm, C.J. MINLP optimisation model for increased power production in small-scale CHP plants. *Appl. Therm. Eng.* **2007**, *27*, 89–99. [CrossRef]

34. Löfberg, J. YALMIP: A toolbox for modeling and optimization in MATLAB. In Proceedings of the 2004 IEEE International Conference on Robotics and Automation (IEEE Cat. No.04CH37508), New Orleans, LA, USA, 2–4 September 2004; pp. 284–289.

35. Djongyang, N.; Tchinda, R.; Njomo, D. Thermal comfort. A review paper. *Renew. Sustain. Energy Rev.* **2010**, *14*, 2626–2640. [CrossRef]

36. Carlucci, S.; Bai, L.; de Dear, R.; Yang, L. Review of adaptive thermal comfort models in built environmental regulatory documents. *Build. Environ.* **2018**, *137*, 73–89. [CrossRef]

MDPI
St. Alban-Anlage 66
4052 Basel
Switzerland
Tel. +41 61 683 77 34
Fax +41 61 302 89 18
www.mdpi.com

Energies Editorial Office
E-mail: energies@mdpi.com
www.mdpi.com/journal/energies